Modern Optics and Photonics

Atoms and Structured Media

Modern Optics and Photonics

Atoms and Structured Media

editors

Gagik Yu. Kryuchkyan
Yerevan State University, Armenia

Gagik G. Gurzadyan
Nanyang Technological University, Singapore

Aram V. Papoyan
Institute for Physical Research of the National Academy of Sciences, Armenia

World Scientific

NEW JERSEY • LONDON • SINGAPORE • BEIJING • SHANGHAI • HONG KONG • TAIPEI • CHENNAI

Published by

World Scientific Publishing Co. Pte. Ltd.
5 Toh Tuck Link, Singapore 596224
USA office: 27 Warren Street, Suite 401-402, Hackensack, NJ 07601
UK office: 57 Shelton Street, Covent Garden, London WC2H 9HE

British Library Cataloguing-in-Publication Data
A catalogue record for this book is available from the British Library.

MODERN OPTICS AND PHOTONICS
Proceedings of the International Advanced Research Workshop

ISBN-13 978-981-4313-26-1
ISBN-10 981-4313-26-2

Printed in Singapore by World Scientific Printers

PREFACE

The present volume includes plenary talks presented at the International Advanced Research Workshop on "Modern Problems in Optics and Photonics", MPOP-2009. The Workshop was held in Armenia, from 27 August to 2 September, 2009. Hosts were Yerevan State University and Institute for Physical Research of National Academy of Sciences. The Workshop was bestowed to the 90-th Anniversary of the Yerevan State University (http://www.ipr.sci.am/mpop/). It provided an opportunity to bring together experienced and young scientists on various facets of modern optics and photonics, to converse and consider the most recent developments in these areas.

Modern laser optics and nanoscience are perhaps the fastest growing areas of material research and technological development. Photonic science is concerned mainly with optical investigations elaborating the new fundamental concepts, the generation and technological utilization of photons. For the further development of optical and laser technologies the quest for new materials and principles for generation, transmission, modulation, signal processing, switching, amplification and manipulation of photons are necessary. When photonics merges with nanoscience and nanotechnology, spatial confinement considerably modifies light propagation and light-matter interaction. It should also be noted that the impetuously growing directions of quantum information and quantum computing are primarily focused on the methods of quantum optics and novel composite materials. Therefore, it is expected that photonic technologies progressively gain importance in the 21st century.

Optics and laser research has a long lasting history in Armenia. In a rapidly changing field like modern optics and photonics it is not so easy exactly to formulate a standpoint of the field. Therefore, we as organizers of the workshop have tried to cover large area of topics and problems in this field. We hope that the workshop provides an impulse of the current activity at the research frontier of the areas.

The 22 papers are the end result of highly selective review process. They represent the state of the art in quantum optics and spectroscopy of the structured nonlinear media. This material is segregated into three chapters: Quantum optics, Structured materials and Photonic processes, Nonlinear optical effects and Spectroscopy.

Chapter I presents articles on quantum optical phenomena. In this chapter, the different aspects of light propagation in atomic and nuclear media, including phase control of light propagation through closed-loop atomic media and coherent control of nuclear forward scattering are discussed. In the other part, high-order harmonic generation with strong laser pulses employing electron ionization-recollision from highly charged ions is investigated in the relativistic regime. Generation of entangled light and superbunched photons via laser-multiparticle interactions as well as entanglement phenomena for matter waves is also studied. The Chapter contains discussion on irreversible photon transfer in an ensemble of -type atoms as well as a photonic diode. The results on applications of the frequency modulated (chirped) laser pulses in atomic physics are presented. This Chapter also elaborates on the investigation of quantum effects in a dissipative chaos where the counterpart of the semiclassical Poincaré section in quantum treatment is discussed. The prospect of utilizing quantum phases of light in a single atom driven by multiple different modes of photons is presented for possible quantum devices. Furthermore, the investigation of quantum adiabatic dynamics, where the external field moves slowly in time and is influenced by system's state (feedback) is presented. The Chapter also includes a meticulous analysis of several versions of Landau-Zener transition in nonlinear quantum systems.

Chapter II consists of papers mainly dedicated to structured nonlinear media, quantum dots and photonic devices. The multiple interactions in superlattice structures of nonlinear second-order material leading to the simultaneous quasi-phase matching of several parametric processes are discussed. A review and original studies of integrated photonic device structures with nanoscale features and sensitive applications are included in this chapter. One and two electronic and excitonic states in quantum dots with the non trivial geometries are discussed in the next article. The nanostructures are considered as the basics of novel photonic devices. In the other part, the properties of planar plasmonic structures and nonlinear metal-dielectric sub wavelength waveguides are investigated. The Chapter also involves material devoted to computer algebra study of structural and symmetry properties of discrete dynamical systems. A complete analysis and characterization of the bound and scattering states of two and three bosonic atoms in a one-dimensional optical lattice are presented. A vital phenomenon for atomic polaritons in a cavity array for the interaction of two-level atoms with quantized optical radiation in the strong coupling regime is also discussed.

Chapter III comprises articles on nonlinear optical effects and spectroscopy. The experimental investigations of formation of spectrally narrow optical resonances based on resonant absorption and fluorescence processes in an extremely thin vapor cells with smoothly variable thickness comparable with the resonant wavelength is presented. The nonlinear magneto-optical resonances in atomic rubidium in extremely thin cells are studied both experimentally and theoretically with high accuracy. The other contribution describes the modern problems of laser separation of isotopes in atomic vapors. The Chapter also incorporates both theoretical and experimental investigations of two-dimensional confined terahertz wave propagation in gap plasmon waveguide formed by two cylindrical surfaces. The complete characterization of the broadband similariton by means of the chirp measurement through the technique of spectral compression and frequency tuning in the sum-frequency generation process is presented in the last article.

We would like to express our sincere gratitude to the authors of the papers. We also thank the sponsors of the Workshop: State Committee of Science of Armenia, National Foundation of Science and Advanced Technologies (NFSAT), International Society of Optics Engineering (SPIE), Optical Society of Armenia (OSA). Sponsorship of EINST Technology, Pte Ltd (Singapore) and Analytical Technologies, Pte Ltd (Singapore) in publication of the present Proceedings is gratefully acknowledged. We would like to mention the valuable role of members of the Organizing and Program Committees, and, in particular, Diana Antonosyan for her generous contributions in organizing the conference and to the preparation of this volume.

Gagik Kryuchkyan
Yerevan State University and Institute for Physical Research, Armenia.
kryuchkyan@ysu.am

Gagik Gurzadyan
Nanyang Technological University, Singapore.
gurzadyan@ntu.edu.sg

Aram Papoyan
Institute for Physical Research, Armenia.
papoyan@ipr.sci.am

1 March 2010

The International Workshop

"Modern Problems in Optics and Photonics-2009"

was organized by

Yerevan State University (*Faculty of Physics*)
National Academy of Sciences of Armenia
(*Institute for Physical Research*)
Russian-Armenian (Slavonic) State University
State Committee of Science of Armenia

Publication of the Proceedings is sponsored by

EINST Technology Pte Ltd

Analytical Technologies Pte Ltd

SPIE

1. Gagik Kryuchkyan — Yerevan State University, Institute for Physical Research of NAS, Armenia. 2. Gagik Gurzadyan — Nanyang Technological University, Singapore; Technical University of Munich, Germany. 3. Aram Papoyan — Institute for Physical Research of NAS, Armenia. 4. Diana Antonosyan — Yerevan State University, Armenia. 5. David Sarkisyan — Institute for Physical Research of NAS, Armenia. 6. Karen Hatsagortsyan — Max-Planck Institute for Nuclear Physics, Germany. 7. Mihal Macovei — Max-Planck Institute for Nuclear Physics, Germany. 8. Jörg Evers — Max-Planck Institute for Nuclear Physics, Germany. 9. Alexander Alodjants — Vladimir State University, Russia. 10. Aranya Bhattacharjee — Max-Planck Für Physik Komplexer Systeme, Dresden, Germany; Department of Physics, University of Delhi, New Delhi, India. 11. Richard De La Rue — University of Glasgow, United Kingdom. 12. Yuriy Gladush — Institute of Spectroscopy, Ras, Russia. 13. Roman Noskov — Institute for Physics of Microstructures of RAS, Nizhny Novgorod, Russia. 14. Maxim Tsarev — Nizhny Novgorod State University, Russia. 15. Robert Sargayan — Institute for Physical Research of NAS, Armenia. 16. David Petrosyan — Institute of Electronic Structure & Laser, Forth, Crete, Greece. 17. Gor Nikoghosyan — University of Kaiserslautern, Germany; Institute for Physical Research of NAS, Armenia. 18. Vahan Nanumyan — Yerevan State University, Armenia. 19. Armen Sargsyan — Institute for Physical Research of NAS, Armenia. 20. Simon Liebing — Institute für Theoretische Physic, Germany. 21. Misnik Kazaryan — P. N. Lebedev Physical Institute of the NAS, Russia. 22. Dariusz Łukaszewski — Warsaw University of Technology, Institute of Micromechanics and Photonics, Poland. 23. Linards Kalvans — Laser Centre of the University of Latvia, Riga, Latvia. 24. Raphael Drampyan — Institute for Physical Research of NAS, Armenia. 25. Behrooz Rezael — University of Tabriz, Research Institute for Applied Physics and Astronomy, Tabriz, Iran. 26. Aryan Hamid Reza — Research Institute for Applied Physics and Astronomy, Tabriz, Iran. 27. Narine Gevorgyan — Institute for Physical Research of NAS, Armenia. 28. Tatyana Buryakina — Institute of Biophotonics, National Yang-Ming University, Taiwan. 29. Emil Gazazyan — Institute for Physical Research of NAS, Armenia. 30. Razmik Unanyan — University of Kaiserslautern, Germany; Institute for Physical Research of NAS, Armenia. 31. Anna Shahinyan — Yerevan State University, Armenia. 32. Mushegh Rafayelyan — Yerevan State University, Armenia. 33. Vardan Gevorgyan — Institute for Physical Research of NAS, Armenia. 34. Anahil Gogyan — Institute for Physical Research of NAS, Armenia; Universite de Bourgogne, France. 35. Levon Yaghiazaryan — Yerevan State University, Armenia. 36. Andranik Hakobyan — Russian Amenian (Slavonic) University, Armenia. 37. Karapet Mkrtchyan — Yerevan Physics Institute, Armenia. 38. Vahagn Petrosyan — Joint Institute for Physical Researches, Russia. 39. Areg Meliksetyan — Yerevan State University, Armenia. 40. Karen Vardanyan — Institute for Physical Research of NAS, Armenia. 41. Hasmik Demirkhanyan — Institute for Physical Research of NAS, Armenia.

CONTENTS

PART A

Quantum Optics

LIGHT PROPAGATION: FROM ATOMIC TO NUCLEAR QUANTUM OPTICS

J. EVERS*, R. FLEISCHHAKER, A. PÁLFFY and C. H. KEITEL

Max Planck Institute for Nuclear Physics
Saupfercheckweg 1, 69117 Heidelberg, Germany
**E-mail: joerg.evers@mpi-hd.mpg.de*
http://www.mpi-hd.mpg.de/~evers/

Coherent control of light propagation in atomic and nuclear media is discussed. The main focus is on light propagation through so-called closed-loop media. We present a calculation technique which allows to evaluate and interpret the light propagation dynamics in closed-loop media. As an application, we discuss the realization of a so-called white-light cavity with artificially enhanced bandwidth. We further study light propagation in media which realize a cross-coupling of the electric and the magnetic probe field component. Finally, we discuss coherent control of nuclear forward scattering, as an example for light propagation in nuclear media.

Keywords: Quantum optics, light propagation, closed-loop media, chiral media, nuclear quantum optics.

1. Introduction

Light propagation through coherently prepared media is a cornerstone of research and applications in quantum optics. In particular, electromagnetically induced transparency (EIT) allows for the propagation with low absorption, controlled velocity and the storage of light pulses, together with other beneficial properties such as high nonlinear response.[1] Consequently, based on the idea of EIT, a great variety of model systems have been analyzed. Here, we will mainly discuss a particular class of level schemes, which are so-called loop systems.[2–7] In these media, the applied laser fields form a closed interaction loop. As a consequence, in such systems a certain initial atomic state is connected to a final atomic state via several different combinations of laser field interactions. On the one hand, this leads to quantum pathway interference, and thus to a dependence of the system properties on the relative phase of the driving fields. This allows for a efficient control

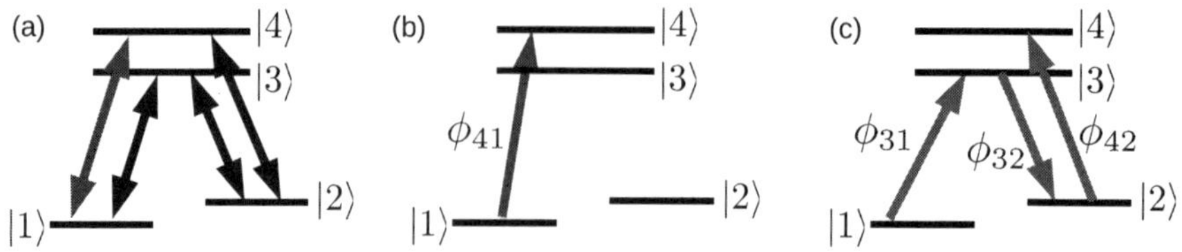

Fig. 1. (a) The four-level double-Λ type scheme with laser fields applied in closed-loop configuration. (b) and (c) are two possible interfering pathways from $|1\rangle$ to $|4\rangle$.

of the optical properties. On the other hand, however, such a system in general does not have a time-independent steady state.

In the following, first, we present a theory to describe light propagation in closed loop media. As an application, we discuss a so-called white light cavity. Next, we address light propagation in media that cross-couple the electric and the magnetic component of a single probe field, which can be seen as an ideal implementation of a closed-loop medium. Finally, we describe possible extensions of light propagation to nuclear systems.

2. Closed-Loop Media

An example for a closed-loop level scheme is given in Fig. 1. Starting from state $|1\rangle$, it is possible to evolve the atomic population by laser fields via states $|3\rangle$, $|2\rangle$ and $|4\rangle$ back to the initial state $|1\rangle$, such that the condition for a closed-loop system is fulfilled. Interference thus can occur, for example, between the pathways $|1\rangle \to |4\rangle$ and $|1\rangle \to |3\rangle \to |2\rangle \to |4\rangle$. Mathematically, the closed-loop property implies that it is impossible to find an interaction picture in which the Hamiltonian does not have an explicit time dependence. In a suitable interaction picture, the Hamiltonian can be written in the simple form V given by

$$\begin{aligned} V = &\hbar(\Delta_{32} - \Delta_{31})\tilde{\rho}_{22} - \hbar\Delta_{31}\tilde{\rho}_{33} + \hbar(\Delta_{32} - \Delta_{31} - \Delta_{42})\tilde{\rho}_{44} \\ &- \hbar\left(g_{31}\tilde{\rho}_{31} + g_{32}\tilde{\rho}_{32} + g_{42}\tilde{\rho}_{42} + g_{41}\tilde{\rho}_{41}e^{-i\Phi} + \text{ h.c.}\right) . \end{aligned} \tag{1}$$

Here, $\tilde{\rho}_{ij}$ is the operator $|i\rangle\langle j|$ in the chosen reference frame ($i, j \in \{1, \ldots, 4\}$). The Rabi frequencies are g_{ij}, and Δ_{ij} are laser field detunings. In this interaction picture, the residual time dependence along with the laser field phases appears only together with the Rabi frequency g_{41}, which we associate to a weak probe field coupled to transition $|1\rangle \leftrightarrow |4\rangle$.

The parameter Φ is given by

$$\Phi = \Delta t - \vec{K}\vec{r} + \phi_0 \,, \qquad \Delta = (\Delta_{32} + \Delta_{41}) - (\Delta_{31} + \Delta_{42}) \,, \tag{2a}$$

$$\vec{K} = (\vec{k}_{32} + \vec{k}_{41}) - (\vec{k}_{31} + \vec{k}_{42}) \,, \quad \phi_0 = (\phi_{32} + \phi_{41}) - (\phi_{31} + \phi_{42}) \,. \tag{2b}$$

The parameters Δ, $\vec{K}$ and ϕ_0 are known as the multiphoton resonance detuning, wave vector mismatch and initial phase difference, respectively. These parameters are a direct consequence of the closed-loop nature. In general it is not possible to find a reference frame where the explicit time dependence due to Δ vanishes from the Hamiltonian, such that for $\Delta \neq 0$ no stationary long-time limit can be expected. Therefore, a time-dependent solution of the density matrix equations is required. Using the notation $\bar{g}_{41} = g_{41}\,\exp[-i\vec{K}\vec{r} + i\phi_0]$, the density matrix equations of motion can be written as

$$\frac{\partial}{\partial t}\tilde{R} + \Sigma = M\tilde{R}, \tag{3}$$

where $\tilde{R}$ is a vector containing the density matrix elements. M is a time-dependent matrix, and Σ a time-dependent vector independent of the density matrix elements which arises from eliminating one of the state populations from the equations of motion via the trace condition $Tr(\tilde{\rho}) = 1$. For a weak probe field, using Floquet's theorem,[8] the solution $\tilde{R}$ can be obtained with a series ansatz

$$\tilde{R} = \tilde{R}_0 + \bar{g}_{41}e^{-i\Delta t}\tilde{R}_1 + \bar{g}_{41}^{*}e^{i\Delta t}\tilde{R}_{-1} + \dots \,, \tag{4}$$

where $\tilde{R}_i$ ($i \in \{0, \pm1, \dots\}$) are time-independent coefficient vectors. Using this ansatz, the time-dependent solution to the density matrix equations of motion can be found, which allows to evaluate all observables.[6,9]

In an interaction picture oscillating in phase with the applied probe field, the probe field transition coherence $\hat{\rho}_{41}$ determining the light pulse propagation through the medium can be written as,

$$\hat{\rho}_{41} = [\tilde{R}_0]_p\, e^{i\Phi} + g_{41}[\tilde{R}_1]_p + g_{41}^{*}[\tilde{R}_{-1}]_p\, e^{2i\Phi} \,, \tag{5}$$

where $[x]_{13}$ denotes the relevant probe field coherence component of the vector x.[6] The different contributions to this result naturally arising from the Floquet analysis correspond to the various involved physical processes and allow to in detail understand the medium response. The first part of Eq. (5) represents the scattering of the driving fields into the probe field mode arises from $[\tilde{R}_0]_p$, as shown in Fig. 2(a). This contribution in general does not oscillate at the probe field frequency, but rather at the combination frequency $\omega_{31} + \omega_{42} - \omega_{32}$ of the three driving fields. This

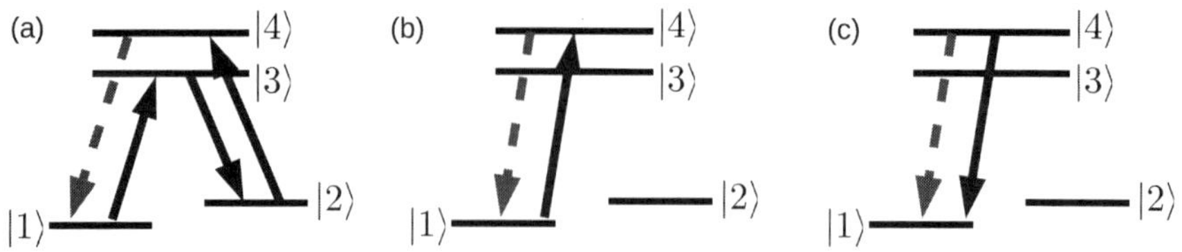

Fig. 2. (Color online) Interpretation of the different contributions to the probe field susceptibility in terms of transition pathways. (a) represents the interaction loop leading to a scattering of the driving fields into the probe field mode. (b) is the direct scattering of the probe field off of the probe transition. (c) shows a counter-rotating term. The solid blue arrows indicate coupling field transitions, the dashed red line is a probe field interaction.

frequency coincides with the probe field frequency only under multiphoton resonance. The contribution proportional to $[\tilde{R}_1]_p$ shown in Fig. 2(b) is in phase with the probe field for all values of Δ, and is independent of the relative field phase. It represents the direct scattering of the probe field off of the probe field transition. The third contribution proportional to $[\tilde{R}_{-1}]_p$ can be interpreted as a counter-rotating term which in the Floquet expansion differs by 2Δ from the probe field frequency, and is depicted in Fig. 2(c).

As an important result, it can be concluded that the phase-dependence of the loop-configuration studied here is restricted to the multiphoton resonance condition $\Delta = 0$, because it arises from the scattering of the coupling fields into the probe field mode. Furthermore, it can be seen from Eq. (5) that all contributions but the direct scattering acquire an additional dependence on the wave vector mismatch $\vec{K}$ together with the dependence on the phase ϕ_0. Therefore, the laser field geometry influences the relevance of these contributions to the detection signal in probe field propagation direction. In general, only the direct scattering contribution can be detected in propagation direction of the probe beam regardless of the separation of detector and the scattering atoms.

The considered system enables one to control the pulse propagation to a great extend.[6] An example is shown in Fig. 3. In (a), around zero probe field detuning, subluminal light propagation with gain could be realized. At $\Delta_{41} \approx \pm 2\gamma_0$, superluminal propagation with small absorption or gain occurs. (b) shows an example of switching between sub- and superluminal light propagation by via the value of one of the coupling field Rabi frequencies. The interesting range around is accompanied by vanishing absorption at $\Delta_{41} = 0$.

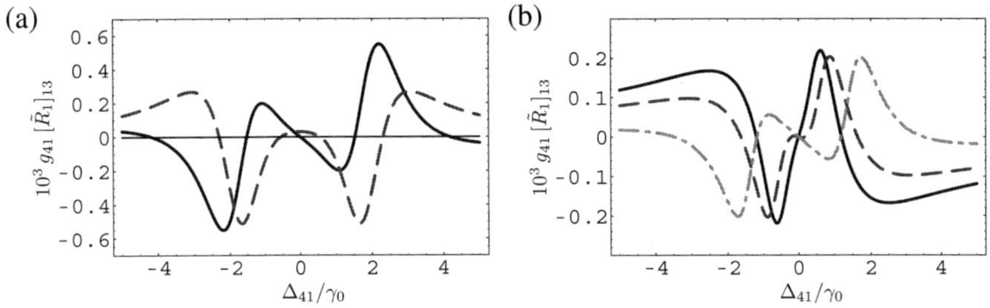

Fig. 3. Results for the susceptibility in the closed-loop system. Both curves show the direct scattering channel indicated in Fig. 2(b). (a) the parameters are $g_{31} = 1.8\gamma_0$, $g_{32} = 0.2\gamma_0$, $g_{42} = 0.5\gamma_0$, and $g_{41} = 0.01\gamma_0$. (b) shows results for different Rabi frequencies of the coupling field g_{31}, namely, $g_{31} = 0.7\gamma_0$ (solid line), $g_{31} = 0.85\gamma_0$ (dashed line), and $g_{31} = 1.5\gamma_0$ (dash-dotted line). The other couplings are $g_{42} = 0.2\gamma_0$, and $g_{41} = 0.01\gamma_0$. In both subfigures, the detunings and decay rates are $\Delta_{31} = \Delta_{32} = \Delta_{42} = 0$, and $2\gamma_{13} = 2\gamma_{14} = 2\gamma_{23} = 2\gamma_{24} = \gamma_0$.

3. An "Open" Closed Loop System

Interesting insight in the dynamics of light propagation can be gained by considering the closed-loop setup shown in Fig. 1 without the driving field coupling to transition $|2\rangle \leftrightarrow |3\rangle$.[9] Then, the driving fields no longer form a closed loop. Also, naively, one would expect strong absorption to occur, for example, since the EIT formed on resonance in the Λ-type subsystem $|1\rangle$, $|4\rangle$ and $|2\rangle$ is constantly perturbed by the driving field between $|1\rangle$ and $|3\rangle$. The absorption of the control fields is symbolically depicted in Fig. 4(a). A more thorough analysis shows that two additional processes need to be considered, as shown in Fig. 4(b,c). Due to a four-wave mixing (4WM) process, an additional field is generated on the undriven transition $|2\rangle \leftrightarrow |3\rangle$ throughout the propagation, as shown in (b). This dynamically generated field then leads to a second 4WM process that scatters the newly created field and the control fields back into the probe field mode. A full numerical solution of the corresponding Maxwell-Schrödinger propagation equations for a probe field pulse reveals that the interplay of these processes in effect leads to a slight gain of the probe field throughout the propagation. The control fields, on the other hand, are substantially absorbed throughout the propagation, while at the same time the amplitude of the 4WM field builds up. Despite this pronounced in-medium dynamics, the effective susceptibility experienced by the probe pulse over the whole propagation can be surprisingly smooth.[9]

As an application, such a medium allows to construct a so-called white-light cavity (WLC).[9–12] A regular optical cavity has a finite bandwidth of

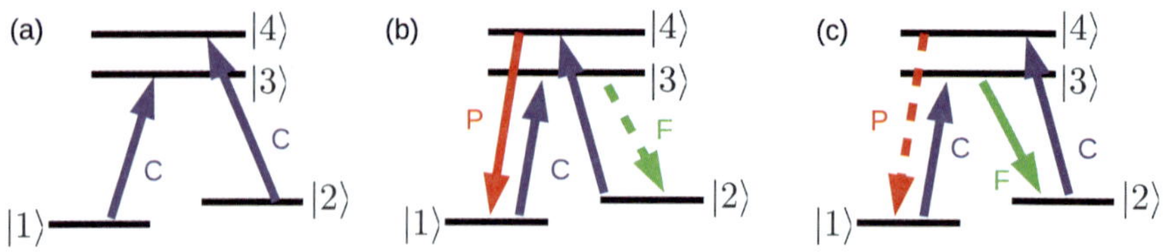

Fig. 4. Main contributions to the in-medium propagation dynamics. (a) absorption of the coupling fields (C). (b) four-wave mixing of coupling fields and probe field (P) to give an additional field (F). (c) "backscattering" of coupling and four-wave-mixing field into the probe field mode.

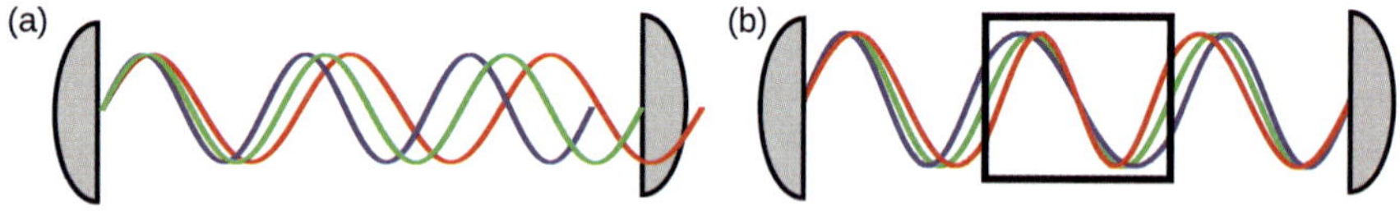

Fig. 5. White light cavity mechanism. In (a), two of the three wavelengths are not supported by the cavity. In (b), the additional dispersive medium placed inside the cavity causes all three wavelengths to be supported by introducing suitable frequency-dependent phase shifts.

supported frequencies, as indicated in Fig. 5(a). Two of the shown waves do not fit into the resonator, and thus would not be supported by the cavity. However, if an additional medium with a suitable (negative) dispersion is introduced into the cavity, then the effective wavelength of the different frequency components may change. In particular, in subfigure (b), it can be seen that a suitable dispersion can transform all three different wavelengths into the supported frequency spectrum. This way, the bandwidth of a cavity can be enhanced, without reducing the maximum intensity buildup. More specific, in one round trip through a cavity of length L, a field with detuning Δ relative to the cavity resonance frequency will acquire a phase $\phi_0 = 2L\Delta/c$. If in addition a medium of length l and index of refraction n is introduced to the cavity, it will lead to an additional phase shift $\phi_1 = 2l\omega_0(n-1)/c$, with ω_0 the resonance frequency. A white light cavity is achieved, if these phase shifts cancel, $\phi_0 + \phi_1 = 0$. This leads to a white-light cavity condition

$$\omega_0 \frac{\partial n(\omega_0)}{\partial \omega} = -\frac{L}{l}, \tag{6}$$

i.e., negative dispersion is required. Based on a full analysis of the above described system, it turns out that for a cavity length of about 59.5 cm, a medium length of 30 cm, and an empty cavity finesse of 1000, a medium

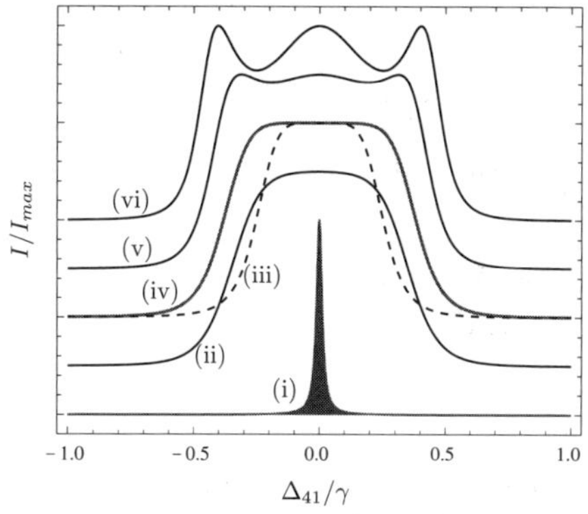

Fig. 6. Intensity profile of the white light cavity. The different cases are discussed in the text. Note that the curves all have the same base line, but are shifted in y direction in the figure for better visibility.

of the type described above driven by suitable coupling fields would allow to enhance the empty cavity bandwidth by a factor of about 30. In Fig. 6, curve (i) shows the empty cavity, and (iv) the corresponding curve with WLC medium. If the WLC-condition Eq. (6) is fulfilled, the cavity has a near-flat intensity distribution across the enhanced resonance width.[9] Interestingly, the pronounced in-medium dynamics leads to an enhancement of the WLC cavity width, as can be seen from curve (iii), which is the result of a calculation with the 4WM processes artificially suppressed. In curves (ii), the WLC condition is overcompensated by 2%, while (vi) and (v) correspond to 2% and 4% overcompensation.

4. Light Propagation in Media with Cross Coupling of Electric and Magnetic Field Component

Motivated not least by ideas related to the realization of materials with a negative index of refraction in atomic gases, in the recent past, atomic level schemes have been considered which lead to a cross-coupling of the electric component **E** and the magnetic component **H** of a single probe laser field. In this case, the medium polarization $\mathbf{P}(t)$ and magnetization $\mathbf{M}(t)$ can be written as,

$$\mathbf{P}(t) = \int \left(\varepsilon_0 \chi_E(\tau)\mathbf{E}(t-\tau) + \frac{1}{c}\xi_{EH}(\tau)\mathbf{H}(t-\tau) \right) \mathrm{d}\tau\,, \tag{7}$$

$$\mathbf{M}(t) = \int \left(\chi_H(\tau)\mathbf{H}(t-\tau) + \frac{1}{c\mu_0}\xi_{HE}(\tau)\mathbf{E}(t-\tau) \right) \mathrm{d}\tau\,. \tag{8}$$

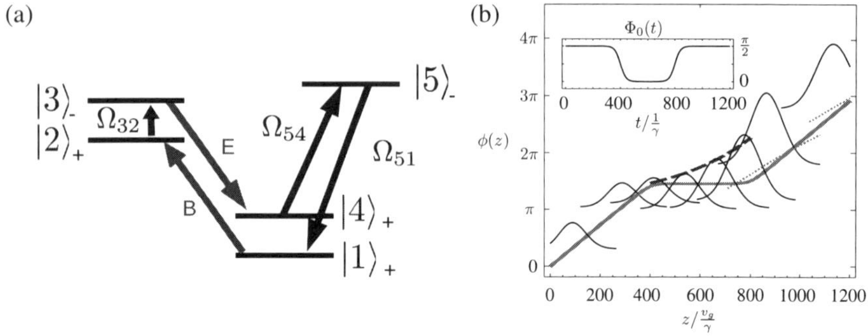

Fig. 7. (a) Level scheme featuring cross coupling of the electric and the magnetic field components of a probe beam. (b) Light propagation in the medium shown in (a). The figure shows the phase $\phi(z)$ of the central pulse component versus propagation distance z. The inset shows the applied closed-loop phase versus propagation time.

Here, χ_E and χ_M are the usual linear response coefficients for electric and magnetic field, respectively, while χ_{HE} and χ_{EH} describe the cross-coupling. It is interesting to note that a setup of this type essentially requires a closed-loop level scheme. A particular example is shown in Fig. 7(a).[7,13] The two probe field components couple to transitions $|1\rangle \leftrightarrow |2\rangle$ and $|3\rangle \leftrightarrow |4\rangle$. Calculating the closed-loop phase for this setup, one obtains

$$\begin{aligned}\phi &= \phi_{\text{probe}} + \phi_{32} - \phi_{\text{probe}} + \phi_{54} - \phi_{51} \\ &= \phi_{32} + \phi_{54} - \phi_{51} \, . \end{aligned} \tag{9}$$

In contrast to regular closed-loop media which couple only to the electric component of the probe field, the closed-loop phase is independent of the applied probe field phase. The reason for this is that in one loop, one absorption and one emission of a probe field photon occur. This is a crucial advantage for potential applications, since then the properties of the probe pulse may not be known. In this sense, media with cross-coupling of electric and magnetic field component are ideal closed-loop media. Fig. 7(b) shows the result of a numerical solution to the full Maxwell-Schrödinger equations describing the pulse propagation through the medium shown in Fig. 7(a). The figure mainly depicts the phase in the center of the propagated pulse versus propagation length z. At the initial stage of the propagation, the phase increases linearly with z, which is due to a real index of refraction. At a certain time throughout the propagation, the closed-loop phase is changed, as shown in the inset. The pulse now propagates with constant phase, but with exponentially increasing pulse amplitude, which corresponds to an imaginary index of refraction with suitable sign. Finally,

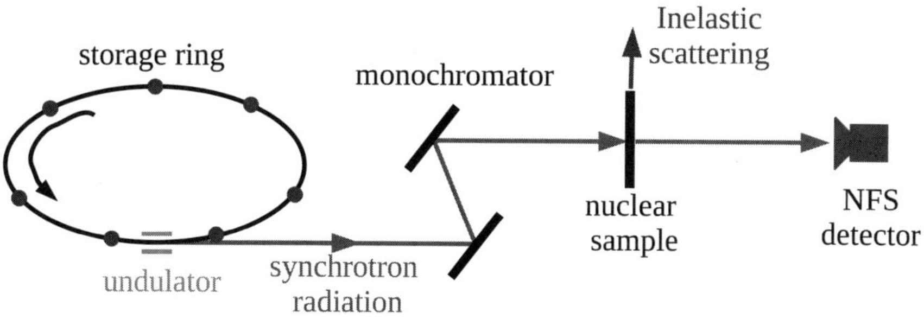

Fig. 8. Typical setup in nuclear forward scattering.

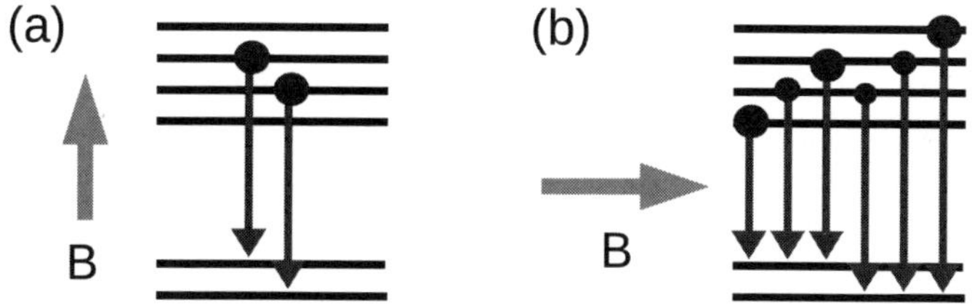

Fig. 9. Level scheme of ^{57}Fe after excitation with linearly polarized light. (a) with magnetic field as during the excitation. (b) with rotated magnetic field. The circles indicate population in the upper states.

at a later time, the closed-loop phase is switched back, and the pulse again propagates with constant amplitude, but with increasing phase. This result demonstrates that it is possible to dynamically influence the propagation dynamics of a probe pulse in a closed loop medium via the control field phases. At the same time, it is interesting to note that the additional coupling of the magnetic field component is crucial to the phase control, even though it is very weak. The reason is that only the magnetic field component allows to close the interaction loop, which becomes significant due to the strong coupling fields contributing to the other transitions in the loop, already at parameters compatible with state-of-the-art light propagation experiments.

5. Light Propagation in Nuclear Media

Coherent control of nuclear dynamics has been a long-term goal over the last few decades, fueled by the desire for a number of fascinating applications such as gamma-ray lasers or clean energy storage. However, not least due to the lack of suitable light sources, a direct transfer of quantum optical schemes to the nuclear case is often not promising. We therefore here pursue a different approach, and study coherent control in a typical experimental setting for light scattering off of nuclei. In coherent nuclear forward scattering, intense high-frequency light such as that from a synchrotron radiation source is monochromatized at a nuclear resonance energy, and then hits a nuclear target, as shown schematically in Fig. 8.[14] A detector placed in forward direction then registers the coherently scattered photons. The most commonly used target consists of ^{57}Fe, which has a magnetic dipole allowed Mößbauer transition from a $J = 1/2$ ground state to a $J = 3/2$ first excited state at 14.4 keV energy. At typical synchrotron source parameters, per incident light pulse at most one nucleus in the sample is excited. This low excitation rate is mostly due to the fact that the nuclear resonances are very narrow ($\sim$ 5 neV) compared to the bandwidth of the light after the monochromator ($\sim$ meV). The coherently scattered light can be described with a wave equation similar to the atomic case,[15]

$$\left(\Delta - \frac{1}{c^2}\frac{\partial^2}{\partial t^2}\right)\vec{E} = \frac{4\pi}{c}\frac{\partial}{\partial t}\vec{I}\,, \tag{10}$$

where $\vec{E}$ is the electric field of the light, and $\vec{I}$ is the nuclear source current. In slowly varying envelope approximation, and for light propagating in z direction, this simplifies to an equation for the envelopes $\mathcal{E}$ and $\mathcal{I}$ given by

$$\frac{\partial}{\partial z}\vec{\mathcal{E}} = -\frac{2\pi}{c}\vec{\mathcal{I}}\,. \tag{11}$$

Calculating the nuclear source current in second order in the interaction of the light and the nuclei, one finally obtains a wave equation

$$\frac{\partial \vec{\mathcal{E}}(z,t)}{\partial z} = -\sum_{\ell} K_\ell \vec{J_\ell}(t) \int_{-\infty}^{t} d\tau \vec{J}_\ell^{\dagger}(\tau)\cdot\vec{\mathcal{E}}(z,\tau) \tag{12}$$

where $\vec{J_\ell}$ are nuclear transition current matrix elements, and ℓ is a summation index over all possible transitions, with properties characterized by K_ℓ.[15] This equation can be iteratively solved starting from an initial synchrotron radiation pulse $E(t) \sim \delta(t)$ which is instantaneous on the time scale of the nuclear dynamics.

A possibility for coherent control arises, if the iron nuclei are embedded in a suitable host material, such as $FeBO_3$. Inside this crystal, the nuclei experience a strong magnetic field of order 30 T, leading to a pronounced Zeeman shift of the magnetic sublevels. Interestingly, the direction of this field can be controlled by a weak external magnetic field (tens of Gauss)

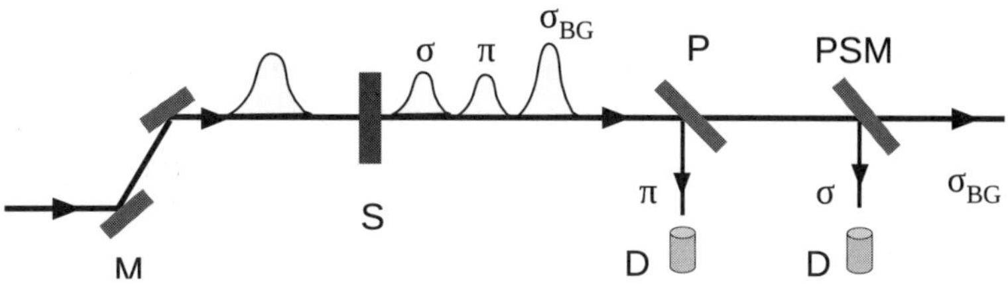

Fig. 10. Schematic setup for keV single photon entanglement generation. σ, π and σ_{BG} indicate the two signal components with σ and π polarization, and the background pulse, respectively. S is the target, and M is a monochromatizer. P and PSM are x-ray optics devices to separate the two signal pulses and to extract the signal from the background.

very rapidly. This feature can be facilitated to control the coherent decay of the nuclei.[16,17] If directly after the excitation of the nuclei via the synchrotron radiation pulse the magnetic field is rotated, the excited upper sublevels are projected onto a new set of states determined by the new quantization axis after the rotation, see Fig. 9. If this switching occurs with a suitable rotation angle of the magnetic field, and at a suitable time, then the transition amplitudes for the coherent decay can destructively interfere, thus suppressing coherent decay. This mechanism is in close analogy to EIT, in which destructive interference of different transition channels prohibits an *excitation* of the atom. It was shown experimentally that this can lead to a storage of the upper state population over durations larger than the natural life time of the nuclei.[16]

Based on these ideas, it is possible to achieve single-photon mode entanglement at x-ray photon energies.[17] Using an advanced switching scheme, the single excitation in the nuclei can be stored after the synchrotron pulse has passed by, and then be released into two subsequent separated pulses. The two pulses are distinct in the emission time and in the polarization of the emitted light. For this, after the excitation, the magnetic field is first rotated at a suitable time in direction parallel to the beam line. Until this point, all background light (the SR pulse and prompt electronic scattering) has left the target in a first pulse. At a suitable second switching time, the magnetic field is rotated back to release the first signal pulse with π-polarization. Then, another rotation in direction along the beam axis and rotation back at definite times leads to an emission of the second σ-polarized signal pulse, see Fig. 10. The obtained state is of the form

$$|\Psi\rangle = \alpha|0\rangle_A|1\rangle_B + \beta|1\rangle_A|0\rangle_B\,, \tag{13}$$

which is known as a single photon mode entangled state.[18] Here, modes

A and B correspond to the different emitted pulses which have different polarizations. Using state-of-the-art x-ray optics, the two signal pulse components can then be extracted from the background, as indicated in Fig. 10 by the polarizing beam splitter (P) and time-variable mirror (PSM). The latter can either be realized using a Bragg-reflecting crystal mounted on a piezoelectric drive, or by facilitating mirror switching devices based on lattice deformations.

Nuclear forward scattering together with the magnetic field switching offers an interesting playground for coherent control of nuclear dynamics. Despite the weak interaction between light and nuclei, which on average leads to less than one excitation in the nuclei per incident pulse, a complete control of the coherent nuclear dynamics is enabled by the switching of the magnetic field. It can be expected that further applications in the coherent control of nuclei will arise from related techniques.

References

1. M. Fleischhauer, A. Imamoglu and J. P. Marangos, *Rev. Mod. Phys.* **77**, p. 633 (2005).
2. O. Kocharovskaya and P. Mandel, *Phys. Rev. A* **42**, p. 523 (1990).
3. C. H. Keitel, O. A. Kocharovskaya, L. M. Narducci, M. O. Scully, S.-Y. Zhu and H. M. Doss, *Phys. Rev. A* **48**, p. 3196 (1993).
4. G. Morigi, S. Franke-Arnold and G.-L. Oppo, *Phys. Rev. A* **66**, p. 053409 (2002).
5. E. A. Korsunsky and D. V. Kosachiov, *Phys. Rev. A* **60**, p. 4996 (1999).
6. M. Mahmoudi and J. Evers, *Phys. Rev. A* **74**, p. 063827 (2006).
7. R. Fleischhaker and J. Evers, *Phys. Rev. A* **80**, p. 063816 (2009).
8. G. Floquet, *Ann. École Norm. Sup.* **12**, p. 47 (1883).
9. R. Fleischhaker and J. Evers, *Phys. Rev. A* **78**, p. 051802 (2008).
10. A. Wicht, K. Danzmann, M. Fleischhauer, M. Scully, G. Müller and R. H. Rinkleff, *Opt. Commun.* **134**, p. 431 (1997).
11. G. S. Pati, M. Salit, K. Salit and M. S. Shahriar, *Phys. Rev. Lett.* **99**, p. 133601 (2007).
12. A. Dogariu, A. Kuzmich and L. J. Wang, *Phys. Rev. A* **63**, p. 053806 (2001).
13. J. Kästel, M. Fleischhauer, S. F. Yelin and R. L. Walsworth, *Phys. Rev. Lett.* **99**, p. 073602 (2007).
14. R. Röhlsberger, *Nuclear Condensed Matter Physics with Synchrotron Radiation: Basic Principles, Methodology and Applications (Springer Tracts in Modern Physics)* (Springer, 2004).
15. Y. V. Shvyd'ko, *Phys. Rev. B* **59**, 9132 (1999).
16. Y. V. Shvyd'ko, T. Hertrich, U. van Bürck, E. Gerdau, O. Leupold, J. Metge, H. D. Rüter, S. Schwendy, G. V. Smirnov, W. Potzel and P. Schindelmann, *Phys. Rev. Lett.* **77**, 3232 (1996).

17. A. Pálffy, C. H. Keitel and J. Evers, *Phys. Rev. Lett.* **103**, p. 017401 (2009).
18. S. J. van Enk, *Phys. Rev. A* **72**, p. 064306 (2005).

RELATIVISTIC HIGH-ORDER HARMONIC GENERATION

M. C. KOHLER and K. Z. HATSAGORTSYAN*

Max-Planck-Institut für Kernphysik
Saupfercheckweg 1, D-69117 Heidelberg, Germany
** E-mail: k.hatsagortsyan@mpi-hd.mpg.de*
http://www.mpi-hd.mpg.de/keitel/

High-order harmonic generation (HHG) with relativistically strong laser pulses employing electron ionization-recollision from highly charged ions is investigated. In the relativistic regime, the magnetically induced drift of an ionized electron severely suppresses the probability of the electron revisiting the ionic core and, consequently, the yield of harmonic photons. We discuss several methods to increase the efficiency of rescattering in the relativistic regime and achieve HHG in the hard x-ray domain. We lay emphasis on a setup with counterpropagating equal-handed circularly polarized laser waves. The setup exhibits the feature that the electric and magnetic fields are aligned. The relativistic quantum mechanical calculations of a single atom response show a significant HHG yield due to frustrated relativistic drift. We also discuss the phase coherence properties of the HHG in this setup.

Keywords: High-Order Harmonic Generation, Relativistic Optics.

1. Introduction

Half a century ago, laser light paved the way to the development of the research field of nonlinear optics, one of the first nonlinear effects being observed was low-order harmonic generation.[1] Decades later, at laser intensities above 10^{14} W/cm^2, the field of extreme nonlinear optics has been entered where so-called high-order harmonic generation (HHG) takes place.[2] The latter means that coherent light of several integer multiples of the fundamental laser frequency is produced. Nowadays, harmonics with orders of several thousands (in the keV range) have been observed.[3,4] The large range of frequencies of HHG allows to synthesize pulses in the attosecond regime.[5] Routinely, coherent laser pulses with very high photon energies of 100 eV in the extreme ultraviolet (XUV) regime and short pulses below 100 as became achievable which opened a door for attosecond control and metrology[6] and for many other applications.[7] The most important application is the pump-probe technique for time-resolved spectroscopy of ultrafast processes. This method has been applied to monitor and control chemical reactions,[8] to investigate the time evolution of molecular dynamics (e.g. molecular fragmentation[9]), to visualize atomic inner-shell processes[10] and was advanced down to the attosecond time resolution and for excitation energies in the XUV regime.[11,12] HHG has also applications in industry, e.g. for lithography and nanoscale spectroscopy.

Over the years, the available laser intensities have been increased and nowadays, lasers with intensities of 10^{22} W/cm^2 are available.[13] The future Extreme Light Infrastructure is aimed even at laser intensities of 10^{26} W/cm^2.[14] Can the strong laser field of infrared (IR) light be used to generate thousends of harmonics and, in this way, to achieve the generation of coherent hard x-rays? In this paper, we will discuss the problems and several possibilities towards this technique. The coherent light with those properties will be vitally important not only for fundamental physics but also for chemistry and life science, in particular, allowing for ultrafast coherent diffraction imaging of single particles, clusters and biomolecules with sub-ångström resolution. In atomic physics, it will permit to track inner-shell electron dynamics and, in nuclear physics, the investigation of time-resolved dynamics of nuclear excitations. The large scale unique x-ray free electron lasers such as the TESLA-XFEL at DESY, Germany,[15] the LCLS at Stanford, USA[16] and SCSS in Japan,[17] will partly fulfill this task but are limited to energies below 10 keV. Moreover, the HHG technique may offer the opportunity of having table-top sources.

In principle, the harmonic photon energy can be increased by using stronger driving laser fields. Because, at higher laser intensities, phase-matching is more difficult to realize, HHG in the non-relativistic regime can be extended up to a kilo-electronvolt photon energy range applying mid-IR driving laser fields.[18] However, progress in this field appears to have reached a limit. Most importantly, the further increase of the driving field intensity transfers the interaction regime into the relativistic domain where the drift motion of the ionized electron in the laser field propagation direction prohibits the recollision and, consequently, suppresses HHG.[19] This limits the non-relativistic HHG to a photon cutoff energy of $\varepsilon_{\text{cutoff}} \approx 10$ keV. The second point hindering HHG at high intensities is the phase matching problem. In ultrastrong laser fields, the atomic outer-shell electrons are rapidly ionized and produce a large free electron background causing phase mismatch between the driving laser wave and the emitted x-rays.

The process of HHG can be intuitively understood within the semiclassical three-step-model introduced by Corkum.[20] In the first step, the strong electric field of the laser field suppresses the Coulomb barrier allowing the atomic electron to tunnel out of the atom. In the second step, the laser field drives the electron in the continuum conveying to it a significant amount of energy and then, in the third step, forces the electron into a recollision with the atomic core, triggering the electron recombination with the ion. As a result, the entire kinetic energy gained by the recombining electron is released in a form of a highly energetic photon. Conventionally, the emission spectrum exhibits a flat plateau with a rapid drop at the cut-toff energy $\varepsilon_{\text{cutoff}} = 3.17\,U_p + I_p$, where the electron ponderomotive energy $U_p = (E_0/2\omega)^2$ is the mean quiver energy of an electron in the laser field, E_0 and ω are the laser field amplitude and frequency, respectively, and I_p the atomic ionization potential. The first analytical quantum mechanical treatment of a HHG process relying on the strong field approximation (SFA)[21,22] was carried out by Lewenstein.[23] The calculation was done within the single active electron (SAE) approach using the dipole approximation (DA) for a linearly polarized plane wave and is valid only for non-relativistic intensities. The model was able to explain the important features of high harmonic generation: the cutoff and the plateau with its interference features.

The increase of the harmonic photon energy that can be achieved by stronger driving laser fields encounters a limit as soon as relativistic intensities are approached. One can explain this limit from the semi-classical model of harmonic generation: For non-relativistic intensities the motion of the electron wave packet is merely along the polarization axis of the fundamental laser field (see Fig. 1(i)). However, if the field is so strong that the electron moves relativistically, the Lorentz force $\mathbf{v}/c\times\mathbf{B}$ causes a drift of the electron into the propagation direction because of the laser magnetic field. Due to this, the electron misses the core and the recombination as well as HHG are suppressed. Therefore, the significance of this effect is not only determined by the ratio $\mathbf{v}/c$ but also by the wave packet size.

The drift becomes crucial as soon as the drift distance exceeds the size of the electron wave packet at the recollision moment.[24,25] The borderline between both regimes can be drawn at intensities of about 4×10^{16} W/cm^2 for IR wavelengths. This intensity is much weaker than the one at which a free electron becomes relativistic in a laser field, i.e. when the relativistic field parameter $\xi \equiv eE_0/mc\omega$ exceeds 1 (this corresponds to a laser intensity of 10^{18} W/cm^2 for IR wavelengths), where e and m are the electron charge and mass, respectively, c the speed of light. One can derive the explicit condition for the significance of the relativistic drift in the tunneling regime:

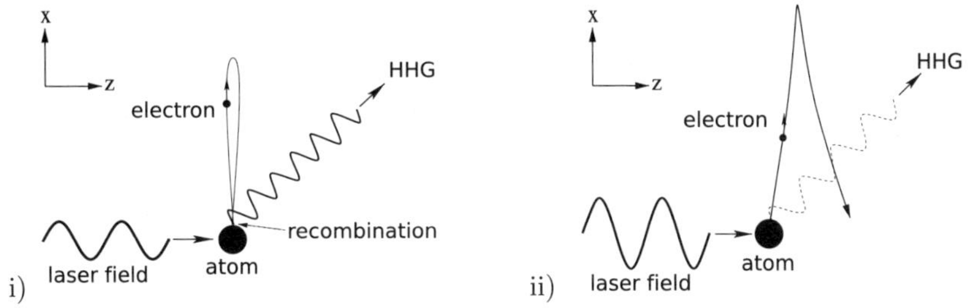

Fig. 1. Sketch of electron trajectories for a linearly polarized wave in the i) non-relativistic and ii) highly relativistic regime. The weaker high harmonic yield caused by the electron's drift in the relativistic case is depicted by a dashed line.

The drift distance in the laser propagation direction can be estimated as $d \sim (c/\omega)\xi^2/2$ (see e.g. Ref. 19) and the electron wave packet size due to its spreading during the excursion in the laser field as $\Delta z \sim v_z \Delta t$, where Δt is the excursion time and v_z the electron velocity along the laser propagation direction. The latter can be related to the energy uncertainty of the electron because of the tunneling: $mv_z^2/2 \sim \hbar/\tau_{tun}$, where $\tau_{tun} \sim \ell_{tun}/v_a$ is the tunneling time, $\ell_{tun} \sim I_p/eE$ the tunneling length, $v_a \sim \sqrt{2I_p/m}$ the bound electron velocity, and $\Delta t \sim 1/\omega$. As a result $\omega \Delta z \sim (eE_0\hbar)^{1/2}/m^{3/4}I_p^{1/4}$. The signature of the drift will be essential when $d > \Delta z$, i.e. at

$$\sqrt{\frac{2I_p}{mc^2}\frac{\xi^3}{16}\frac{mc^2}{\hbar\omega}} > 1\,. \tag{1}$$

Another point of view on the relativistic recollision is also useful: Due to the relativistic drift, the ionized electron misses the ionic core when it is ionized with zero momentum. Nevertheless, the recollision will occur if the electron is ionized with a certain initial momentum ($p_d \approx mc\xi^2/4$[26]) opposite to the laser propagation direction. However, the probability of this process is exponentially damped due to the non-zero momentum:

$$W(p_d) \sim \exp\left(-\frac{2(2mI_p + p_d^2)^{3/2}}{3m\hbar eE}\right) \approx \exp\left(-\frac{2(2mI_p)^{3/2}}{3m\hbar eE} - \frac{\sqrt{2mI_p}}{3m\hbar eE}p_d^2\right), \tag{2}$$

The drift term in the exponent will be important if $\sqrt{2mI_p}p_d^2/(3m\hbar eE) > 1$, which again yields the condition of Eq.(1).

The structure of the paper is the following. In Section 2, the existing proposals are presented and their advantages and disadvantages are discussed. Thereafter, the emission rate of one of these proposals is calculated

(Section 3) and the phase matching problem is discussed. The conclusion follows in Section 4.

2. Existing Proposals for Relativistic HHG

There are different proposals how to circumvent the relativistic drift. However, no universal solution has been found, each method has its drawbacks. On the one hand, the properties of the atomic system can be modified to this end. Thus, positronium atoms could be employed instead of ordinary atoms.[27] Both constituents would experience an identical drift and recollide eventually due to their equal mass and same amount of charge. Antisymmetrical molecular orbits[28] could also provide the electron with a certain initial momentum to compensate the drift. For instance, atomic p-orbitals or antisymmetrical molecular π-orbitals could be used for this scheme employing experimentally well-established methods to align those molecules. However, the method works only in the weakly relativistic regime since the delivered molecular momenta are small. Furthermore, multiply-charged ions moving against the laser propagation direction[29] may reduce the drift due to a Doppler-shifted laser frequency in the center of mass system of the ion.

On the other hand, the laser field can be modified to suppress the relativistic drift. Thus, a tightly focused laser beam could be employed yielding a drift compensation in a certain part of the focal volume due to the lon-

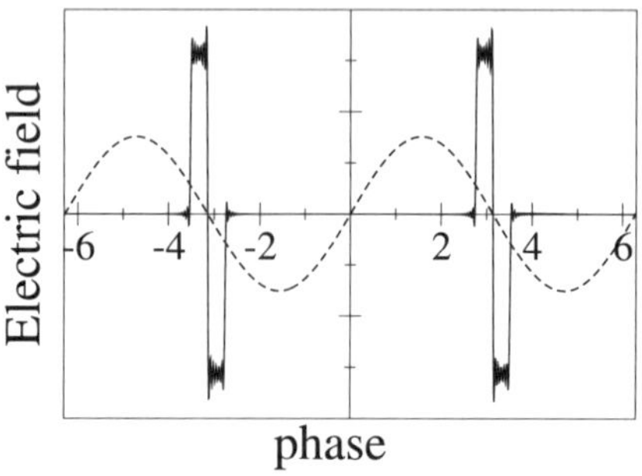

Fig. 2. The optimized pulse shape of a laser field for relativistic HHG. The pulse consists of 100 harmonics of a IR laser field. A wave packet which is freed at the end of one pulse moves away from the core with its initial momentum. Within the field free region it is not affected by any magnetic field. Finally, the recollision is caused by the next pulse where the only drift takes place.

gitudinal component of the laser field.[30] Another possibility is to apply two linearly polarized counter-propagating laser beams[31] instead of only one linearly polarized laser field. Then, the Lorentz-force is eliminated in a small area near the antinodes of the standing wave. In the weakly relativistic regime where the relativistic drift velocity is not very high, the Lorentz-force can also be compensated by a second weak laser beam which has its polarization in the strong beam propagation direction.[32] By varying the time delay between the laser pulses, it is then possible to achieve a drift suppression for selected harmonics. A different approach was chosen in Refs. 33 and 34 where it is investigated whether the drift can be reduced by a special tailoring of the laser pulse shape. Figure 2 shows such an optimized pulse shape. The recipe is that the laser field consists of a very strong first pulse of which the electric field strength becomes quickly negligible. Then, a long field free time span follows and a second strong pulse arrives driving the electron back. In this field configuration, the electron is ionized by the first pulse also obtaining some initial momentum. However, as soon as it is ionized, the pulse is gone and the electron moves away from the parent ion without any magnetic drift. A relativistic drift, though reduced, arises only during the third short stage of the process of acceleration back to the parent ion. Thus, the tailored laser pulse scheme shows that strong attosecond pulse trains (APTs) employed as a driving field for HHG can be very useful to suppress the relativistic drift. The further development of this idea brought other setups for relativistic HHG: A proposal is to use two counterpropagating attosecond pulses. The first pulse ionizes the elec-

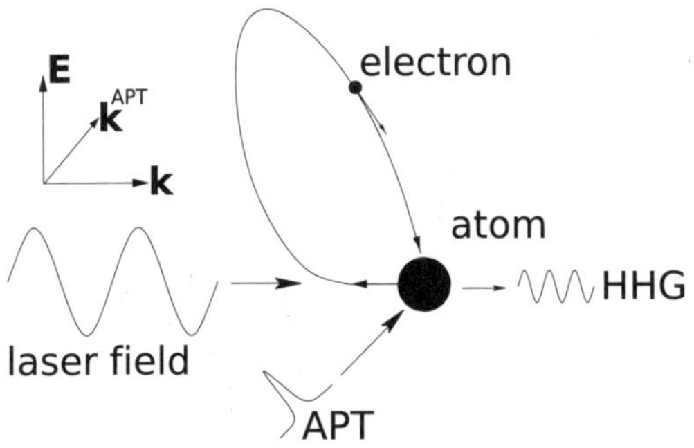

Fig. 3. Relativistic HHG setup where an XUV field of moderate intensity assists a super-strong laser field to produce harmonics in the hard x-ray region. The electron is one-photon-ionized by a XUV photon and obtains initial momentum to compensate the relativistic drift.

tron and drives it away from the ion. The electron undergoes a relativistic drift thereby. The second counterpropagating pulse reverts the drift and the electron recollides with the core.[25] At last, we mention a recent proposal[35] which seems to be not very difficult to realize experimentally. In this setup, a moderate XUV field with a photon energy of several hundreds of electron-volt assists the super-strong laser field to produce relativistic harmonics in the hard x-ray region. Here, the three step process begins with a single XUV photon ionization of an atom. Then, the electron moves in the strong IR laser field acquiring a large amount of energy. The important point is that the single XUV energy is chosen such that it also provides a large initial momentum to the ionized electron, which can be sufficient to compensate the further relativistic drift of the electron in the IR laser field if the polarization of the XUV field is along the propagation direction of the strong laser wave (see Fig. 3). With the compensated drift, the recollision becomes probable and the HHG yield is enhanced.

Proposal of a Standing Wave Configuration. Another candidate for entering HHG in the relativistic regime is a standing wave configuration consisting of two equal-handed circularly polarized counterpropagating waves (see Fig. 4). Originally, it was proposed by Milosevic *et al.*[36] In this field configuration, the electric and magnetic fields are aligned in each point in space. Thus, the Lorentz force vanishes, i.e. the usual relativistic drift

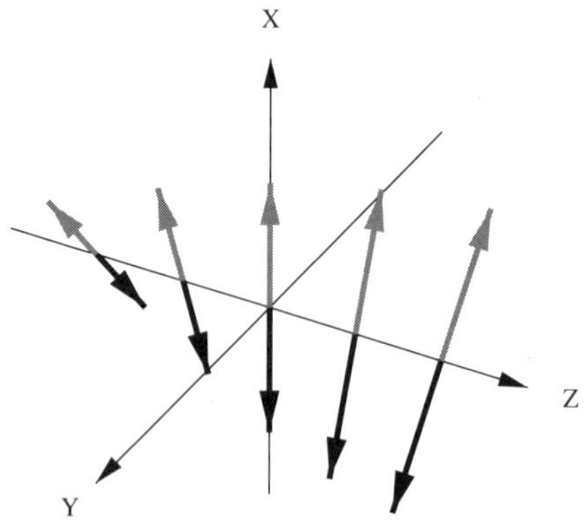

Fig. 4. The electric (black) and magnetic (gray) fields are shown for two equal-handed counterpropagating circularly polarized light waves. The magnetic field is always antiparallel to the electric field. A particle moving along the electric field experiences no Lorentz force or is focused additionally. It is plotted at the time where $\sin(\omega t) = \cos(\omega t)$.

along the laser propagation direction is absent since the accelerated electron moves along the magnetic field lines. Moreover, the Lorentz force acts as a focusing force because of the electron initial transversal momentum. It minimizes the spreading of the wave packet and increases the recollision probability. Therefore, this field configuration shows no drift and allows for the recollision of the ionized electron with parent ion without limitations in terms of laser intensity and can be applied to initiate high-energy processes with a single atom. The recolliding electron with MeV energy, for example, can induce nuclear reactions[36] or the electron and positron from a positronium atom driven by this field can induce muon production at the recollision.[37] However, applicability of this scheme to HHG was not clear. Firstly, so far, there is no fully quantum mechanical description of the problem of HHG in this field configuration and, secondly, possibilities of phase matching are not analyzed. The second part of this paper is devoted to the investigation of HHG in this setup employing the field of two equal handed circularly polarized counterpropagating waves.

3. HHG in a Circularly Polarized Standing Wave

In this Section, the HHG response of an atom to a circularly polarized standing wave with the vector potential

$$\boldsymbol{A}_L(\boldsymbol{r},t) = -E_0\frac{c}{\omega}\sin(\omega t)\Big(\cos(kz)\hat{\boldsymbol{x}} - \sin(kz)\hat{\boldsymbol{y}}\Big) \tag{3}$$

is calculated. We adopt the SAE approach well-justified in IR strong field physics.[38] A fully quantum mechanical calculation is analytically treatable within SFA.[22] The latter assumes that after ionizationF, the electron is only affected by the laser field and that the influence of the laser field on the bound state of the atom is negligible. Usually, for the description of the ionized electron motion in the laser field, the Volkov solution is applied.[39] Unfortunately, an exact solution of the Klein-Gordon or Dirac equation does not exist for an electron in a standing laser wave. Therefore, in order to calculate the evolution of the wave packet in the laser field, an expansion of the laser field in multipole moments is applied also taking into account the mass shift in a $1/c^2$ correction. This is applicable in the weakly relativistic regime restricting the relativistic parameter of the laser field to values below $\xi < 0.3$. The amplitude of HHG in SFA is given by:[40]

$$M = -i\int_{-\infty}^{\infty} dt \int_{-\infty}^{t} dt' \int d^3\boldsymbol{r} \int d^3\boldsymbol{r}'\Phi_0^*(\boldsymbol{r},t)\tilde{V}_H G_f(\boldsymbol{r},t;\boldsymbol{r}',t')V\Phi_0(\boldsymbol{r}',t'), \tag{4}$$

where the wave function $\Phi_0(\boldsymbol{r},t) = \phi_0(\boldsymbol{r})\exp[i(I_p t - \boldsymbol{r}\boldsymbol{A}(0,t)/c)]$ is chosen to be an eigenstate of the physical energy operator as we employ the gauge-invariant SFA[41] (see also Ref. 42); $\phi_0(\boldsymbol{r})$ is the stationary ground-state wave function of the electron in a zero range-potential (three-dimensional δ-potential) $V = \frac{2\pi}{\kappa}\delta(\boldsymbol{r})\partial_r r$,[43] with $\kappa = \sqrt{2I_p}$; $V_H(\boldsymbol{r},t) = \frac{\boldsymbol{A}_H(\boldsymbol{r},t)}{c}\Big(-i\nabla + \frac{\mathbf{A}(0,t)}{c}\Big)$ is the Hamiltonian of the harmonic field interaction with the electron, with a vector potential $\boldsymbol{A}_H(\boldsymbol{r},t) = c\sqrt{\frac{2\pi}{\omega_H}}\hat{\boldsymbol{e}}_H\exp(i\omega_H t - i\boldsymbol{k}_H\boldsymbol{r})$, and $G_f(\boldsymbol{r},t;\boldsymbol{r}',t')$ is the Green function of the electron in the standing laser field. We use the momentum representation of the ground state wave function $\tilde{\phi}_0(\boldsymbol{p}) = \langle\phi_0|\boldsymbol{p}\rangle = \frac{\sqrt{\kappa/\pi}}{\kappa^2+p^2}$ as well as the matrix element of the binding potential in momentum space $\langle\boldsymbol{p}|V|\phi_0\rangle = -\frac{\sqrt{\kappa}}{2\pi}$.[34] The Green function of the electron in the laser field is given by

$$G_f(\boldsymbol{r},t;\boldsymbol{r}',t') = -i\Theta(t-t')\int d^3\boldsymbol{p}\Psi^f_{\boldsymbol{p}}(\boldsymbol{r},t)(\Psi^f_{\boldsymbol{p}})^*(\boldsymbol{r}',t'), \tag{5}$$

where $\Psi^f_{\boldsymbol{p}}(\boldsymbol{r},t)$ is the solution of the Schrödinger equation in the laser field with the non-dipole interaction Hamiltonian including relativistic corrections

$$H^f = \frac{1}{2}\Big(-i\nabla + \frac{\boldsymbol{A}(\boldsymbol{r},t)}{c}\Big)^2 - \frac{1}{8c^2}\Big(-i\nabla + \frac{\boldsymbol{A}(\boldsymbol{r},t)}{c}\Big)^4. \tag{6}$$

In the weakly relativistic regime, the vector potential can be safely expanded about the parent ion since the potential drift distance is small. This renders an analytic solution of the Schrödinger equation possible. The expansion is carried out up to an order of $1/c^2$ and, consequently, the relativistic mass shift is also accounted for since it has the same order of magnitude. After the expansion, the Hamiltonian reads

$$\begin{aligned} H^f = &-\frac{1}{2}\triangle - i\frac{A(0,t)}{c}\Big((1-\frac{1}{2}k^2z^2)\frac{\partial}{\partial x} - kz\frac{\partial}{\partial y}\Big) \\ &+ \frac{A^2(\omega t)}{2c^2} - \frac{1}{8c^2}\Big(-i\nabla + \frac{\boldsymbol{A}(0,t)}{c}\Big)^4. \end{aligned} \tag{7}$$

In order to solve the corresponding Schrödinger equation, the following ansatz is used

$$\Psi^f_{\boldsymbol{p}}(\boldsymbol{r},t) = \frac{1}{(2\pi)^{(3/2)}}\exp i\Big(\boldsymbol{p}\boldsymbol{r} + w(t) + g(t)kz + h(t)k^2z^2\Big). \tag{8}$$

The term $\boldsymbol{pr}$ expresses the plane wave properties of the wave function, whereas the Taylor expansion shall cover the impact of the vector potential on the Hamiltonian. Therefore, the expansion of the vector potential and the exponent of (8) are of the same order. It is expected that the functions

$w(t)$, $g(t)$ and $h(t)$ vanish if the electric field is adiabatically switched off, as in this case, the plane wave shall be recovered. Inserting the ansatz into the Schrödinger equation and solving each order of z^n $(n = 0, 1, 2)$ independently, leads to

$$\begin{aligned} w(t) \approx & -\frac{\boldsymbol{p}^2}{2}t + \frac{\boldsymbol{p}^4}{8c^2}t - \frac{E_0 p_x \cos(\omega t)}{\omega^2} - \frac{E_0^2 t}{4\omega^2} + \frac{E_0^2 \sin(2\omega t)}{8\omega^3} \\ & - \frac{E_0 p_y p_z \sin(\omega t)}{c\omega^2} - \frac{3E_0^2 p_x^2 \sin(2\omega t)}{16c^2\omega^3} - \frac{3E_0^2 p_y^2 \sin(2\omega t)}{16c^2\omega^3} \\ & - \frac{E_0^2 p_z^2 \sin(2\omega t)}{16c^2\omega^3} + \frac{3E_0^2 p_x^2 t}{8c^2\omega^2} - \frac{E_0^2 p_y^2 t}{8c^2\omega^2} + \frac{E_0^2 p_z^2 t}{8c^2\omega^2} + \frac{3E_0^3 p_x \cos(\omega t)}{8c^2\omega^4} \\ & - \frac{E_0^3 p_x \cos(3\omega t)}{24c^2\omega^4} + \frac{E_0 p_x^3 \cos(\omega t)}{2c^2\omega^2} + \frac{E_0 p_x p_y^2 \cos(\omega t)}{2c^2\omega^2} \\ & - \frac{E_0 p_x p_z^2 \cos(\omega t)}{2c^2\omega^2} + \frac{3E_0^4 t}{64c^2\omega^4} - \frac{E_0^4 \sin(2\omega t)}{32c^2\omega^5} + \frac{E_0^4 \sin(4\omega t)}{256c^2\omega^5} \end{aligned} \tag{9}$$

$$g(t) = -\frac{E_0 p_x p_z \sin(\omega t)}{c\omega^2} + \frac{E_0 p_y \cos(\omega t)}{\omega^2} \tag{10}$$

$$h(t) = \frac{E_0 p_x \cos(\omega t)}{2\omega^2}. \tag{11}$$

From the matrix element (4), the harmonic emission rate per solid angle is found from the Fermi's golden rule. The probability of harmonic photon emission per unit time is given by

$$\frac{dw_n}{d\Omega} = \frac{1}{(2\pi)^3} n\omega \frac{\omega^2}{c^3} |M_n|^2, \tag{12}$$

where the amplitude of the emission reads

$$M_n = \int_{-\pi}^{\pi} d\eta \int_{-\infty}^{\eta} d\eta' \int d^3\boldsymbol{p}\, \tilde{m}(\boldsymbol{p}, \eta, \eta') e^{i(-S(\boldsymbol{p}, \omega t, \omega t') + n\eta)}, \tag{13}$$

with $\eta = \omega t$ and

$$\begin{aligned} \tilde{m}(\boldsymbol{p}, \eta, \eta') = & -\frac{1}{\omega^2}(p_x + A(0, \eta/\omega)/c) \langle \phi_0(\eta) | \boldsymbol{p} + g(\eta/\omega) k\hat{z} \\ & + \frac{A(0, \eta/\omega)\hat{x}}{c} - \boldsymbol{k}_H \rangle \\ & \times \left\langle \boldsymbol{p} + g(\eta'/\omega) k\hat{z} + \frac{A(0, \eta'/\omega)\hat{x}}{c} \middle| V \middle| \phi_0(\eta') \right\rangle. \end{aligned} \tag{14}$$

3.1. *Saddle point integration*

As long as we are in the low frequency tunneling regime $\omega \ll I_p \ll U_p$, the integral (13) can be calculated by means of the saddle point integration (SPI). In fact, for the employed parameters: $\omega = 0.05$ a.u., the ionization potential of a Li^{2+} ion $I_p = 4.51$ a.u., the low frequency tunneling regime is valid. The SPI approximation assumes that the main contributions to the integral are around the saddle points $(\boldsymbol{p}_s, \eta_s, \eta'_s)$ which are the extremum points of the exponent: $\partial_i S(\boldsymbol{p}_s, \eta_s, \eta'_s) = 0$, where i can be p_x, p_y, p_z, η or η'. The SPI yields the following expression

$$M \approx \sum_s \sqrt{\frac{(2\pi i)^5}{\det(\partial_i \partial_j S(\boldsymbol{p}_s, \eta_s, \eta'_s))}} \tilde{m}(\boldsymbol{p}, \eta, \eta') e^{i(-S(\boldsymbol{p},\omega t,\omega t')+n\eta)}, \tag{15}$$

where $\det \partial_i \partial_j S(\boldsymbol{p}_s, \eta_s, \eta'_s)$ is the functional determinant of the classical action. The SPI means the application of the WKB approximation assuming that the electron wave packet evolution in the continuum takes place along the classical trajectories. The two saddle point equations in phase have to be solved numerically, whereas the momentum saddle point equations can be solved by a perturbation expansion with the ansatz

$$\boldsymbol{p}_s = \begin{pmatrix} p_{x,0} + p_{x,1}/c + p_{x,2}/c^2 + O(1/c^4) \\ p_{y,0} + p_{y,1}/c + p_{y,2}/c^2 + O(1/c^4) \\ p_{z,0} + p_{z,1}/c + p_{z,2}/c^2 + O(1/c^4) \end{pmatrix}. \tag{16}$$

Inserting this into the saddle point equations $\partial_i S(\boldsymbol{p}_s, \eta_s, \eta'_s) = 0$ and solving each order of $1/c$ independently gives

$$\begin{aligned} p_{x,s} = {} & \frac{E_0(\cos(\eta') - \cos(\eta))}{(\eta - \eta')\omega} + \frac{E_0^3(\cos(\eta') - \cos(\eta))}{24c^2(\eta - \eta')^3\omega^3} \\ & \times \Big((2((\eta - \eta')^2 - 6)\cos(2\eta) + 2(5(\eta - \eta')^2 + 24\cos(\eta)\cos(\eta') \\ & - 6\cos(2\eta') - 12) + (\eta - \eta')(2(\eta - \eta')(2\cos(\eta)\cos(\eta') + \cos(2\eta')) \\ & - 9\sin(2\eta) + 9\sin(2\eta')))\Big) \end{aligned} \tag{17}$$

and $p_{y,s} = p_{z,s} = 0$. The vanishing saddle points $p_{y,s}$ and $p_{z,s}$ prove that the relativistic drift plays no role.

3.2. *Spectra*

The resulting spectral emission rates are shown in Fig. 5 (red curve) for different fundamental laser intensities. Additionally, spectra for the linearly polarized traveling wave (green) and the dipole approximation (black) are shown for comparison.

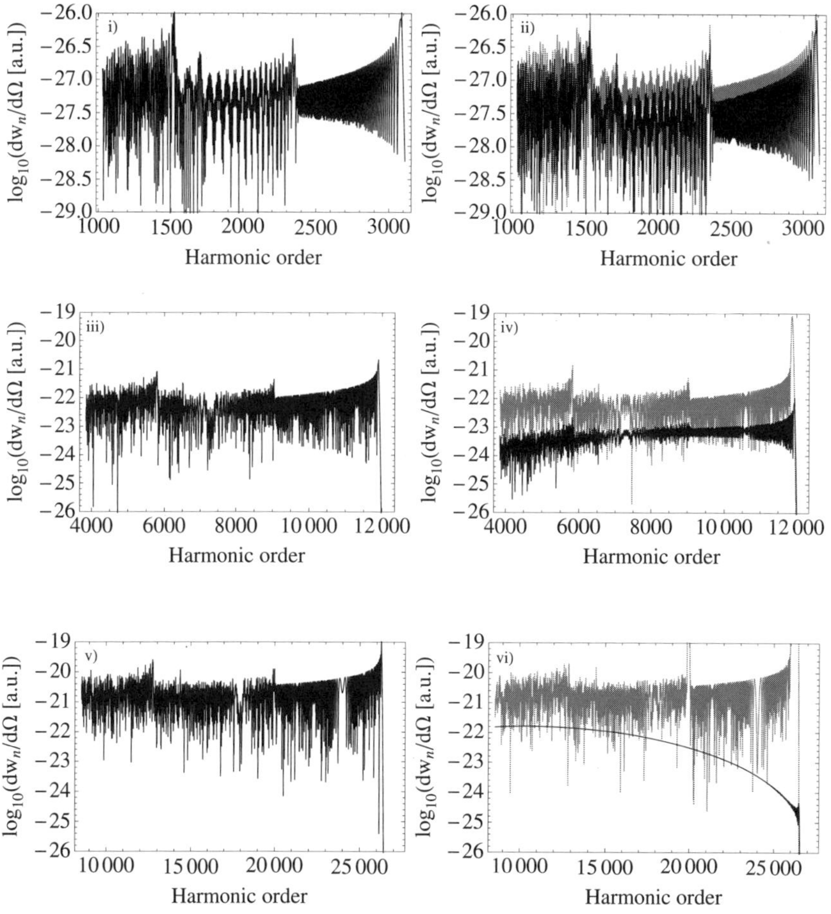

Fig. 5. HHG spectra for different fundamental laser intensities: i) and ii) $\xi = 0.1$, $I = 1.6 \times 10^{16} \frac{\mathrm{W}}{\mathrm{cm}^2}$, iii) and iv) $\xi = 0.2$, $I = 6.6 \times 10^{16} \frac{\mathrm{W}}{\mathrm{cm}^2}$, v) and vi) $\xi = 0.3$, $I = 1.5 \times 10^{17} \frac{\mathrm{W}}{\mathrm{cm}^2}$. The left panel shows the emission spectra for the counterpropagating circularly polarized laser fields beyond the DA for different laser intensities. For comparison, in the right panel HHG spectra for a conventional linearly polarized travelling wave (black) with the relativistic mass shift and non-dipole corrections and (gray) the same within the DA are shown. The emission rate of the conventional field gets steadily suppressed compared to the standing wave as the laser intensity increases. The absolute yields increase with ξ because of the increased tunneling probability, however, limited by depletion of the quantum state.

Most importantly, one can see that the plateaus of the black and red curves have roughly the same height. This means that higher-order multipole moments have no influence on the harmonic emission rate on the plateau for this field configuration. Only a change in the cutoff energy is observed which is expected from the $1/c^2$ correction of the saddle point $p_{x,s}$. Moreover, in all plots, there are several cutoffs visible below the final cutoff at the maximum energy. They are characterized by a sudden ascent and a step down afterwards and correspond to trajectories which visited the vicinity of the core several times. The three graphs show the pathway from the non-relativistic regime to the weakly relativistic regime. This can be observed from the green curve. Due to the magnetic drift, the harmonic plateau gets suppressed with increasing intensity. Fig. 5 iii) also indicates that the interference between different trajectories makes a contribution even at higher laser intensities in the standing wave configuration. This is not the case for the travelling plane wave (green) since the recombination probability of the short and long trajectories become too different. We see that at high intensities of $I = 1.5 \times 10^{17} \frac{\mathrm{W}}{\mathrm{cm}^2}$ about 25000 harmonics of the driving field can be produced. The cutoff photon energy is equal to a 34 keV which is about one order higher than current experimental achievements.

3.3. *Phase matching*

So far, harmonic radiation from single atoms has been considered. Due to the driftless electron motion in this field configuration, the single-atom response with respect to HHG is significant in the relativistic regime. In the experiment, harmonic radiation is emitted from a gas target situated in the laser beam. Therefore, the overall harmonic yield is not only determined by the single-atom response but also by the way harmonics from different atoms interfere with each other. From this point of view, the setup is problematic: The electric field triggering the HHG process has the same strength in the whole medium at any time. Only the polarization directions is tilted. Consequently, the harmonics are emitted in the whole medium at the same time but from different spatial points and, therefore, with different phases $\eta = \omega t - kz$. Nevertheless, the coherence length is non-vanishing since the harmonic polarization directions also spin around the propagation axis and allows for emission from a quarter laser period if the medium density is sharply modulated with precision of the harmonic wavelength. The number of coherently emitting atoms can be estimated to be $\rho L w_0^2 \lambda_H / \lambda$, where ρ is the density of atoms, L the interaction length along the z-axis, λ_H and λ the wavelengths of the harmonic and the laser fields, respectively. In this field configuration, the role of the free electron background due to ionized electrons is not crucial.

4. Conclusion

High harmonic generation (HHG) from gases is a state-of-the-art source of short-wavelength coherent radiation. On the road to shorter harmonic wavelengths, it is crucial to overcome the magnetic drift which arises once relativistic fundamental laser intensities are reached. The drift is caused by the strong magnetic field and strongly suppresses the recollision probability. An auspicious laser geometry circumventing this is a standing laser wave configuration consisting of equal-handed circularly polarized waves. Here, the electric and magnetic field are parallel and hence from the classical point of view, no drift is expected.

In the weakly relativistic regime, it was shown that this configuration overcomes the suppression. Due to the symmetry, it is clear that no drift of the wave packet center would occur. Additional wave packet spreading effects due to the multipole moments do not appear in this regime. It has been proven that the only consequence is a slightly lower cutoff compared to the dipole approximation (DA). However, one has to note that the single-atom response is only the first step in studying the feasibility. The harmonic efficiency is not only determined by this but also by the way light from different atom interferes with each other. Macroscopic HHG can only be expected with a proper phase matching which is problematic in this setup. From this point of view, one can conclude that for the phase matched HHG in the relativistic regime the setups employing counterpropagating APTs[25] or the XUV photon-assisted HHG in a strong laser field[35] are more promising.

References

1. P. Franken, A. Hill, C. Peters and G. Weinreich, Phys. Rev. Lett. **7**, 118 (1961).
2. A. McPherson *et al.*, J. Opt. Soc. Am. B **4**, 595 (1987).
3. J. Seres, *et al.*, Nature **433**, 596 (2005).
4. P. Agostini and L. F. DiMauro, Rep. Prog. Phys. **67**, 813 (2004).
5. Y. Mairesse *et al.*, Science **28**, 302 (2003); E. Goulielmakis *et al.*, Science **320**, 1614 (2008).
6. F. Krausz and M. Ivanov, Rev. Mod. Phys. **81**, 163 (2009).
7. A. L'Huillier, *et al.*, Eur. Phys. J. D **26**, 91 (2003).
8. A. Mokhtari, P. Cong, J. L. Herek, and A. H. Zewail, Nature **348**, 225 (1990).
9. Th. Ergler, *et al.*, Phys. Rev. Lett **95**, 093001 (2005).

10. M. Drescher *et al.*, Nature **419**, 803 (2002).
11. M. Hentschel *et al.*, Nature **414**, 509 (2001).
12. R. Kienberger *et al.*, Nature **427**, 817 (2004).
13. V. Yanovsky, *et al.*, Opt. Express **16**, 2109 (2008).
14. `http://www.extreme-light-infrastructure.eu/`
15. `http://tesla.desy.de/new_pages/TDR_CD/start.html`
16. `http://lcls.slac.stanford.edu/`
17. `http://www-xfel.spring8.or.jp/`
18. T. Popmintchev *et al.*, Proc. Nat. Acad. Sci. USA **106**, 10516 (2009).
19. Y. I. Salamin, *et al.*, Phys. Reports **427**, 41 (2006).
20. P. B. Corkum, Phys. Rev. Lett. **71**, 1994 (1993).
21. L. V. Keldysh, Sov. JETP **20**, 1307 (1964); F. H. M. Faisal, J. Phys. B **6**, L89 (1973); H. R. Reiss, Phys. Rev. A **22**, 1786 (1980).
22. H. R. Reiss, Phys. Rev. A **42**, 1476 (1990); J. Opt. Soc. Am. B **7**, 574 (1990).
23. M. Lewenstein, *et al.*, Phys. Rev. A. **49**, 2117 (1994).
24. S. Palaniyappan, *et al.*, Phys. Rev. A **74**, 033403 (2006).
25. K. Z. Hatsagortsyan, *et al.*, J. Opt. Soc. Am. B **25**, 93 (2008).
26. M. W. Walser, *et al.*, Phys. Rev. Lett. **85**, 5082 (2000).
27. B. Henrich, K. Z. Hatsagortsyan and C. H. Keitel, Phys. Rev. Lett. **93**, 013601 (2004).
28. R. Fischer, M. Lein, and C. H. Keitel, Phys. Rev. Lett. **97**, 143901 (2006); J. Phys. B **40**, F113 (2007); J. Mod. Opt. **54**, 1911 (2007).
29. G. Mocken, and C. H. Keitel, J. Phys. B **37**, L275 (2004); C. C. Chirilă, *et al.*, Phys. Rev. Lett. **93**, 243603 (2004).
30. Q. Lin, S. Li, and W. Becker, Opt. Lett. **31**, 2163-2165 (2006).
31. N. J. Kylstra, *et al.*, Phys. Rev. Lett. **85**, 1835 (2000); V. D. Taranukhin, Laser Phys. **10**, 330-336 (2000); V. D. Taranukhin and N. Yu. Shubin, J. Opt. Soc. Am. B **19**, 1132 (2002).
32. C. C. Chirilă, *et al.*, Phys. Rev. A **66**, 063411 (2002).
33. M. Klaiber, K. Z. Hatsagortsyan, and C. H. Keitel, Phys. Rev. A **74** 051803 (2006).
34. M. Klaiber, K. Z. Hatsagortsyan, and C. H. Keitel, Phys. Rev. A **75**, 063413 (2007).
35. M. Klaiber, *et al.*, Opt. Lett. **33**, 411 (2008).
36. N. Milosevic, P. B. Corkum, and T. Brabec, Phys. Rev. Lett. **92**, 013002 (2004).
37. K. Z. Hatsagortsyan, C. Müller and C. H. Keitel, Europhys. Letters **76**, 29 (2006); C. Müller, K. Z. Hatsagortsyan and C. H. Keitel, Physics Letters B **659**, 209 (2008).
38. K. C. Kulander, Phys. Rev. A **38**, 778 (1988); G. G. Paulus, *et al.*, Phys. Rev. Lett. **72**, 2851(1994).
39. D. M. Volkov, Z. Physik **94**, 250(1935); see also in V. B. Berestetskii, L. P. Pitaevskii, E.M. Lifshitz, *Quantum Electrodynamics* (Butterworth-Heinemann, Oxford, 1999) p.148.
40. D. B. Milošević, S. Hu, and W. Becker, Phys. Rev. A **63**, 011403(R) (2001); Laser Phys. **12**, 389 (2002).

41. M. Klaiber, K. Z.Hatsagortsyan, C. H. Keitel, Phys. Rev. A **73** 053411 (2006).
42. F. H. M Faisal, J. Phys. B **40** (2007) F145; Phys. Rev. A **75**, 063412 (2007).
43. W. Becker, S. Long, J. K. McIver, Phys Rev. A **50**, 1540 (1994).

ENTANGLED LIGHT AND MATTER WAVES VIA NON-LINEAR INTERACTIONS

M. MACOVEI*, G. YU. KRYUCHKYAN* and G.-X. LI**

*Max-Planck-Institut für Kernphysik, Saupfercheckweg 1, D-69117 Heidelberg, Germany

*Yerevan State University, A. Manookyan 1, 375025 Yerevan, Armenia

**Department of Physics, Huazhong Normal University, 430079 Wuhan, China

We discuss different schemes on matter or light generation with unusual properties. First, we shall demonstrate the generation of two-particle entanglement via external field pumping and damping through a squeezed electromagnetic field reservoir. Then, continuous variable entanglement of the generated field of a pumped multi-atom sample will be reported. Further, in a different setup, a particular attention will be devoted to spatial entanglement of matter waves. Various entangled interference patterns will be described and analyzed.

Keywords: Entanglement, multi-particle systems, matter waves.

1. Introduction

Non-linear interactions play an essential role to entangle light or matter waves modes, for instance, via spontaneous parametric down-conversion processes, multi-wave mixing phenomena or very intense laser-vacuum interactions. Potential applications of entanglement in quantum information processing makes it attractive and it is widely investigated. However, this topic is difficult due to the fact that a general criterion quantifying the entanglement in such systems does not exist. One may consider the state that cannot be represented as a factorized product of individual states describing each particle separately as an entangled multi-particle state. As a consequence, the condition for entanglement of formation of an arbitrary state of two-qubits[1] and an inseparability criterion for continuous variable systems[2] were proposed. Other criteria are, respectively, the entanglement of distillation,[3] the relative entropy of entanglement[4] or the so-called negativity.[5] Based on these formulations, numerous investigations were performed in order to describe the entanglement in various systems. In particular, the electromagnetically induced transparency has been used to generate narrow-band entangled photons[6] while a four-wave parametric interaction in a two-level system was shown to generate highly non-classical light.[7] The Kerr non-linearity was exploited to generate two-mode entangled light in an optical fiber[8] or cavities,[9] respectively. Entanglement induced by a single-mode heat environment was shown to occur in,[10] and entanglement between two atoms can be effectively created in an overdamped cavity injected with squeezed vacuum.[11] Two initially entangled and afterward not interacting qubits can become completely disentangled in a finite time.[12] Also, dark periods and revivals of entanglement in a two-qubit system were shown to occur due to the coupling to environmental vacuum modes.[13] Further, the motion of $2N$ ions trapped in two separate single mode cavities can be entangled via a two-mode squeezed vacuum field.[14] Schemes to produce entangled motional states for the two trapped ions in one or two cavities were proposed, too.[15] The entanglement of two particles inside an atomic cloud was demonstrated in Ref.[16] while spatially entangled atomic deflections in twin-photon light beams were shown to occur in.[17] An experimental demonstration of continuous variable entanglement using cold atoms in a high-finesse optical cavity was presented in,[18] and experimental evidence of quantum entanglement of a large number of photons was found in Ref.[19] Also, the entanglement of formation for an arbitrary two-mode Gaussian state and entangled light via non-linear vacuum-multi-particle interactions were demonstrated in[20] and,[21] respectively. Note, however, that in spite of a large amount of contributions to the subject there are still a lot of discussions about it, particularly on quantifying entanglement.

Here, we investigate the possibility to generate entanglement in various systems. We start our investigations with a simple Dicke-like two-atom system. We demonstrate the generation of two-particle entanglement via external field pumping and damping through a squeezed electromagnetic field reservoir by calculating the concurrence. Then, continuous variable entanglement of the generated field of a pumped multi-atom sample in a two-mode resonator will be analyzed. Further, in a different setup, a particular attention will be devoted to spatial entanglement of matter waves. Various entangled interference patterns will be described and analyzed.

2. Two-Particle Quantum Entanglement

We consider a pair (a and b) of two-state ($|e\rangle$ and $|g\rangle$) sample pumped with a coherent laser field and damped via its interaction with the environmental squeezed electromagnetic field reservoir. The atomic frequencies are identical and equal to ω_0, while the laser frequency is ω_L and the carrier frequency of the squeezed bath is ω_s. The particles are placed in a volume with dimensions small compared to the emission wavelength. Such a model can be described in the Born-Markov, the rotating-wave and the electric-dipole approximations by the master equation approach:

$$\begin{aligned}\frac{d}{dt}\rho &= -i[H_0,\rho] - \gamma(\bar{N}+1)\{[S^+,S^-\rho]+[\rho S^+,S^-]\}\\ &- \gamma\bar{N}\{[S^-,S^+\rho]+[\rho S^-,S^+]\}\\ &- \gamma\bar{M}e^{2i\Delta_p t}\{[S^-,S^-\rho]+[\rho S^-,S^-]\}\\ &- \gamma\bar{M}^*e^{-2i\Delta_p t}\{[S^+,S^+\rho]+[\rho S^+,S^+]\},\end{aligned} \tag{1}$$

where

$$H_0 = \tilde{\Delta}S_z - \delta S^+S^- + \Omega(S^+ + S^-), \tag{2}$$

with $\tilde{\Delta} = \omega_0 - \omega_L + \delta \equiv \Delta + \delta$ and $\Delta_p = \omega_0 - \omega_s$. Here, Ω is the Rabi frequency and γ is the single-particle spontaneous emission decay rate. δ represents the dipole-dipole potential. $\bar{N}$ is the mean photon number in the squeezed reservoir while $\bar{M} = |\bar{M}|e^{i(\phi_s - 2\phi_L)}$ describes the two-particle excitation and de-excitation by the squeezed bath. ϕ_s and ϕ_L are the phases of the squeezed reservoir and laser field, respectively. The collective angular momentum operators, $S^\pm$ and S_z, are defined in the following way:

$$S^+ = \sum_{j\in a,b}\sigma_j^+, \quad S^- = (S^+)^\dagger, \quad S_z = \sum_{j\in a,b}\sigma_{zj}/2, \tag{3}$$

where the raising and lowering operators for each qubit are denoted by $\sigma_j^+ = |e\rangle_{jj}\langle g|$ and $\sigma_j^- = (\sigma_j^+)^\dagger$, while $\sigma_{zj} = |e\rangle_{jj}\langle e| - |g\rangle_{jj}\langle g|$. The collective operators obey the commutation relations for su(2) algebra, i.e. $[S_z, S^\pm] = \pm S^\pm$ and $[S^+, S^-] = 2S_z$.

Since our main interest here is the entanglement, we shall discuss the concurrence for a mixed state of qubits $\{a,b\}$ with density matrix ρ_{ab}, where the concurrence C is defined as

$$C = \max\{0, \lambda_1 - \lambda_2 - \lambda_3 - \lambda_4\}, \tag{4}$$

where the quantities λ_i $(i \in \{1,2,3,4\})$ are the square roots of the eigenvalues of the matrix product

$$R = \rho_{ab}(\sigma_{ay} \otimes \sigma_{by})\rho^*_{ab}(\sigma_{ay} \otimes \sigma_{by}), \tag{5}$$

in descending order. Here, ρ^*_{ab} denotes complex conjugation of ρ_{ab}, and σ_{iy} are Pauli matrices for the two-level systems $\big(i \in \{a, b\}\big)$. The values of the concurrence range from zero for an unentangled state to unity for a maximally entangled two-particle state.[1]

The state of two qubits which is symmetric under the exchange of the sub-systems can be represented in the basis $\{|e_a e_b\rangle, |e_a g_b\rangle, |g_a e_b\rangle, |g_a g_b\rangle\}$ as follows·

$$\rho_{ab} = \begin{pmatrix} \rho_{11} & \rho_{12} & \rho_{13} & \rho_{14} \\ \rho_{21} & \rho_{22} & \rho_{23} & \rho_{24} \\ \rho_{31} & \rho_{32} & \rho_{33} & \rho_{34} \\ \rho_{41} & \rho_{42} & \rho_{43} & \rho_{44} \end{pmatrix}. \tag{6}$$

The matrix elements of ρ_{ab} can be determined by:[16]

$$\begin{aligned}
\rho_{11} &= \frac{N^2 - 2N + 4\langle S_z^2\rangle + 4(N-1)\langle S_z\rangle}{4N(N-1)}, \\
\rho_{12} = \rho_{13} &= \frac{N\langle S^+\rangle + 2\langle S^+ S_z\rangle}{2N(N-1)}, \\
\rho_{14} &= \frac{\langle (S^+)^2\rangle}{N(N-1)}, \quad \rho_{21} = (\rho_{12})^*, \\
\rho_{22} = \rho_{23} &= \frac{N^2 - 4\langle S_z^2\rangle}{4N(N-1)}, \\
\rho_{24} &= \frac{\langle S^+\rangle(N-2) - 2\langle S^+ S_z\rangle}{2N(N-1)}, \\
\rho_{31} &= (\rho_{13})^*, \quad \rho_{32} = (\rho_{23})^*, \quad \rho_{33} = \rho_{22}, \\
\rho_{34} &= \rho_{24}, \quad \rho_{41} = (\rho_{14})^*, \quad \rho_{42} = (\rho_{24})^*, \\
\rho_{43} &= (\rho_{34})^*, \\
\rho_{44} &= \frac{N^2 - 2N + 4\langle S_z^2\rangle - 4(N-1)\langle S_z\rangle}{4N(N-1)},
\end{aligned} \tag{7}$$

with $N = 2$ in our case, and

$$S_z^2 + (S^+S^- + S^-S^+)/2 = 2. \tag{8}$$

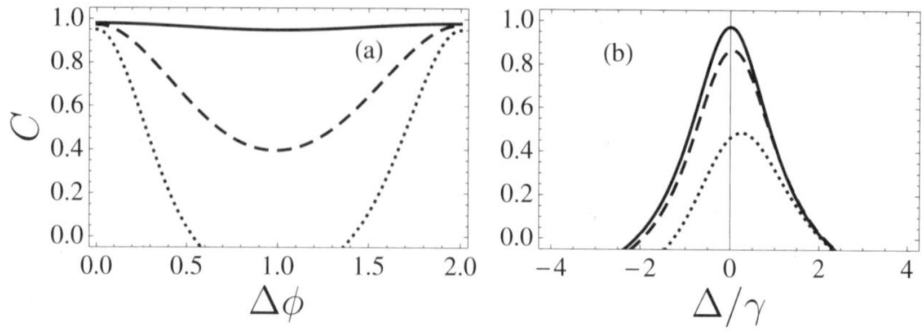

Fig. 1. The steady-state two-particle concurrence. (a) Here, $\Delta\phi = \phi_s - 2\phi_L$ and $\Delta/\gamma = 0$. The solid curve stands for $\Omega/\gamma = 0.1$, the dashed one for $\Omega/\gamma = 0.5$ while the dotted line corresponds to $\Omega/\gamma = 1$. (b) Here, $\Delta\phi = 0$. The solid curve is for $\Omega/\gamma = 0.5$, the dashed one stands for $\Omega/\gamma = 2$ and the dotted line depicts the case of $\Omega/\gamma = 5$. Other parameters are: $\delta/\gamma = 5$, $\bar{N} = 2$, and $|\bar{M}| = \sqrt{\bar{N}(\bar{N}+1)}$.

Further, we present the equations of motion for the atomic variables that enter in Eq. (7) which follow from the master equation (1):

$$\begin{aligned}
\frac{d}{dt}\langle S_z\rangle &= i\Omega(\langle S^-\rangle - \langle S^+\rangle) - 2\gamma\langle S^+S^-\rangle - 4\gamma\bar{N}S_z, \\
\frac{d}{dt}\langle S^+\rangle &= (i\tilde{\Delta} - 2\gamma\bar{N})\langle S^+\rangle + 2(\gamma + i\delta)\langle S^+S_z\rangle - 2i\Omega\langle S_z\rangle + 2\gamma\bar{M}\langle S^-\rangle, \\
\frac{d}{dt}\langle S^+S^-\rangle &= 2i\Omega(\langle S^+S_z\rangle - \langle S_zS^-\rangle) - 8\gamma(1 + 3\bar{N}/2)\langle S^+S^-\rangle \\
&\quad + 8\gamma(1+\bar{N})\langle S_z\rangle + 2\gamma\bar{M}\langle S^{-^2}\rangle + 2\gamma\bar{M}^*\langle S^{+^2}\rangle + 8\gamma(1+2\bar{N}), \\
\frac{d}{dt}\langle S_zS^-\rangle &= -(i\tilde{\Delta} - 2i\delta + 6\gamma(1 + 5\bar{N}/3))\langle S_zS^-\rangle + 2\gamma\bar{M}^*(3\langle S^+S_z\rangle + \langle S^+\rangle) \\
&\quad - 4\gamma(1+\bar{N})\langle S^-\rangle + i\Omega(\langle S^{-^2}\rangle + 4 + 2\langle S_z\rangle - 3\langle S^+S^-\rangle), \\
\frac{d}{dt}\langle S^{+^2}\rangle &= 2(i\tilde{\Delta} - i\delta - \gamma(1+2\bar{N}))\langle S^{+^2}\rangle - 2i\Omega(\langle S^+\rangle + 2\langle S^+S_z\rangle) \\
&\quad - 4\gamma\bar{M}(3\langle S_z\rangle - 3\langle S^+S^-\rangle + 4),
\end{aligned} \tag{9}$$

where $\Delta_p = 0$ and $\langle S^-\rangle = [\langle S^+\rangle]^\dagger$, $\langle S^+S_z\rangle = [\langle S_zS^-\rangle]^\dagger$, and $\langle S^{-^2}\rangle = [\langle S^{+^2}\rangle]^\dagger$. The relations (4)-(9) will help us to estimate the entanglement, i.e. the concurrence of the atomic pair.

Fig. (2) shows the steady-state behaviors of two-particle concurrence. Entanglement occurs for a variety of parameters. In particular, for weak laser fields, the concurrence is maximal and less sensitive to phase difference $\Delta\phi$ (see Fig. 2a). Furthermore, increasing the laser strength for $\Delta\phi = 0$ the concurrence decreases (see Fig. 2b). The asymmetry observed in Fig. (2)

is due to dipole-dipole interactions. Thus, here, we have shown efficient entanglement of a particle pair driven by a coherent source and damped through the environmental squeezed bath. The next Section, however, deals with the continuous variable entanglement of the generated electromagnetic field.

3. Light Entanglement via Non-linear Four-Wave Mixing Processes

Here, we investigate the generation of a strongly entangled electromagnetic field through laser pumping a collection of N two-level atoms, possesing the transition frequency ω_0, in a two-mode optical resonator (ω_1, ω_2) via a four-wave mixing process. The excited particles may decay spontaneously during the transitions $|e\rangle \leftrightarrow |g\rangle$, with a decay rate γ, due to interaction with the environmental vacuum modes. The collisional damping rate of atoms which alter the phase of the atomic state but not its population is given by γ_c. We shall describe the laser-atom system in the dressed-state representation of a single particle: $|e\rangle = \cos\theta|+\rangle - \sin\theta|-\rangle$ and $|g\rangle = \cos\theta|-\rangle + \sin\theta|+\rangle$ where $\cot 2\theta = \Delta/(2\Omega_0)$. Here, $\Delta = \omega_0 - \omega_L$ is the detuning of the laser frequency ω_L from the atomic transition frequency ω_0 while Ω_0 is the Rabi frequency describing the strength of the atom-laser interaction.

In the mean-field, dipole, Born-Markov and secular approximations the system is characterized by the master equation (1) with $\bar{N} = \bar{M} = 0$, which in the dressed-state picture takes the form:

$$\dot{\rho} + \frac{i}{\hbar}[H, \rho] = -\gamma_0[R_z, R_z\rho] - \gamma_+[R_{+-}, R_{-+}\rho] \\ - \gamma_-[R_{-+}, R_{+-}\rho] - \sum_{i\in\{1,2\}} \kappa_i[a_i^\dagger, a_i\rho] + H.c., \tag{10}$$

where $H = H_f + H_a + H_{int}$ with $H_f = \hbar\Delta_2 a_2^\dagger a_2 - \hbar\Delta_1 a_1^\dagger a_1$ being the free electromagnetic field modes Hamiltonian, $H_a = \hbar\Omega R_z$ is the free Hamiltonian of the laser-dressed atoms and

$$H_{int} = \{(F\cos^2\theta - F^\dagger \sin^2\theta)R_{+-} + H.c.\} \\ + (F + F^\dagger)R_z \sin 2\theta/2, \tag{11}$$

describes the interaction of the cavity modes with the dressed atomic sample. Here, $\Delta_1 = \omega_L - \omega_1$, $\Delta_2 = \omega_2 - \omega_L$, $\Omega = \sqrt{\Omega_0^2 + (\Delta/2)^2}$ and $F = \hbar(g_1 a_1 + g_2 a_2)$, where g_i are the atom-cavity couplings, while a_i $[a_i^\dagger]$ are the cavity photon annihilation [creation] operators, and $[a_i, a_l^\dagger] = \delta_{il}$,

$[a_i, a_l] = [a_i^\dagger, a_l^\dagger] = 0$, $\{i, l \in 1, 2\}$. The collective dressed-state atomic operators $R_{\alpha\beta} = \sum_{j=1}^{N} R_{\alpha\beta}^{(j)} = \sum_{j=1}^{N} |\alpha_j\rangle\langle\beta_j|$ describe the transition between the dressed-states $|\beta\rangle$ and $|\alpha\rangle$ for $\alpha \neq \beta$ and population for $\alpha = \beta$ $\{\alpha, \beta \in +, -\}$, and obey the standard commutation relations of su(2) algebra. $R_z = R_{++} - R_{--}$ is the dressed-state inversion operator. The quantum dissipations due to spontaneous emission into surrounding electromagnetic field modes as well as collisional damping are described by the terms proportional to $\gamma_0 = (\gamma \sin^2 2\theta + \gamma_c \cos^2 2\theta)/4$, $\gamma_+ = \gamma \cos^4 \theta + \gamma_c \sin^2 2\theta/4$ and $\gamma_- = \gamma \sin^4 \theta + \gamma_c \sin^2 2\theta/4$, respectively. The last term in Eq. (10) characterizes the damping of the cavity modes with κ_i being the field decay rates of the mode $i \in \{1, 2\}$.

We assume an intense pumping field, i.e. $\Omega \gg \{N\gamma, g_{1,2}\sqrt{N}\}$, and a high quality cavity such that $\gamma N \gg \kappa_{1,2}$. In this case, the atomic subsystem achieves its steady-state on a time scale faster than the cavity field and, thus, the atomic variables can be eliminated to arrive at a master equation for the cavity field modes alone:[21]

$$\begin{aligned}
\dot{\rho} - i\delta \sum_{i\in\{1,2\}} [a_i^\dagger a_i, \rho] = &- \sum_{i\in\{1,2\}} [A_i(\rho a_i a_i^\dagger - a_i^\dagger \rho a_i) \\
&+ \bar{B}_i(a_i^\dagger a_i \rho - a_i \rho a_i^\dagger)] + \sum_{i\neq j\in\{1,2\}} [C_i(a_j^\dagger a_i^\dagger \rho - a_i^\dagger \rho a_j^\dagger) \\
&+ D_i(\rho a_i^\dagger a_j^\dagger - a_j^\dagger \rho a_i^\dagger)] + H.c.
\end{aligned} \tag{12}$$

Here $\delta = (\Delta_2 - \Delta_1)/2$, $\bar{B}_i = B_i + \kappa_i$ $\{i \in 1, 2\}$ and

$$\begin{aligned}
A_1 &= \frac{Ng_1^2}{\gamma}[\chi_0(\bar{\Delta}_1) + \sin^4\theta\chi_-(\bar{\Delta}_1) + \cos^4\theta\chi_+(\bar{\Delta}_1)], \\
B_1 &= \frac{Ng_1^2}{\gamma}[\chi_0(\bar{\Delta}_1) + \cos^4\theta\chi_-^*(-\bar{\Delta}_1) + \sin^4\theta\chi_+^*(-\bar{\Delta}_1)], \\
C_1 &= \frac{Ng_1g_2}{4\gamma}[\sin^2 2\theta(\chi_-^*(\bar{\Delta}_1) + \chi_+^*(\bar{\Delta}_1)) - 4\chi_0(-\bar{\Delta}_1)], \\
D_1 &= \frac{Ng_1g_2}{4\gamma}[\sin^2 2\theta(\chi_-(-\bar{\Delta}_1) + \chi_+(-\bar{\Delta}_1)) - 4\chi_0(-\bar{\Delta}_1)].
\end{aligned} \tag{13}$$

Further, A_2, B_2, C_2 and D_2 can be obtained from A_1, B_1, C_1 and D_1 by replacing $\bar{\Delta}_1$ with $-\bar{\Delta}_2$, $\{\Delta_1, \Delta_2 \neq 0\}$, and $g_1 \leftrightarrow g_2$, respectively. Other parameters are:

$$\chi_{\mp}(z) = \frac{\langle R_{\mp\pm}R_{\pm\mp}\rangle_s/N^2}{\bar{\Gamma}_{\perp} \mp i(2\bar{\Omega} \mp z)}, \quad \chi_0(z) = \frac{\langle R_z^2\rangle_s/4N^2}{\bar{\Gamma}_{||} + iz}\sin^2 2\theta,$$

$\bar{\Gamma}_m = (\Gamma_m^{(s)} - \cos 2\theta\langle R_z\rangle_s)/N$ $\{m \in ||, \perp\}$, with $\Gamma_{||}^{(s)}$=$1 + \cos^2 2\theta + \gamma_c \sin^2 2\theta/\gamma$ and $\Gamma_{\perp}^{(s)} = 1 + \sin^2 2\theta/2 + \gamma_c(1 + \cos^2 2\theta)/(2\gamma)$ being the corresponding single-particle decay rates, and scaled parameters were introduced, i.e. $\bar{\Omega} = \Omega/(N\gamma)$, $\bar{\Delta}_i = \Delta_i/(N\gamma)$ $\{i \in 1, 2\}$. Note that to obtain Eq. (12), we decoupled the involved multi-particle correlators - an approximation valid for larger N, i.e. $N \gg 1$.[21] The corresponding equation for $N = 1$ (or independent atoms) is identical to Eq. (12) but with single-atom decay rates $\{\Gamma_{||}^{(s)}, \Gamma_{\perp}^{(s)}\}$ instead of collective ones. The steady-state expectation values for the atomic correlators entering into the above expressions can be estimated from the steady-state solution of the master equation describing the strongly pumped atoms in the absence of the cavity, i.e.

$$\rho_s^{(0)} = Z^{-1}\mathrm{Exp}[-\xi \mathrm{R_z}], \tag{14}$$

where $2\xi = \ln(\gamma_+/\gamma_-)$ and Z is chosen such that $\mathrm{Tr}\{\rho_s^{(0)}\} = 1$. Considering an atomic coherent state $|n\rangle$ denoting a symmetrized N-atom state in which $N - n$ particles are in the lower dressed state $|-\rangle$ and n atoms are excited to the upper dressed state $|+\rangle$, and that $R_{-+}|n\rangle = \sqrt{n(N - n + 1)}|n - 1\rangle$, $R_{+-}|n\rangle = \sqrt{(N - n)(n + 1)}|n + 1\rangle$, and $R_z|n\rangle = (2n - N)|n\rangle$ one can calculate the expectation values of any atomic correlators of interest[16,21] as soon as $\Omega \gg \{N\gamma, g_{1,2}\sqrt{N}\}$ and $N\gamma \gg \kappa_{1,2}$.

The physical meaning of the parameters in Eq. (12) is as follows. A ($\bar{B}$) describes the process of increasing (decreasing) of the photon number into cavity modes as well as the Stark shift. On the other hand, C and D characterize the creation and annihilation of a photon into each cavity mode as well as their correlations induced by the four-wave mixing effect. Finally, continuous variable entanglement between the two modes occurs if:

$$E = \sqrt{\langle a_1^{\dagger}a_1\rangle\langle a_2^{\dagger}a_2\rangle} - |\langle a_1 a_2\rangle| < 0. \tag{15}$$

To elucidate the role the collectivity plays to entangle the two-field modes, we shall consider further that $g_1 = g_2 \equiv g$, $\kappa_1 = \kappa_2 \equiv \kappa$ and $|\Delta_1| \approx |\Delta_2| \gg g\sqrt{N}$ so that $|\Delta_1 - \Delta_2| \sim \kappa$. We also assume that $|\bar{\Delta}_{1(2)}| \gg \bar{\Gamma}_{||}$ and $|2\bar{\Omega} \pm \bar{\Delta}_{1(2)}| \gg \bar{\Gamma}_{\perp}$. Then, the steady-state expectation values for the field correlators in Eqs. (15) are obtained from Eq. (12):

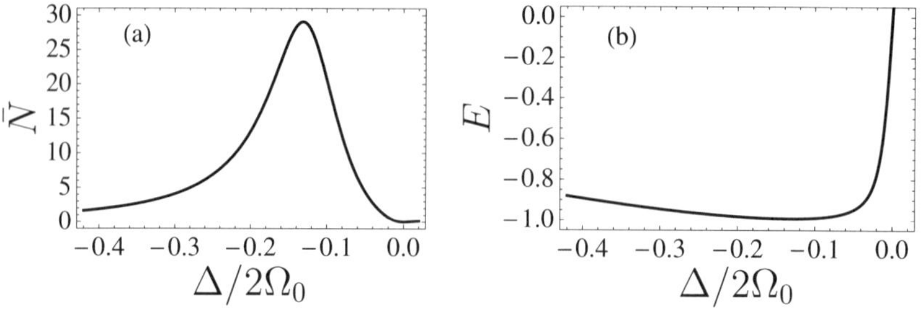

Fig. 2. (a) Mean number of entangled photons $\bar{N}$ and (b) entanglement parameter E as function of $\Delta/2\Omega_0$. Here $|\Delta_1| \approx |\Delta_2| = \Omega_0$, $Ng^2/\kappa\Omega_0 = 3.2$, $\gamma_c/\gamma = 0.2$, $\delta/\kappa = -1$ and $N = 75$.

$$\langle a_1^\dagger a_1\rangle = \langle a_2^\dagger a_2\rangle = \frac{(\bar{g}/\sqrt{2}\kappa)^2}{1 + [(\delta+\delta_s)/\kappa]^2 - (\bar{g}/\kappa)^2},$$
$$\langle a_1 a_2\rangle = [\langle a_1^\dagger a_2^\dagger\rangle]^* = \frac{\bar{g}/2\kappa}{1 + [(\delta+\delta_s)/\kappa]^2 - (\bar{g}/\kappa)^2} \times [(\delta+\delta_s)/\kappa + i]. \tag{16}$$

Here, the effective coupling of the dressed atomic system to the cavity modes, i.e. $\bar{g}$, and the frequency shift δ_s due to the dispersive atom-cavity interaction are represented as follows

$$\bar{g} = \frac{g^2\bar{\Omega}\langle R_z\rangle_s/(N\gamma)}{4\bar{\Omega}^2 - \bar{\Delta}_1^2}\sin^2 2\theta,$$
$$\delta_s = \frac{g^2\bar{\Omega}\langle R_z\rangle_s/(N\gamma)}{4\bar{\Omega}^2 - \bar{\Delta}_1^2}(2 - \sin^2 2\theta). \tag{17}$$

To reach the cavity-field steady-state, these parameters must satisfy the inequality $(\bar{g}/\kappa)^2 < 1 + [(\delta+\delta_s)/\kappa]^2$. The frequency shift δ_s can be compensated by slightly modifying δ, i.e. $(\delta+\delta_s)/\kappa = 0$, to arrive at $\bar{g}/\kappa < 1$. Next, we shall focus on the quantum properties of the cavity electromagnetic fields. The expression (15) characterizing the continuous variable field entanglement is given now by

$$E = \frac{2(\bar{g}/\kappa)\left(\bar{g}/\kappa - \sqrt{1 + [(\delta+\delta_s)/\kappa]^2}\right)}{1 + [(\delta+\delta_s)/\kappa]^2 - (\bar{g}/\kappa)^2}. \tag{18}$$

If $(\delta+\delta_s)/\kappa = 0$, then $E \to -1$ when $|\bar{g}/\kappa|$ approaches unity.

Figure (3a) shows the generated average cavity photon number, $\bar{N} = \langle a_1^\dagger a_1\rangle + \langle a_2^\dagger a_2\rangle$, as function of laser detuning according to Eq. (16). Below threshold, one has a steady-state output with $\bar{N} \gg 1$. Figure (3b) depicts the steady-state behavior of the continuous variable entanglement characteristics E. One can observe here that the generated two-mode electromagnetic field is highly correlated, i.e., entangled. The entanglement is due to a four-wave mixing process, i.e., the atoms absorb two photons from the pump field and re-distribute them in the two cavity modes.

The following Section describes the spatial entangled atomic deflections in a different setup via twin-photon light fields.

4. Spatially Entangled Atomic Deflections in Photon-Number Correlated Light Beams

Here, we shall consider interesting effects resulting from the combination of atomic optics and cavity QED. For this purpose, we investigate the deflection of a $V-$level atomic beam in a two-mode quantized cavity field. The two-mode field is in a twin-photon state or photon-number correlated state (PNCS) which can be generated by parametric down-conversion processes. The role of photon-number correlations is important in the proposed scheme of atomic deflection. Indeed, we shall show that the atomic deflection patterns, in the transverse plane to the direction of the center of atomic mass motion, are essentially different for the case of independent and correlated modes. In particular, an atomic deflection pattern for twin photons displays spatial correlations between atomic position variables in the transverse directions monitored in the position distribution of the atomic pattern. These effects are understood as the spatial entanglement in the deflected pattern. Note that PNCS has attracted an increasing interest in recent years. Several implementations regarding the generation of PNCS have already been reported. For instance, experiments on quantum correlations among the intensities of twin beams were performed in.[22] Photon-number correlated states have been investigated in the spontaneous parametric down-conversion processes.[23]

We proceed by considering the quantum dynamics of a three-level atom with a $V-$type configuration of energy levels moving along the z-direction and passing through a cavity electromagnetic field.[17] The field consists of two crossed one-mode standing waves. The atomic flux is adjusted so that collective interactions among particles are negligible. Our aim here is to investigate the position patterns of deflected atoms in the x-y plane. We shall assume off-resonant interactions with detunings of same or opposite sign. The cavity modes are in a photon-number correlated state, i.e. in a twin-photon state. Suppose that initially, i.e. at $t = 0$, each atom is prepared in its ground state, and the quantum state vector $|\Psi(t = 0)\rangle$ of the combined system is a direct product of the atomic and electromagnetic field mode states, i.e $|\Psi(t=0)\rangle = |\text{atom}\rangle \otimes |\text{field}\rangle$. The couplings of the atom to the two modes are determined by the spatial mode functions $\Omega_3 \equiv g_3 \sin(k_3 x)$ and $\Omega_2 \equiv g_2 \sin(k_2 y)$, where $\{k_3 = k_x, k_2 = k_y\}$, with $k_x = k_y = \omega/c$, while $g_2 = E_0 \vec{e}_+ \langle 1|\vec{d}|2\rangle$ and $g_3 = E_0 \vec{e}_- \langle 1|\vec{d}|3\rangle$. $E_0 = \sqrt{\omega/2v}$ is the so-called electric field per photon. $\vec{e}_+$ and $\vec{e}_-$ are the polarization vectors while $\langle 1|\vec{d}|2\rangle$ and $\langle 1|\vec{d}|3\rangle$ are the dipole moments of the corresponding atomic transitions.

We consider the transverse atomic motion in the x-y plane to be given in the position representation, so that $|\text{atom}\rangle = \int dxdy f(x,y)|x\rangle|y\rangle|1\rangle$, where $f(x,y)$ is the atomic wave function describing the transverse motion of the atom as it enters the resonator. We also assume that $f(x,y) = f(x)f(y)$, which means that the initial transverse distribution of atoms does not display any correlations in the x-y direction. Also, the position distribution of the atom before it enters the cavity is assumed to be Gaussian, i.e. $|f(x,y)|^2 = (2\pi\Delta x\Delta y)^{-1} e^{-(x-\langle x\rangle)^2/2(\Delta x)^2} e^{-(y-\langle y\rangle)^2/2(\Delta y)^2}$, with widths $\Delta z = \sqrt{\langle (z - \langle z\rangle)^2\rangle}$ centered at the nodes of both waves ($z = x, y$). We shall consider the cases of narrow initial position distribution of atoms, assuming that the width of this distribution is smaller than the wavelength of the cavity modes, that is $\{\Delta x, \Delta y\} \ll \lambda$. The electromagnetic field is described by two standing quantized modes: $|\text{field}\rangle = \sum_{n_i,n_j} C(n_i,n_j)|n_i\rangle_2|n_j\rangle_3$, where $|n_i\rangle_2$, $|n_j\rangle_3$ are the respective photon number states at the frequencies corresponding to the transitions $|1\rangle \leftrightarrows |2\rangle$ and $|1\rangle \rightleftarrows |3\rangle$, while $C(n_i,n_j)$ is the probability amplitude of finding two photons with opposite circular polarizations in the two channels. For the case of independent modes, the amplitude is factorized as $C(n_i,n_j) = C_2(n_i)C_3(n_j)$, so that $|\text{field}\rangle = (\sum_{n_i} C_2(n_i)|n_i\rangle_2)(\sum_{n_j} C_3(n_j)|n_j\rangle_3)$, while for photon-number correlated modes, we choose $|\text{field}\rangle = \sum_n A_n|n\rangle_2|n\rangle_3$, where A_n is the probability amplitude of twin photons, i.e. the amplitude of finding n-photon pairs in the light fields with $\sum_n |A_n|^2 = 1$. The foregoing article is based on the Raman-Nath approximation, i.e. we neglected the transverse kinetic energy in comparison with the atom-field interaction energy.

Further, we consider the limiting case of a large atom-cavity detuning, i.e. $\Delta_i \gg g_i^2\langle n_i\rangle$, $(i = 2,3)$, with $\langle n_i\rangle$ being the mean photon number of the ith cavity mode and $\Delta_i = \omega_{i1} - \omega$ the corresponding detunings of the excited atomic level frequencies $\{\omega_{31}, \omega_{21}\}$ from the cavity frequency ω. In this case, there is no exchange of energy between the field and the atom, i.e. the interaction does not change the internal atomic state. An atom initially in the ground state will remain there and spontaneous atomic processes can be neglected for small atom-field interaction time durations. Thus, we can eliminate adiabatically the upper states to arrive at an effective interaction Hamiltonian

$$H_{eff} = -\hbar \sum_{i\in\{2,3\}} \frac{\Omega_i^2}{\Delta_i} a_i^\dagger a_i |1\rangle\langle 1|, \tag{19}$$

where a_i and $a_i^\dagger$ are the annihilation and creation operators of the ith mode, while the quantum state of the combined system after the interaction will be calculated in the following way:

$$|\Psi(t)\rangle = e^{-iH_{eff}t/\hbar}|\Psi(t=0)\rangle. \tag{20}$$

Note that here the combined atom-field system can be described by the effective Hamiltonian (19) if the system parameters and the interaction time t are such that $|g_i\langle n_i\rangle/\Delta_i| \ll 1$, and $|\Delta_i t \left(g_i^2\langle n_i\rangle/\Delta_i^2\right)^2| \ll 1$, for $i = 2, 3$. After projecting the states $|x\rangle$ and $|y\rangle$ onto the state vector $|\Psi(t)\rangle$, one obtains

$$\langle y, x|\Psi(t)\rangle = f(x)f(y)\sum_{n_i n_j} C(n_i, n_j) \times e^{i\left[\chi_2 n_i \sin^2(k_2 y) + \chi_3 n_j \sin^2(k_3 x)\right]}|1\rangle|n_i\rangle_2\,|n_j\rangle_3 \tag{21}$$

where we have introduced the interaction parameters $\chi_2 = g_2^2 t/\Delta_2$ and $\chi_3 = g_3^2 t/\Delta_3$. Both χ_2 and χ_3 may be either positive or negative, depending on the signs of Δ_2 and Δ_3, respectively.

The corresponding probability distribution $|\langle x|\langle y|\Psi(t)\rangle|^2$ depends on the photon-number states and hence ignores the field phases. Another possibility is to consider the conditional probability amplitude to measure the atomic position, assuming that the cavity modes are in a given reference state. Therefore, for the reference state, we use the phase states for each of the modes $|\Psi_R\rangle = |\varphi_2\rangle|\varphi_3\rangle = (2\pi)^{-2}\sum_{n_i}\sum_{n_j} e^{i(n_i\varphi_2 + n_j\varphi_3)}|n_i\rangle_2|n_j\rangle_3$.

In this case, for the joint position distribution $W(x, y, \varphi_2, \varphi_3)$ to find the atom at positions $\{x, y\}$ and phases $\{\varphi_2, \varphi_3\}$ of the electromagnetic field modes, one obtains

$$W(x, y, \varphi_2, \varphi_3) = \frac{|f(x)|^2|f(y)|^2}{(2\pi)^2}\left|\sum_{n_2,n_3} C(n_2, n_3) \times e^{-i(n_2\varphi_2 + n_3\varphi_3)} e^{i\left[\chi_2 n_2 \sin^2(k_2 y) + \chi_3 n_3 \sin^2(k_3 x)\right]}\right|^2. \tag{22}$$

Further, for independent modes the probability amplitude of having two photons factorizes, i.e. it is represented as a product $W(x,y) = W(x)W(y)$ of two independent probability distributions for x and y, respectively. For the case of a photon-number correlated state, the situation is different from the usual ones with two independent standing waves because here, each number of photon pairs contributes to the scattering amplitude separately. In particular, the state vector $\langle y,x||\Psi(t)\rangle$, i.e. Eq. (21), and hence the probability distribution (22) is not a factorisable function, i.e. $W(x,y) \neq W(x)W(y)$, and, thus, this could be understood as spatial entanglement of the deflected particles.

Let us investigate the influence of photon-number correlations on the atomic distribution patterns. The photon-pair distribution $P_n = |A_n|^2$ has been theoretically and experimentally investigated mainly for different regimes of parametric down-conversion processes. In the absence of the initial photons in the generated sub-harmonic modes, the non-degenerate spontaneous down-conversion process produces the well-known two-mode squeezed vacuum state, $|\Psi_V\rangle = \exp[\lambda_p a_2^\dagger a_3^\dagger + \lambda_p^* a_2 a_3]|0,0\rangle = (\cosh\lambda_p)^{-1}\sum_n(\tanh\lambda_p)^n|n,n\rangle$, where λ_p is proportional to the pump amplitude. The photon-pair distribution in this regime is thermal. However, in our case, we should operate with two-mode states of bright light produced, for instance, by an optical parametric oscillator (OPO). So far, there is no complete theory of the OPO state in the Schrödinger picture. A useful approximation is the so-called pair-coherent state with an equal photon number in both the modes, $(a_2^\dagger a_2 - a_3^\dagger a_3)|\xi\rangle = 0$, described as eigenstates of the product of the annihilation operators of the two radiation modes $a_2 a_3|\xi\rangle = \xi|\xi\rangle$. This state reads as: $|\xi\rangle = N_0\sum_n(\xi^n/n!)|n,n\rangle$, where $N_0 = 1/\sqrt{I_0(2|\xi|)}$, with I_0 being the modified Bessel function of the zero order and the parameter ξ being the amplitude of the pair coherent state. Note that the expansion coefficients $A_n = N_0\xi^n/n!$ in this case are different from the analogous coefficients of the squeezed vacuum state, $A_n = (\cosh\lambda_p)^{-1}(\tanh\lambda_p)^n$, and these coefficients decrease quickly with increasing $n!$. Nevertheless, here, we shall use a quasi-phenomenological approach to describe these states relying on both theoretical and experimental considerations, i.e., for concrete calculations, we shall use the PNCS with a Poissonian distribution.[24]

We now discuss the position patterns due to twin-photon number modes. The distribution (22) can be rewritten in a simplified form:

$$W(x,y,\varphi) = \frac{|f(x)|^2|f(y)|^2}{(2\pi)^2}\left[1 + 2\mathrm{Re}\{\sum_{n_2>n_3} A_{n_2}A^*_{n_3}\right.$$
$$\left.\times\, e^{i(n_2-n_3)(\chi(x,y)-\varphi)}\}\right], \qquad (23)$$

where $\varphi = \varphi_2 + \varphi_3$ and $\chi(x,y) = \chi_2 \sin^2(k_2 y) + \chi_3 \sin^2(k_3 x)$, which clearly demonstrates that the interfering probability amplitudes A_n of finding n twin-photon pairs in the crossed light beams contribute to $W(x,y)$. We perform concrete calculations based on the PNCS with the Poissonian photon-pair distribution assuming that $A_n = e^{-N/2}\alpha^n/\sqrt{n!}$ and hence $P_n = N^n e^{-N}/n!$, where $N = |\alpha|^2$ is the averaged value of the total number of photon pairs generated by a single pump pulse. As $\alpha = \sqrt{N}e^{i\theta}$, we choose $\varphi = \theta$, which gives the best localization. If we take the width of the atomic wave packet to be much smaller than the atomic transition wavelength and redirect it at a node of the field, we can then replace the sinusoidal potentials in Eq. (23) with its linear expansion, i.e. $\sin(k_i z) \approx k_i z$. Then, according to the formula (23), the maximal interference is realized for

$$|\chi_2|(k_2 y)^2 \pm |\chi_3|(k_3 x)^2 = 2\pi m, \qquad (24)$$

with $k_2 y \ll 1$, $k_3 x \ll 1$ and $m \in \{0, 1, 2, \cdots\}$. The upper sign stands for identical detunings, that is $(\Delta_2, \Delta_3) > 0$ or $(\Delta_2, \Delta_3) < 0$ while the lower one for opposite detunings, i.e. $\Delta_2 > 0, \Delta_3 < 0$ or vice versa. It is not difficult to realize that the maxima of the deflection pattern, which is characterized by $W(x,y)$, adopt the form of a sequence of circles in the x-y plane for the same detunings and of crossed lines for opposite detunings.

Fig. (3a) and Fig. (3b) depict the corresponding probability distributions $W(x,y)$ for initially Gaussian atomic wave packets. In particular,

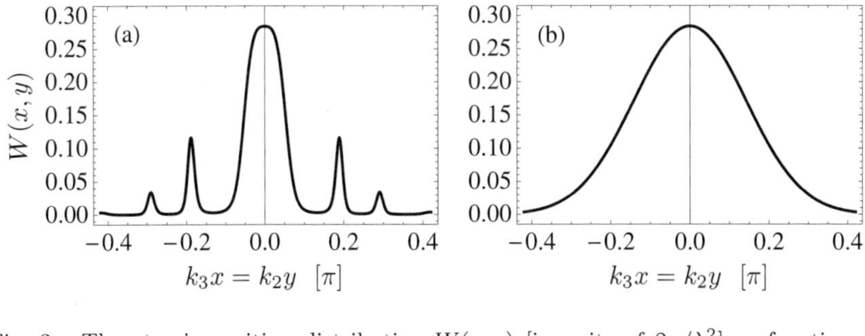

Fig. 3. The atomic position distribution $W(x,y)$ [in units of $2\pi/\lambda^2$] as function of $k_3x/\pi = k_2y/\pi$ for correlated modes with $\varphi = \theta$. Here $N = 1$, $|\chi_2| = |\chi_3| = 11$, and $\Delta x = \Delta y = 0.1\lambda$ while (a) $\{\chi_2, \chi_3\} > 0$ and (b) $\{\chi_2 > 0, \chi_3 < 0\}$.

Fig. (3a) shows the corresponding deflection pattern with correlated beams that are identically detuned from the resonance while Fig. (3b) describes the respective case with opposite detunings. By increasing the interaction parameter χ for both modes, one can obtain smaller structures. Thus, this scheme provides the mechanism for two-dimensional localization of atoms.

In summary, we discussed different schemes entangling light fields or matter waves. We demonstrated the generation of two-particle entanglement and its phase control via external field pumping and damping through a squeezed electromagnetic field reservoir. Continuous variable entanglement of the generated field of a pumped multi-atom sample was reported as well. In a different setup, we showed spatial entanglement of matter waves.

References

1. S. Hill, W. K. Wootters, Phys. Rev. Lett. **78**, 5022 (1997); W. K. Wootters, Phys. Rev. Lett. **80**, 2245 (1998).
2. Lu-Ming Duan, G. Giedke, J. I. Cirac, and P. Zoller, Phys. Rev. Lett. **84**, 2722 (2000); R. Simon, Phys. Rev. Lett. **84**, 2726 (2000).
3. Charles H. Bennett, Gilles Brassard, Sandu Popescu, Benjamin Schumacher, John A. Smolin, and William K. Wootters, Phys. Rev. Lett. **76**, 722 (1996).
4. V. Vedral, M. B. Plenio, K. Jacobs, and P. L. Knight, Phys. Rev. A **56**, 4452 (1997).
5. A. Peres, Phys. Rev. Lett. **77**, 1413 (1996); P. Horodecki, Phys. Lett. A **232**, 333 (1997).
6. C. H. van der Wal et al., Science **301**, 196 (2003); A. Kuzmich et al., Nature (London) **423**, 731 (2003).
7. S. Du, J. Wen, M. H. Rubin, and G. Y. Yin, Phys. Rev. Lett. **98**, 053601 (2007).
8. C. Silberhorn, P. K. Lam, O. Weiss, F. Konig, N. Korolkova, and G. Leuchs, Phys. Rev. Lett. **86**, 4267 (2001).
9. M. K. Olsen, Phys. Rev. A **73**, 053806 (2006).
10. M. S. Kim, Jinhyoung Lee, D. Ahn, and P. L. Knight, Phys. Rev. A **65**, 040101(R) (2002).
11. G.-x. Li, K. Allaart, D. Lenstra, Phys. Rev. A **69**, 055802 (2004).
12. Ting Yu, J. H. Eberly, Phys. Rev. Lett. **93**, 140404 (2004).
13. Z. Ficek, R. Tanas, Phys. Rev. A **74**, 024304 (2006).
14. D. T. Pope and G. J. Milburn, Phys. Rev. A **67**, 052107 (2003).
15. A. S. Parkins, and E. Larsabal, Phys. Rev. A **63**, 012304 (2000).
16. M. Macovei, J. Evers, and C. H. Keitel, Jr. of Modern Optics (in print, 2010).
17. G. Yu. Kryuchkyan, M. Macovei, and C. H. Keitel, Phys. Rev. A **77**, 035603 (2008).
18. V. Josse, A. Dantan, A. Bramati, M. Pinard, and E. Giacobino, Phys. Rev. Lett. **92**, 123601 (2004).

19. H. S. Eisenberg, G. Khoury, G. A. Durkin, C. Simon, and D. Bouwmeester, Phys. Rev. Lett. **93**, 193901 (2004).
20. P. Marian, T. A. Marian, Phys. Rev. Lett. **101**, 220403 (2008).
21. M. Macovei, G.-x. Li, Phys. Rev. A **76**, 023818 (2007).
22. S. Feng, and O. Pfister, Phys. Rev. Lett. **92**, 203601 (2004).
23. J. Laurat, T. Coudreau, N. Treps, A. Maître, and C. Fabre, Phys. Rev. Lett. **91**, 213601 (2003).
24. E. Jeffrey, N. A. Peters, and P. G. Kwiat, New J. Phys. **6**, 100 (2004).

IRREVERSIBLE PHOTON TRANSFER IN AN ENSEMBLE OF Λ-TYPE ATOMS AND PHOTON DIODE

G. NIKOGHOSYAN[1,2] and M. FLEISCHHAUER[1]

[1] *Dept. of Physics and research center OPTIMAS, University of Kaiserslautern, D-67663 Kaiserslautern, Germany*

[2] *Institute of Physical Research, 0203, Ashtarak-2, Armenia*

We show that a pair of quantized cavity modes interacting with a spectrally broadened ensemble of Λ-type atoms is analogous to an ensemble of two level systems coupled to a bosonic reservoir. This provides the possibility for an irreversible photon transfer between photon modes. The density of states as well as the quantum state of the reservoir can be manipulated allowing the observation of effects such as the quantum Zeno- and anti-Zeno effect, the destructive interference of decay channels and the decay in a squeezed vacuum. As a particular application we discuss a photon diode, i.e. a device which directs a single photon from anyone of two input ports to a common output port.

Keywords: Quantum optics, Non-Markovian processes.

1. Introduction

In the present paper (see also[1]) we propose a photonic analogue of the irreversible decay of an ensemble of two-level systems coupled to a bosonic reservoir.[2] In particular we consider a pair of quantized cavity modes interacting with a spectrally broadened ensemble of Λ-type atoms. The two cavity modes replace the collective states of the ensemble of two-level systems and the Λ-type atoms form the modes of the reservoir. In contrast e.g. to the quantized radiation field as a reservoir, the atomic ensemble can be easily modified and controlled dynamically. E.g. the density of states of the reservoir can be tailored by application of inhomogeneous magnetic or electric fields and thus it should be possible to implement e.g. the quantum Zeno[3] and anti-Zeno[4,5] effects, which are otherwize difficult to realize. Moreover the reservoir of Λ atoms can be prepared in different initial quantum states. E.g. coherent ensemble states can be created by using electromagnetically induced transparency[6] or methods of adiabatic population transfer.[7] Also non-classical states can be prepared[8–11] which can be used to simulate a squeezed vacuum reservoir.[12–14] If the atomic ensemble is prepared in only one internal state, serving as the vacuum of reservoir excitations, the analogue of spontaneous decay can be observed, where the photons of one cavity mode are transferred irreversibly, i.e. non-unitarily, to the second mode. This effect can have a large variety of applications; e.g. creation of new quantum states, transfer of photons of optical frequency to the microwave domain and vice versa, or a photon diode, i.e. a device where a single photon injected into anyone of two inputs ports leads to a single photon emission from the same output port. The system considered here can be easily constructed with current technology and is available in several labs. In a number of groups experiments were implemented where a Bose-Einstein condensate of atoms[15–17] or a cold atomic cloud[18,19] was strongly coupled to a cavity mode. As the two modes orthogonal polarizations of the same frequency can be used and the required spectral broadening of atomic ensemble can be achieved e.g. by application of inhomogeneous magnetic field.

Let us consider the interaction of two light modes described by the annihilation operators $\hat{a}_1$ and $\hat{a}_2$ with an ensemble of three level Λ-atoms (see Fig. 1). $\hat{a}_1$ and $\hat{a}_2$ couple the ground-state $|g\rangle$, respectively a meta-stable lower state $|s\rangle$ to a common excited state $|e\rangle$ in a Raman transition. The two-photon transition between states $|g\rangle$ and $|s\rangle$ is assumed to be inhomogeneoulsy broadened, as indicated in Fig. 1. For simplicity we assume a discrete spectrum consisting of m energy levels. In each spectral class there are N atoms, so that the total number of atoms is mN.

To describe the quantum properties of the medium, we use collective atomic operators for each spectral component $\hat{\sigma}^n_{ij} = \frac{1}{\sqrt{N}} \sum\limits_{k=1}^{N} |i\rangle^{nn}_{kk} \langle j|$, where $i, j \in e, g, s$, and k labels the atom. The dynamics of the system is then governed by the Hamiltonian

$$\hat{H} = \hbar\sqrt{N} \sum_{n=1}^{m} \left[\Delta\hat{\sigma}^n_{ee} + \delta_n\hat{\sigma}^n_{ss} + \left(g_1\hat{\sigma}^n_{eg}\hat{a}_1 + g_2\hat{\sigma}^n_{es}\hat{a}_2 + H.c. \right) \right] \tag{1}$$

where Δ is the one-photon detuning of the cavity field 1, and δ_n is the two-photon detuning corresponding to the nth spectral class. $g_{1,2}$ are the coupling strength of both modes, which are assumed to be the same for all spectral components for simplicity. The factor $\sqrt{N}$ is due to the collective coupling of the atoms of each spectral class to the cavity modes.[20,21] If the vacuum Rabi frequencies $g_{1,2}$ and the two-photon detunings δ_n are significantly smaller than Δ, we can adiabatically eliminate the upper level

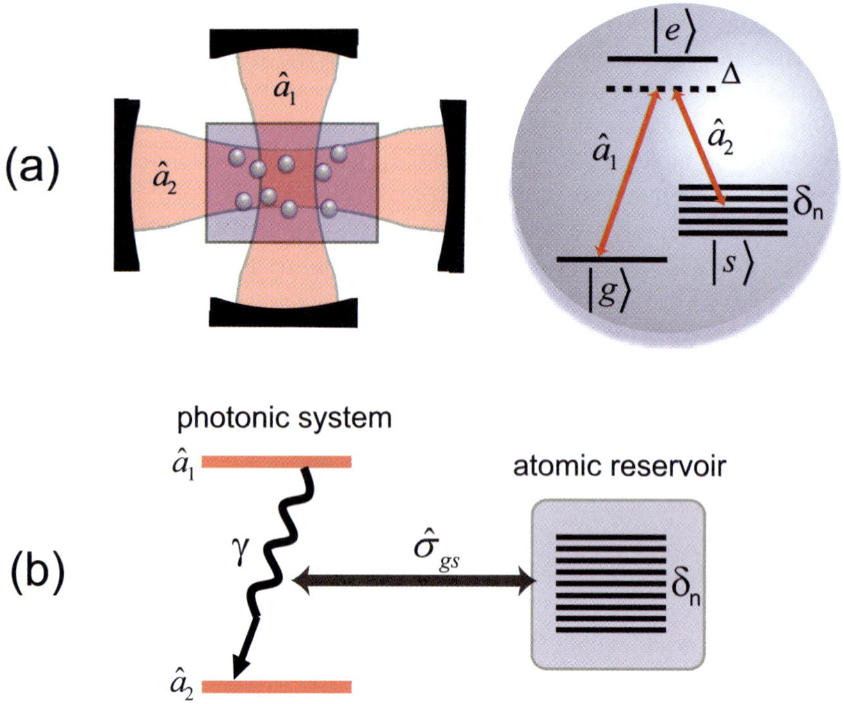

Fig. 1. (a) Schematic setup: Two quantized cavity modes $\hat{a}_1$ and $\hat{a}_2$ interact with an ensemble of three-level Λ type atoms with inhomogeneously broadened two-photon transition $|g\rangle - |s\rangle$. (b) For a large number of three-level atoms and a sufficiently large spectral width of the Raman transition the system resembles a collection of two-level systems coupled to a bosonic reservoir.

$|e\rangle$ by standard techniques. Let us further assume that the atomic ensemble is prepared in a state which is close to the collective ground state and that the total number of photons is much less than the total number of atoms. Then the population of state $|s\rangle$ remains small and the atomic operators $\hat{\sigma}_{gs}^n$ and $\hat{\sigma}_{sg}^n$ obey aproximately bosonic commutation relations $\left[\hat{\sigma}_{sg}^n, \hat{\sigma}_{sg}^m\right] \approx 0$, $\left[\hat{\sigma}_{gs}^n, \hat{\sigma}_{sg}^m\right] \approx \delta_{m,n}$, i.e. we can set $\hat{\sigma}_{gs}^n = \hat{\beta}_n$, $\hat{\sigma}_{sg}^n = \hat{\beta}_n^\dagger$. Thus passing to an interaction picture we arrive at an effective Hamiltonian

$$\hat{H}_{\text{eff}} = -\hbar\sqrt{N}\sum_{n=1}^{m}\left\{\eta\hat{\beta}_n\hat{a}_1^\dagger\hat{a}_2\,\mathrm{e}^{i\omega_n t} + H.c.\right\} \tag{2}$$

where $\eta = \dfrac{(g_1)^* g_2}{\Delta}$ and $\omega_n = \delta_n + \dfrac{|g_1|^2}{\Delta}$ is an effective detuning containing ac-Stark shift contributions.

$\hat{H}_{\text{eff}}$ is similar to the Hamiltonian that describes the interaction of an ensemble of two level systems consisting of states $|1\rangle$ and $|2\rangle$ with a reservoir of bosonic modes $\hat{\beta}_n$. $\hat{a}_1^\dagger \hat{a}_2$ destroys an atom in state $|2\rangle$ in the ensemble and creates an atom in state $|1\rangle$.

The dynamics of a two-level system that interacts with a reservoir becomes irreversible when the number of modes of the bath tends to infinity. A well-known consequence of this sort of dynamics is spontaneous decay and a similar dynamics can be obtained here for photons. To see this we derive an effective equation of motion for the photon modes by tracing out the atomic degrees of freedom using standard second order perturbation theory. This yields for the density operator ρ of the photon modes

$$\begin{aligned}\dot{\hat{\rho}}(t) = -\int_0^t \mathrm{d}\tau f(t-\tau)\Big[&\hat{a}_1^\dagger(t)\hat{a}_2(t)\hat{a}_1(\tau)\hat{a}_2^\dagger(\tau)\hat{\rho}(\tau) \\ &-\hat{a}_1(\tau)\hat{a}_2^\dagger(\tau)\hat{\rho}(\tau)\hat{a}_1^\dagger(t)\hat{a}_2(t)\Big] + H.c.\end{aligned}$$

where $f(t) = N\sum_{n=1}^{m} |\eta|^2 e^{i\omega_n t}$. The behavior of the system depends on the reservoir response function $f(t)$ which is determined by the effective light–field coupling constants η and the resonance frequencies ω_n of the atomic ensemble. For simplicity let us assume that the ensemble has equidistant spectral lines with $\omega_n = \varepsilon_{\text{max}}(2n-1-m)/m$, where $2\varepsilon_{\text{max}}$ is the spectral width of the atomic ensemble. If the number of spectral lines is large, and the spectrum of the atomic ensemble is broad, $f(t)$ can be approximated as a delta function $f(t) \approx \pi(Nm/\varepsilon_{\text{max}})\delta(t)|\eta|^2$, which constitites the Born-Markov approximation. In this limit we obtain

$$\frac{\partial\hat{\rho}}{\partial t} = -\frac{\gamma}{2}\left(\hat{a}_1^\dagger\hat{a}_1\hat{a}_2\hat{a}_2^\dagger\hat{\rho} + \hat{\rho}\hat{a}_1^\dagger\hat{a}_1\hat{a}_2\hat{a}_2^\dagger - 2\hat{a}_1\hat{a}_2^\dagger\hat{\rho}\hat{a}_1^\dagger\hat{a}_2\right) \tag{3}$$

with $\gamma = \dfrac{\pi N m |\eta|^2}{\varepsilon_{\text{max}}}$.

The Lindblad-equation (3) describes the irreversible transfer of excitations from mode 1 to 2 with rate γ. If there are many photons in the system the decay will be enhanced due to stimulated Raman emission into mode 2. The atomic analogue of this process is the collective decay of atoms (superfluorescence).[21]

2. Irreversible Photon Transfer

Let us first consider the irreversible transfer of photons from mode 1 to mode 2 for the case when mode 2 is initially not populated and the initial state of the system is pure, i.e. $\hat{\rho}_0 = |\phi_0\rangle\langle\phi_0|$ with $|\phi_0\rangle = \sum_n \alpha_n |n\rangle_1 |0\rangle_2$. where $|n\rangle_1 |m\rangle_2$ are the Fock state of the modes 1 and 2 respectively. This state will evolve to the mixed state $\hat{\rho}_{\text{fin}} = \sum_n |\alpha_n|^2 |0\rangle_{11}\langle 0| \otimes |n\rangle_{22}\langle n|$. It is pure only if mode 1 is initially in a single Fock state.

More interesting is the general case of an initially mixed state of both modes

$$\hat{\rho}_0 = \sum_{m,m',n,n'} A^{n,n'}_{m,m'} |n\rangle_{11} \langle n'| \otimes |m\rangle_{22} \langle m'| \tag{4}$$

In this case the final state reads

$$\hat{\rho}_{\text{fin}} = \sum_{l,l'} B_{l,l'} |0\rangle_{11}\langle 0| \otimes |l\rangle_{22} \langle l'| \tag{5}$$

where $B_{l,l'} = \sum_{q=0} A^{q,q}_{l-q,l'-q}$. This state is a pure state $|\phi\rangle_2 = \sum_m \beta_m |m\rangle_2$ if $B_{l,l'} = \beta^*_l \beta_{l'}$. This implies that all initial states with $A^{m,m'}_{n,n'} = \beta_{m+n}\beta^*_{m'+n}\alpha_n\delta_{n,n'}$, where α_n is an arbitrary function of n, evolve into a pure state. E.g., assume the initial state of the light is completely mixed and contains only one photon

$$\hat{\rho}_0 = \frac{1}{2} |1\rangle_{11} \langle 1| \otimes |0\rangle_{22} \langle 0| + \frac{1}{2} |0\rangle_{11} \langle 0| \otimes |1\rangle_{22} \langle 1| .$$

Then the final state will be a pure single photon Fock state

$$\hat{\rho}_{\text{fin}} = |0\rangle_{11}\langle 0| \otimes |1\rangle_{22} \langle 1| .$$

I.e. an initially mixed states of light can be purified with *conservation* of photon number. Another related example is the case when initially we have Fock states

$$\hat{\rho}_0 = |n\rangle_{11} \langle n| \otimes |m\rangle_{22} \langle m|$$

in both modes. The final state is again a pure Fock state

$$\hat{\rho}_{\text{fin}} = |0\rangle_{11}\langle 0| \otimes |n+m\rangle_{22} \langle n+m|$$

that contains the sum of initial photon numbers. The latter process can be realized also by unitary operations, but for that it is necessary to know the exact number of photons in each mode,[22,23] while the process which we describe is irreversible and can be realized without any information about the initial photon numbers in each mode. It is enough to know, that the initial states are Fock states.

3. Reservoir Tailoring

As originally formulated by Mishra and Sudarshan a decaying quantum system that is continously observed in a specific state does not decay, which they called the quantum Zeno effect.[3] In practice a continuous observation is approximated by a periodic sequence of measurements. In order to observe the decay suppression the period of the repeated projections has to be shorter than a characteristic time determined by the spectral structure of the reservoir coupling. The latter makes the observation of the effect in the decay of a two-level atom to the free-space electromagnetic vacuum rather difficult. In this case the spectrum of the reservoir coupling is flat and the frequency of measurements has to be comparable to the transition frequency in order to see the quantum Zeno suppression of decay. If the reservoir spectrum is structured also the opposite effect, called anti-Zeno effect is possible.[4] The periodic interaction between system and measurement device shifts the effective resonance frequency of the system and moves it to a different part of the reservoir spectrum. This can increase (decrease) the system–reservoir coupling, and thus lead to an increase (decrease) of the decay rate. Similar effects can be observed by keeping the resonant frequency of the system unperturbed but shifting the spectrum of the reservoir. The latter can be realized by the measurement of the reservoir.[5] In our system the reservoir is atomic thus one can easily tailor the reservoir spectrum and measure it by application of an electromagnetic field. Depending on the reservoir response function $f(t)$ the measurement will either accelerate (anti-Zeno effect) or slow down (Zeno effect) the spontaneous photon transfer from one mode into another mode.

Another potential application of a tailored reservoir spectrum is the possibility to observe destructive interference of decay channels. The radiative decay in three-level V-type systems with nearly degenerate dipole transitions can exhibit destructive interference and thus suppression of decay if the dipole moments of the transitions from the two excited states to the common ground state are parallel.[24] Unfortunately in atomic systems degenerate transitions with parallel dipole moments do not exist. Generalizing the present discussion to multi-mode photonic systems provides however a means to observe such a destructive interference of decay channels. To this end consider a system where in addition to mode $\hat{a}_1$ a second cavity mode $\hat{a}_1'$, e.g. of orthogonal polarization, couples the same atomic transition $|g\rangle - |e\rangle$. In this case the "excited state" $\hat{a}_1$ in Fig. 1 b is replaced by two "excited states" $\hat{a}_1$ and $\hat{a}_1'$ coupling to the common atomic reservoir. Then in the Lindblad Eq. (3) $\hat{a}_1$ is replaced by $(\hat{a}_1 + \hat{a}_1')$. As a consequence there are non-decaying states, as e.g. the anti-symmetric single-photon excitation $(|0\rangle|1\rangle - |1\rangle|0\rangle)/\sqrt{2}$.

Furthermore it is also possible to prepare the reservoir in a certain quantum state, e.g. in multi-mode squeezed states.[8–11] With this it should be possible to study the decay in a squeezed vacuum.[12–14] The latter has been proposed and analyzed theoretically for atoms coupled to a squeezed reservoir of radiation modes. An experimental verification is however extremely difficult due to the requirement of a broad-band and isotropic squeezed-vacuum radiation field.

4. Photon Diode

In the last part of this paper we discuss an interesting application of the irreversible photon transfer, the realization of a diode for photons, i.e. a four-port device where single photon pulses injected into any of the two input ports will be directed to the same output port. To model the input–output processes we introduce a continuum of free-space modes with field operators $\hat{b}_{1q}$ and $\hat{b}_{2q}$ which are coupled to the cavity modes $\hat{a}_1$ and $\hat{a}_2$ respectively. For simplicity we assume that the coupling constants κ_1 and κ_2 are the same for all relevant modes. The corresponding interaction is described by the following Hamiltonians $(m = 1, 2)$

$$\hat{V}_m = \hbar\kappa_m \sum_q \left(\hat{a}_m^\dagger \hat{b}_{mq} + h.c.\right) + \sum_q \hbar\Delta_m^q \hat{b}_{mq}^\dagger \hat{b}_{mq}.$$

Here Δ_m^q are the detunings of free field modes from the cavity resonance. Let us consider input fields in a single-photon state $|\psi_{in}\rangle_m = \sum_q P_q^{in}(t)\, \hat{b}_{mq}^\dagger |0\rangle$. All properties of the fields are then described by the single-photon wave-function $\Phi_{in}^m(z,t) = \sum_q \langle 0| \hat{b}_{mq} e^{iqz} |\psi_{in}\rangle_m$. Since all atoms of the atomic ensemble are initially in the ground state, the field in state $|\psi_{in}\rangle_2$ sees an empty cavity and will be reflected from it with some time delay corresponding to the bare cavity ring down time. In the following we want to prove that $|\psi_{in}\rangle_1$ is transferred to $|\psi_{out}\rangle_2$, i.e. to a state where the excitation is in the orthogonal output channel. In general the state of the system can be written in the following form

$$|\psi(t)\rangle = \Big(\sum_q P_q(t)\,\hat{b}^\dagger_{1q} + Q(t)\,\hat{a}^\dagger_1 + \sum_{n=1}^{m} R_n(t)\,\hat{a}^\dagger_2\hat{\beta}^\dagger_n$$
$$+\sum_{n=1}^{m}\sum_q S_{qn}(t)\,\hat{b}^\dagger_{2q}\hat{\beta}^\dagger_n\Big)\,|\mathbf{gs}\rangle\,,$$

with $|\mathbf{gs}\rangle$ denoting the ground state of the atomic ensemble, where all atoms are in state $|g\rangle$. Since initially only mode 1 is excited and all atoms are in the ground state $Q(t) = R_n(t) = S_{qn}(t) = 0$ if $t < 0$, where $t = 0$ is the beginning of interaction. The evolution of the system is described by the Schrödinger equation

$$\dot{P}_q(t) = -i\Delta_1^q P_q(t) - i\kappa_1 Q(t) \tag{6}$$

$$\dot{Q}(t) = -i\kappa_1 \sum_q P_q(t) + i\eta\sqrt{N}\sum_{n=1}^{m} e^{i\omega_n t}R_n(t) \tag{7}$$

$$\dot{R}_n(t) = i\eta^*\sqrt{N}e^{-i\omega_n t}Q(t) - i\kappa_2\sum_q S_{qn}(t) \tag{8}$$

$$\dot{S}_{qn}(t) = -i\Delta_2^q S_{qn}(t) - i\kappa_2 R_n(t) \tag{9}$$

Substituting the formal solution of Eq. (9) into (8) and assuming the Markov limit yields

$$\dot{R}_n(t) = i\eta^*\sqrt{N}e^{-i\omega_n t}Q(t) - \frac{\gamma_2}{2}R_n(t) \tag{10}$$

where $\gamma_2 = \dfrac{\kappa_2^2 L}{c}$ is the cavity loss rate of mode 2 and L is the quantization length of $\hat{b}$ modes. (Note that $\kappa_m \sim 1/\sqrt{L}$, so that the dependence on L drops.) Furthermore substituting the formal solutions of Eqs. (10) and (6) into (7) we find again using the Markov approximation

$$\dot{Q}(t) = -\frac{(\gamma+\gamma_1)}{2}Q(t) - i\kappa_1\sum_q P_q(0)\,e^{-i\Delta_1^q t} \tag{11}$$

where we have used the photon decay rate γ and introduced the cavity loss rate of mode 1 $\gamma_1 = \dfrac{\kappa_1^2 L}{c}$. Upon integrating Eq. (11) we finally find the input-output relation for port 1, i.e. for the modes $\hat{b}_1^q$.

$$\Phi^1_{out}(t) = \Phi^1_{in}(t) - \gamma_1\int_0^t \mathrm{d}\tau\,\Phi^1_{in}(\tau)\,e^{-\frac{(\gamma+\gamma_1)}{2}(t-\tau)} \tag{12}$$

In order to achieve a maximum transfer of free-field photons into the cavity, the outgoing component should be minimized. According to Eq. (12) this can be realized by requiring impedance matching. If $\gamma_1 = \gamma$ and the pulse is much longer than the relaxation rates the two terms on the right hand side of Eq. (12) cancel each other and there is no output into the modes $\hat{b}_{1q}$.

Due to the dissipative nature of the coupling between the two cavity modes the output field will be in general in a mixed state when tracing out the degrees of freedom of the atomic ensemble. Only if the input is an eigenstate of the total excitation number, the cavity output can be in a pure state. We will show now that in this case the final state indeed factorizes into a single photon distributed over many modes $\hat{b}_{2q}$ and a single collective excitations in the atomic ensemble. Following similar steps than above we find for the output wave function for the modes $\hat{b}_{2q}$ when the atomic excitation is in mode n, $\Phi^{2,n}_{\text{out}} \equiv \sum_q \langle 0|\langle 1_n|\hat{b}_{2q}|\psi\rangle_{\text{out}}$ where $|1_n\rangle$ denotes a single excitation in the nth spectral class of the atomic ensemble,

$$\Phi^{2,n}_{out}(t) = -i\frac{\kappa_1}{\kappa_2}\gamma_2\eta^*\sqrt{N}\int_0^t d\tau_1\, e^{-i\omega_n\tau_1}e^{-\frac{\gamma_2}{2}(t-\tau_1)} \times \int_0^{\tau_1} d\tau_2\, \Phi^1_{in}(\tau_2)\, e^{-\frac{(\gamma+\gamma_1)}{2}(\tau_1-\tau_2)}. \tag{13}$$

Thus the probability of having a photon in output port 2, obtained by tracing out the atomic part, reads

$$\rho_{out}(t) = \sum_n \left|\Phi^{2,n}_{out}\right|^2 = \gamma_1\gamma_2\gamma\int_0^t d\tau_1 e^{-\gamma_2(t-\tau_1)} \times \left|\int_0^{\tau_1} d\tau_2\, \Phi^1_{in}(\tau_2)\, e^{-\frac{(\gamma+\gamma_1)}{2}(\tau_1-\tau_2)}\right|^2 \tag{14}$$

If the cavity decay rate of mode 2, γ_2, is much larger than the relaxation rates γ and γ_1 the integral over τ_1 can easily be evaluated. In this case (14) can be expressed as a product of two single photon wavepackets $\rho_{out}(t) = \left|\Phi^2_{out}(t)\right|^2$, where

$$\Phi^2_{out}(t) = \sqrt{2\gamma_1\gamma}\int_0^t d\tau_1\, \Phi^1_{in}(\tau_2)\, e^{-\frac{(2\gamma+\gamma_1)}{2}(t-\tau_1)}.$$

Thus indeed a single-photon input wavepacket results into an asymptotic final state which is a product of a collective atomic excitation and a single-photon wavepacket in the output mode 2. It should be noted that since the physical mechanism that leads to the diode function is irreversible, input superposition states will in general not be mapped to pure output states.

In Fig. 2 we have illustrated the performance of the photon diode. Shown are two cases with a single-photon wavepacket injected into input port 1 (modes $\hat{b}_{1q}$) and input port 2 (modes $\hat{b}_{2q}$). In both cases the output in modes 1 is negligible, while the output in modes 2 is a nearly perfect single-photon wavepacket.

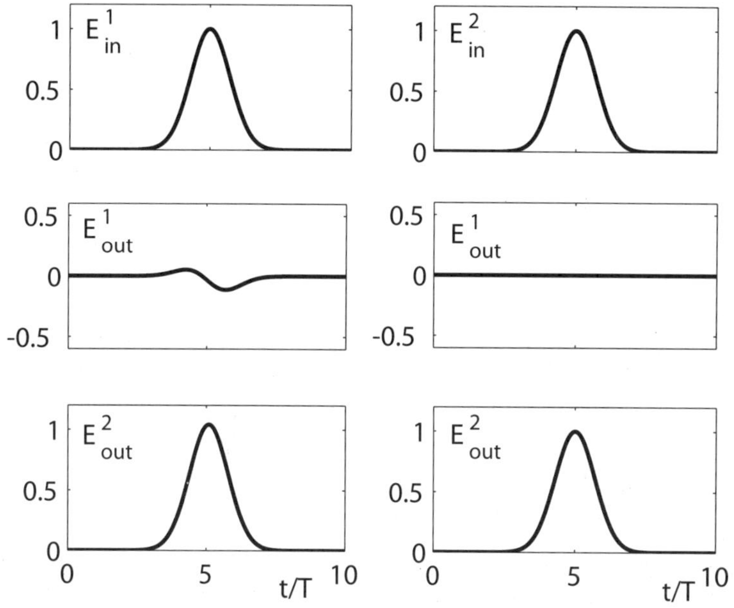

Fig. 2. Left (right) column corresponds to a single photon input in chanel 1 (2). The photon injected into either of the two input modes is transfered to the same output port 2. The amplitude of the field in output channel 1 is negligible.

5. Conclusions

We have shown that a two-mode system interacting with a spectrally broadened ensemble of Λ-type atoms behaves as a collection of two-level systems interacting with a bosonic reservoir. The analogy between these two systems allows the observation and simulations of several interesting phenomena of dissipative processes in tailored reservoirs. In particular we have shown that similarly to spontaneous decay of atoms one can irreversibly transfer photons from one mode to another. The possibility to tailor the reservoir spectrum and to prepare collective quantum states of the ensemble of Λ atoms can be used to observe the quantum Zeno- and anti-Zeno effects, to study the decay in a squeezed reservoir and to observe destructive interference of spontaneous decay channels. As a particular application we have discussed in more detail a photon diode, i.e. a device where a single-photon input in any of the two input ports is always emitted in only one of the two outputs. Besides being intersting in its own right the diode may be used for the implemention of a classical logical OR.

Acknowledgments

The work of G.N. was supported by Alexander von Humboldt Foundation and Armenian National Science and Education Foundation.

References

1. G. Nikoghosyan and M. Fleischhauer, Phys. Rev. Lett. **103**, 163603 (2009).
2. C.W. Gardiner, P.W Zoller, *Quantum Noise* (2nd edition) (Springer-Verlag, Berlin, 2000)
3. B. Misra and E.C.G. Sudarshan, J. Math. Phys. **18**, 756 (1977).
4. A.G. Kofman and G. Kurizki, Nature **405**, 546 (2000).
5. A.G. Kofman and G. Kurizki Phys. Rev. Lett. **87**, 270405 (2001).
6. M. Fleischhauer, A. Imamoglu, and J.P. Marangos, Rev. Mod. Phys. **77**, 633 (2005).
7. K. Bergmann, H. Theuer, and B.W. Shore, Rev. Mod. Phys. **70**, 1003 (1998).
8. B. Julsgaard, A. Kozhekin, and E.S. Polzik, Nature (London) **413**, 400 (2001).
9. B. Julsgaard, J. Sherson, J.I. Cirac, J. Fiurasek, and E.S. Polzik, Nature (London) **432**, 482 (2004).
10. K. Hammerer, E.S. Polzik, and J.I. Cirac, Phys. Rev. A **72**, 052313 (2005).
11. J. Esteve, C. Gross, A. Weller, S. Giovanazzi, and M.K. Oberthaler, Nature **455**, 1216 (2008).
12. M.J. Collet and C.W. Gardiner, Phys. Rev. A **30**, 1386 (1984).
13. C.W. Gardiner, Phys. Rev. Lett. **56**, 1917 (1986).

14. J. Gea-Banachloche, R.R. Schlicher, and M.S. Zubairy, Phys. Rev. A **38**, 3514 (1988).
15. F. Brennecke, T. Donner, S. Ritter, T. Bourdel, M. Köhl and T. Esslinger, Nature **540**, 268 (2007).
16. Y. Colombe, T. Steinmetz, G. Dubois, F. Linke, D. Hunger and J. Reichel, Nature **450**, 272 (2007).
17. S. Gupta, K. L. Moore, K. W. Murch, and D. M. Stamper-Kurn, Phys. Rev. Lett. **99**, 213601 (2007).
18. K. Hammerer, M. Aspelmeyer, E.S. Polzik, P. Zoller Phys. Rev. Lett. **102**, 020501 (2008).
19. K. Jaehne, C. Genes, K. Hammerer, M. Wallquist, E.S. Polzik, P. Zoller, Phys. Rev. A **79** 063819 (2009).
20. M. Tavis and F.W. Cummings, Phys. Rev. **170**, 379 (1968).
21. R.H. Dicke, Phys. Rev. **93**, 99 (1954).
22. J.M. Raimond, M. Brune, S. Haroche, Rev. Mod. Phys. **73**, 565 (2001).
23. H. Walther, B.T.H. Varcoe, B.G. Englert and Th. Becker, Rep. Prog. Phys. **69**, 1325 (2006).
24. S.Y. Zhu, L.M. Narducci, and M.O.Scully, Phys. Rev. A **52**, 4791 (1995).

DISSIPATIVE CHAOS IN QUANTUM DISTRIBUTIONS

T. V. GEVORGYAN [1,*], S. B. MANVELYAN [1]

[1] *Institute for Physical Researches, National Academy of Sciences, Ashtarak-2, 0203, Ashtarak, Armenia*
[*] E-mail: t_gevorgyan@ysu.am
www.ipr.sci.am

A. R. SHAHINYAN[2,**], G. Yu. KRYUCHKYAN[1,2,***]

[2] *Yerevan State University, A. Manoogian 1, 0025, Yerevan, Armenia*
[**] E-mail: anna_shahinyan@ysu.am,
[***] E-mail: kryuchkyan@ysu.am
www.ysu.am

We discuss some problems of dissipative chaos for open quantum systems in the framework of semiclassical and quantum distributions. For this goal, we propose a driven nonlinear oscillator with time-dependent coefficients, i.e. with time-dependent Kerr-nonlinearity and time-modulated driving field. This model showing both regular and chaotic dynamics in the classical limit is realized in several experimental schemes. Quantum dissipative chaos is analyzed on the base of numerical method of quantum trajectories. Three quantities are studied: the Wigner function of oscillatory mode from the point of view of quantum-assemble theory and both semiclassical Poincaré section and quantum Poincaré section calculated on a single quantum trajectory. The comparatively analysis of these distributions for various operational chaotic regimes of the models is performed, as well as scaling invariance in dissipative chaos and quantum interference effects assisted by chaos are discussed.

Keywords: Quantum optics, Nonlinear dynamics.

1. Introduction

Quantum nonlinear systems with chaotic classical counterparts have received much attention in the last two decades. This field of investigation is sometimes called quantum chaos.[1] The majority of studies of quantum chaos for isolated or so-called Hamiltonian systems, the classical counterparts of which are chaotic, focus on static properties such as spectral statistics of energy levels and transition probabilities between eigenstates of the system. A variety of studies have also been carried out to understand the features of time-dependent chaotic systems. In contrast to that very little work has been done to investigate quantum chaos for open nonlinear systems. The beginning of study of an open chaotic system can be dated back to the papers[2] where the authors have analyzed the kicked rotor and similar systems with discrete time interacting with a heat bath. Quite generally, chaos in classical conservative and dissipative systems with noise has completely different properties, e.g., strange attractors can appear only in dissipative systems. For dissipative systems Poincaré section has form of strange attractor in phase space while for Hamiltonian systems it has form of close contours with separatrices.

In classical mechanics a standard characterization of chaos might be given in terms of the unpredictability of phase-space trajectories or Poincaré section that consists of non-localized distributed points. However, the most important characteristic of classical chaotic systems-exponential divergence of trajectories, starting at arbitrarily close initial points in phase space does not have quantum counterpart because of the Hiesenberg uncertainty principle. The question has been posed of what constitutes the quantum mechanical equivalent of chaos. Many criteria have been suggested to define chaos in quantum systems, varying in their emphasis and domain of application.[1,3] As yet, there is no universally accepted definition of quantum chaos.

In recent years much effort has been expended, both theoretically and experimentally, to explore the role of quantum fluctuations and noise in the order-to-chaos transition for open systems. It is obvious that the investigations in this area are connected with the quantum-classical correspondence problem, in general, and with the environment induced decoherence and dissipation, in particular. Recently, this topic has been the focus of theoretical investigations. As a part of these studies, it has been recognized[4] that the decoherence has rather unique properties for systems classical analogs of which are chaotic. In particular, the formation of sub-Planck structure in phase space has been discussed for chaotic system.[5] The connection between quantum and classical treatments of chaos was also realized by means of comparison between strange attractors on the classical Poincaré section and the contour plots of the Wigner functions.[10]

In this paper we investigate some problems of dissipative chaos for open quantum systems in the framework of quantum distributions. We use the traditional ensemble description of Markovian open systems, based on the master equation. This equation is presented in quantum trajectories in the framework of the quantum state diffusion approach.[6] Recently, it was shown how quantum state diffusion can be used to model dissipative chaotic systems on individual quantum trajectories.[7,8] In contrast with these papers, here we show that it is possible to describe quantum chaos using also a statistical ensemble of trajectories, which is actually realized in nature.

The requirement in realization of this study is to have a proper quantum model showing both regular and chaotic dynamics in the classical limit. We propose a driven nonlinear oscillator with time-dependent coefficients for this goal. This model proposed to study the quantum chaos in the series of papers[9–13] allow us to examine challenging problems of quantum dissipative chaos, including the problem of the quantum counterpart of a strange attractor.

The other problem of our interest in this paper relates to the quantum effects in systems with chaotic dynamics. Particularly, in a recent paper, sub-Poissonian statistics of oscillatory excitations numbers was established for chaotic dynamics of nonlinear oscillator.[11] It was shown that quantum-interference phenomena can be realized for the dissipative nonlinear systems exhibiting hysteresis-cycle behavior and quantum chaos.[10,13] The study of these phenomena provides a fundamental understanding of quantum fluctuations in quantum chaos and opens a way for new experimental studies of the quantum dissipative chaos in the field of quantum optics.

The outline of this paper is as follows. In Sec. II we describe both the model and the method of calculations. In Sec. III we analyze correspondence between Poincaré sections and Wigner functions for the chaotic dynamics. In Sec. IV we analyze scaling invariance for quantum system and discuss quantum chaos for the regimes in which the classical chaos is lost. In Sec. V we present results on quantum interference phenomena for chaotic dynamics. We summarize our results in Sec. VI.

2. Driven Nonlinear Oscillator as an Open Quantum System

In this section the systems and the methods of calculations are presented. We treat the Duffing oscillator as an open quantum system and assume that its time evolution is described by Markovian dynamics in terms of the Lindblad master equation for the reduced density matrix ρ. In the interaction picture that corresponds to the transformation $\rho \longrightarrow e^{-i\omega a^+ at}\rho e^{i\omega a^+ at}$, where a^+ and a are the Bose annihilation and creation operators of the oscillatory mode and ω is the driving frequency, this equation reads as

$$\frac{d\rho}{dt} = \frac{-i}{\hbar}[H_0 + H_{int}, \rho] + \sum_{i=1,2}\left(L_i\rho L_i^+ - \frac{1}{2}L_i^+ L_i\rho - \frac{1}{2}\rho L_i^+ L_i\right). \tag{1}$$

The Hamiltonians are

$$H_0 = \hbar\Delta a^+ a,$$
$$H_{int} = \hbar\chi(t)(a^+ a)^2 + \hbar(f(t)a^+ + f(t)^* a), \tag{2}$$

where $\chi(t)$ and $f(t)$, which may or may not depend on time, represent, respectively, the strength of the nonlinearity and amplitude of the force, ω_0 is the resonant frequency, $\Delta = \omega_0 - \omega$ is the detuning. The dissipative and decoherence effects, losses, and thermal noise are included in the last part of the master equation, where L_i are the Lindblad operators:

$$L_1 = \sqrt{(N+1)\gamma}a, \qquad L_2 = \sqrt{N\gamma}a^+, \tag{3}$$

γ is the spontaneous decay rate of the dissipation process and N denotes the mean number of quanta of a heat bath. Here we focus on the pure quantum effects and assume $N = 0$.

This model seems experimentally feasible and can be realized in several experimental schemes. In fact, a single mode e.m. field is well described in terms of an anharmonic oscillator (AHO), and the nonlinear medium could be an optical fiber or a $\chi(3)$ crystal, placed in a cavity. The anharmonicity of mode dynamics comes from the self-phase modulation due to the photon-photon interaction in the $\chi(3)$ medium. In this case, it is possible to realize time modulation of the strength of the nonlinearity by using a media with periodic variation of the $\chi(3)$ susceptibility.

On the other side, the Hamiltonian described by Eq. (2) describes a single nanomechanical resonator with a^+ and a raising and lowering operators related to the position and momentum operators of a mode quantum motion

$$x = \sqrt{\frac{\hbar}{2m\omega_0}}(a + a^+), \; p = -i\sqrt{2\hbar m\omega_0}(a - a^+),$$

where m is the effective mass of the nanomechanical resonator, ω_0 is the linear resonator frequency and χ proportional to the Duffing nonlinearity. One of the variants of nano-oscillators is based on a double-clamped platinum beam[14] for which the nonlinearity parameter equals to $\chi = \hbar/4\sqrt{3}Qma_c^2$, where a_c is the critical amplitude at which the resonance amplitude has an infinite slope as a function of the driving frequency, Q is the mechanical quality factor of the resonator. In this case, the giant nonlinearity $\chi \cong 3.4 \cdot 10^{-4} s^{-1}$ was realized. Note, that details of this resonator, including the expression for the parameter a_c, are presented in.[15] On decreasing nanomechanical resonator mass, its resonance frequency increases, exceeding 1 GHz in recent experiments.[16,17] It is possible to reach a quantum regime for such frequencies, i.e., to cool down the temperatures for which thermal energy will be comparable to the energy of oscillatory quanta. The recent investigations in this direction are devoted to classical to quantum transition of a driven nanomechanical oscillator,[18] generation of Fock states,[19] nonlinear dynamics, and stochastic resonance.[20] Note, that in the last few years there has been rapid progress in the construction and manipulation of such nanomechanical oscillators with giant $\chi(3)$-Kerr nonlinearity. The nanomechanical resonator with a significant fourth-order nonlinearity in the elastic potential energy has been experimentally demonstrated.[21] It has also been shown that this system is dynamically equivalent to the Duffing oscillator with varied driving force.[22] This scheme is widely employed for a large variety of applications as well as the other schemes of micro- and nanomechanical oscillators, more commonly as sensors or actuators in integrated electrical, optical, and optoelectrical systems.[16,23]

Cyclotron oscillations of a single electron in a Penning trap with a magnetic field are another realization of the quantum version of the Duffing oscillator.[24–26] In this case the anharmonicity comes from the nonlinear effect that is caused by the relativistic motion of an electron in a trap, while the dissipation effects arise from the spontaneous emission of the synchrotron radiation and thermal fluctuations of the cyclotron motion. Note that a one-electron oscillator allows one to achieve a relatively strong cubic nonlinearity, $\chi/\gamma \lesssim 1$.

In recent years the study of quantum dynamics of oscillators with time-dependent parameters has been focus of considerable attention. This interest is justified by many applications in different contexts. Particulary, one application concerns to the center of mass motion of a laser cooled and trapped ion in a Paul trap.[27] The quantum dynamics of an AHO with time dependent modulation of its frequency and nonlinearity parameters has been investigated in applications to macroscopic superposition of quantum states.[28]

It is well assessed that in the case of unitary dynamics, without any losses, an anharmonic oscillator leads to sub-Poissonian statistics of oscillatory excitation number, quadratic squeezing and superposition of macroscopically distinguishable coherent states. For dissipative dynamics the important parameter responsible for production of nonclassical states via $\chi(3)$ materials is the ratio between nonlinearity and damping. Therefore, the practical realization of such quantum effects requires a high nonlinearity with respect to dissipation. In this direction the largest nonlinear interaction was proposed in many papers, particularly, in terms of electromagnetically induced transparency[29] and by using the Purcell effect,[30] and in cavity QED.[31] The significant nonlinearity has also been observed for nanomechanical resonators.[14] These methods can lead to $\chi(3)$ nonlinearity of several orders of magnitude higher than natural optical self-Kerr interactions. Note, that high $\chi(3)$ nonlinear oscillators generate also a lot of interest recently due to their applications in areas of quantum computing.[32]

In the case of nonlinear dissipative $\chi(3)$ interaction stimulated by coherent driving force, the time evolution cannot be solved analytically for arbitrary evolution times and suitable numerical methods have to be used. Nevertheless, with dissipation included a driven AHO model has been solved exactly in the steady-state regime in terms of the Fokker-Planck equation in the complex P representation.[33] Analogous solution has been obtained for a combined driven parametric oscillator with Kerr nonlinearity.[34] The Wigner functions for both these models have been obtained using these solutions.[33,35]

The investigation of quantum dynamics of a driven dissipative nonlinear oscillator for non-stationary cases is much more complicated and only a few papers have been done in this field up to now. More recently, the quantum version of dissipative AHO or the Duffing oscillator with time-modulated driving force has been studied in the series of the papers[9–11] in the context of a stochastic resonance,[9] quantum-to-classical transition and investigation of quantum dissipative chaos.[10,11]

For the constant parameters $\chi(t) = \chi$ and $f(t) = f$ the equations (1) and (2) describe the model of a driven dissipative AHO that was introduced long ago in quantum optics to describe bistability due to a Kerr nonlinear medium.[36] For the case of time-dependent parameters $\chi(t)$ and $f(t)$ the dynamics of the AHO exhibits a rich phase-space structure, including regimes of regular, bistable and chaotic motion. We perform our calculations for regular and chaotic regimes concerning two models of time-modulated AHO corresponding to two physical situations: (i) $\chi = \chi(t) = \chi_0 + \chi_1 sin(\Omega t)$ and $f(t) = const = f$; (ii) $f = f(t) = f_0 + f_1 \exp(\delta t)$ and $\chi(t) = const = \chi$ with $\delta \ll \omega$ and $\Omega \ll \omega$ are the modulation frequencies.

We analyze the master equation numerically using quantum state diffusion method (QSD).[6] According to this method, the reduced density operator is calculated as an ensemble mean

$$\rho(t) = M(|\psi_\xi(t)\rangle\langle\psi_\xi(t)|) = \lim_{N\to\infty}\frac{1}{N}\sum_{\xi}^{N}|\psi_\xi(t)\rangle\langle\psi_\xi(t)| \tag{4}$$

over the stochastic pure states $|\psi_\xi(t)\rangle$ describing evolution along a quantum trajectory. The stochastic equation for the state $|\psi_\xi(t)\rangle$ involves both Hamiltonian described by Eq. (2) and the Linblad operators described by Eq. (3) and reads as:

$$|d\Psi_\xi\rangle = -\frac{i}{\hbar}H|\Psi_\xi\rangle dt - \frac{1}{2}\sum_{i=1,2}(L_i^+L_i - 2 < L_i^+ > L_i$$

$$+ < L_i >< L_i^+ >)|\Psi_\xi\rangle dt + \sum_{i=1,2}(L_i - < L_i >)|\Psi_\xi\rangle d\xi, \tag{5}$$

where ξ is the generated complex gaussian noise that satisfies the following conditions:

$$M(d\xi_i) = 0, M(d\xi_i d\xi_j) = 0, M(d\xi_i d\xi_j^*) = \delta_{ij}dt. \tag{6}$$

We calculate the density operator using an expansion of the state vector $|\psi_\xi\rangle$ in a truncated basis of Fock's number states of a harmonic oscillator

$$|\psi_\xi(t)\rangle = \sum_n a_n^\xi(t)|n\rangle. \tag{7}$$

3. Wigner Function and Poincaré Section

In this section we shortly examine correspondence between the Poincaré section and the Wigner function of the oscillatory mode considering the model (ii) with time-dependent driving amplitude.[10,11] In semiclassical treatment a chaotic operational regime is analyzed on phase space of dimensionless position and momentum $x = Re(\alpha)$ and $y = Im(\alpha)$, where $\alpha =< a >$ is the oscillatory complex amplitude (see, Eq.(11)). Choosing x_0 and y_0 as an arbitrary initial phase-space point of the system at the time t_0, we define a constant phase map in the plane by the sequence of points at $t_n = t_0 + (2\pi/\delta)n$, where $n = 0, 1, 2....$ This means that for any $t = t_n$ the system is at one of the points of the Poincaré section.

The analysis show that for time scales exceeding the damping time, $t \gg \gamma^{-1}$, the asymptotic dynamics of the system is regular in the limits of small and large values of the modulation frequency, i.e., $\delta \gg \gamma$, $\delta \ll \gamma$ and small or large amplitude of driving field $f_1 \ll f_0$, $f_1 \gg f_0$. Fig. 1(c) shows the results of numerical calculations of the classical maps, for the parameters chosen in the range of chaos, i.e. $f_0 \approx f_1$ and $\delta \geq \gamma$. As we see, the figure clearly indicate the classical strange attractor with fractal structure that is typical for a chaotic dynamics.

It should be mentioned, that the Wigner function is one of the quantities that allows to observe chaos in quantum theory. The nonstationary Wigner function is written as

$$W(\alpha,t) = \frac{2}{\pi^2}\exp(-2|\alpha|^2)\int d^2\beta\langle-\beta|\psi(t)\rangle\langle\psi(t)|\beta\rangle\exp\left(-2(\beta\alpha^* - \beta^*\alpha)\right) \tag{8}$$

for the state vector $|\psi(t)\rangle$. We apply the QSD to determine Wigner functions for the quantum states of a driven anharmonic oscillator during time evolution. For this, we use the well-known expression for the Wigner function in terms of the matrix elements $\rho_{nm} = \langle n|\rho|m\rangle$ of the density operator in the Fock state representation:

$$W(r,\theta) = \sum_{n,m}\rho_{nm}(t)W_{mn}(r,\theta). \tag{9}$$

Here: (r,θ) are the polar coordinates in the complex phase-space plane, $x = r\cos\theta$, $y = r\sin\theta$, while the coefficients $W_{mn}(r,\theta)$ are the Fourier transform of matrix elements of the Wigner characteristic function:

$$W_{mn}(r,\theta) = \begin{cases} \frac{2}{\pi}(-1)^n\sqrt{\frac{n!}{m!}}\exp i(m-n)\theta(2r)^{m-n} \\ \qquad \exp(-2r^2)L_n^{m-n}(4r^2), \ m \geq n \\ \frac{2}{\pi}(-1)^m\sqrt{\frac{m!}{n!}}\exp i(m-n)\theta(2r)^{n-m} \\ \qquad \exp(-2r^2)L_m^{n-m}(4r^2), \ n \geq m. \end{cases}, \tag{10}$$

It is remarkable that there is a correspondence between contour plots of the Wigner function and the Poincaré section. This point is illustrated on the Figs. 1, where the Wigner function Fig. 1(a), its contour plot, Fig. 1(b) and Poincaré section, Fig. 1(c) are presented for the same parameters. Note, that the Wigner function is a quasidistribution in phase space averaging an ensemble of quantum trajectories obtained for a define time moment while Poincaré section is the distribution for time intervals: it is constructed by fixing points in phase space at a sequence of periodic moments. What we can conclude from their correspondence is the fact, that the quantum

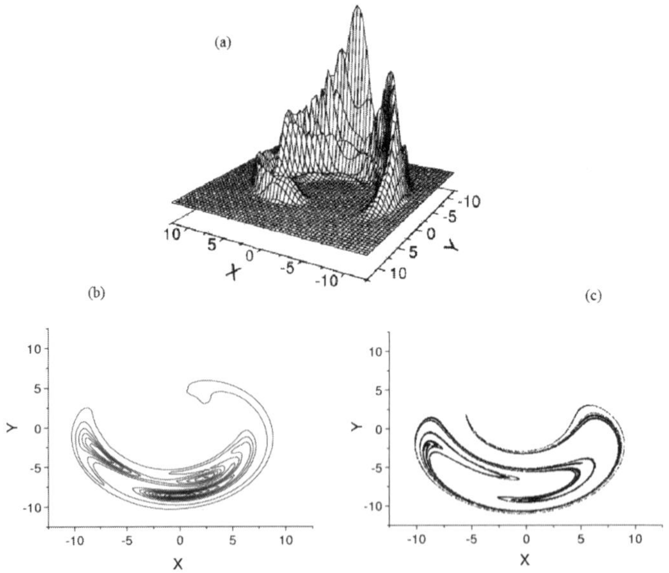

Fig. 1. The Wigner function (a) and its contour plot (b) averaged over 3000 trajectories. The Poincaré section (c) (20000 points plotted at times of the constant phase for the same parameters). In both cases, the dimensionless parameters are in the range of chaos, i.e. $\chi/\gamma = 0.1$, $\Delta/\gamma = -15$, $f_0/\gamma = f_1/\gamma = 27$, $\delta/\gamma = 5$.

distribution in phase space corresponds on the form to the semiclassical distribution but for big numbers of time intervals. As it is seen, the Poincaré section has fine fractal structure while Wigner function contour plot has not. This is due to the Hiesenberg uncertainty relations which prevent sub-Plank structures in phase space.

4. Scaling Invariance in Dissipative Chaos

In this section the scaling invariance for the case of chaotic dynamics is considered on the base of anharmonic oscillator with time-dependent driven force. In the classical limit the system is described by the following equation of motion for the dimensionless amplitude:

$$\frac{d\alpha}{dt} = -\frac{\gamma}{2}\alpha - i(\Delta + \chi(1 + 2|\alpha|^2))\alpha - i(f_0 + f_1 \exp(-\delta t)). \qquad (11)$$

This equation is invariant for the scaling transformation of complex amplitude $\alpha = \lambda\alpha$ if the other parameters are transformed like: $\Delta \to \Delta' = \Delta + \chi(1 - 1/\lambda^2)$, $\chi \to \chi' = \chi/\lambda^2$, $f \to f' = \lambda f$, $\gamma \to \gamma' = \gamma$. This scaling property of the classical equation for the chaotic dynamics leads to the symmetry of strange attractors: they have the same form in the phase space and differ from each other only in scale. This fact is demonstrated in the Figs. 2.

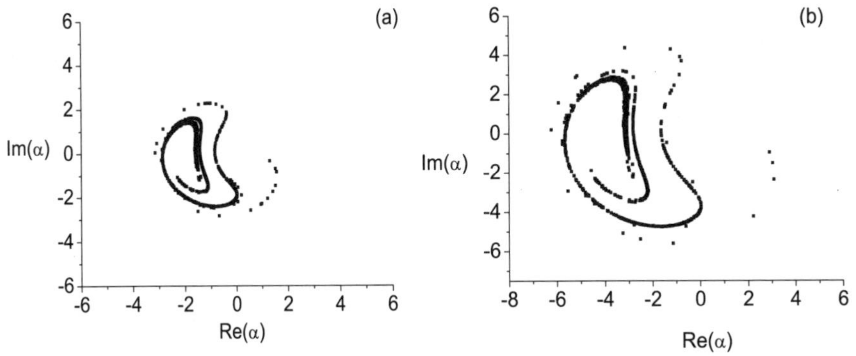

Fig. 2. The Poincaré section for the following parameters: $\Delta/\gamma = -15$, $\chi/\gamma = 2$, $f_0/\gamma = 5.8$, $f_1/\gamma = 4.9$, $\delta/\gamma = 2$, (a); for the scaled $\lambda = 2$ parameters (b). The Poincaré sections are generated for time evaluation 5000 force periods.

Strictly speaking the quantum system does not obey the same scaling invariance as classical one. An analysis of scaling invariance from the point of view of quantum-statistical theory has been performed.[12] It was shown that such parameter scaling occurs for wider ranges of the parameters, but for not large values of the parameter χ/γ, where system displays strong quantum properties.

Now we use the scaling arguments considering stochastic dynamic of a single trajectory. In this way, to investigate chaos we use Poincaré section based on evolution of a single quantum trajectory, Eq.5. The Poincaré section is obtained by recording $(Re(\alpha_\xi), Im(\alpha_\xi))$ at time intervals of $2\pi/\delta$, where ξ indicates the stochastic variable and α_ξ is obtained from Eq. 5. Our goal is to analyze the scaling invariance on the base of this quantity. On the other side we present results that relies on quantum-classical correspondence.

In this direction, we consider the specific parameters for which DAO is in the vicinity of chaotic behavior determined classically. That means, if the parameters are slightly tuned in this range the transition from chaotic to regular dynamics might be realized. As we have realized above the dynamics of the system is chaotic in the range of parameters $f_0 \simeq f_1$ and $\delta \geq \gamma$. Particularly, it is demonstrated on the base of the semiclassical Poincaré section the system exhibits chaotic dynamics for $f_0/\gamma = f_1/\gamma = 5.8$, $\Delta/\gamma = -15$ and $\delta/\gamma = 2$ in classical treatment. As analysis shows for the considered parameters the system dynamics continues to be chaotic till $f_1/\gamma = 4.9$ (Fig. 2), while chaotic dynamics becomes regular for $f_1/\gamma = 4.8$. Thus, we examine Poincaré section in quantum trajectory started from the regular regime, for the parameter $f_1/\gamma = 4.8$. The results of calculations are presented in Fig. 3 for the parameters: $\Delta/\gamma = -15$, $\chi/\gamma = 2$, $f_0/\gamma = 5.8$, $\delta/\gamma = 2$, $f_1/\gamma = 4.8$. As we see, for the scaled parameter $\lambda = 1$ the Poincaré section is distributed stochastically but approximately exhibits localization in two ranges (Fig. 3(a)); its shape does not coincide with Poincaré section (Fig. 2(c)). For the case of $\lambda = 2$, points of the Poincaré section on a trajectory are distributed chaotically, however, unlike to the previous case $\lambda = 1$, here a correspondence between both shapes takes place. This likeness become more obvious for the case of $\lambda = 3$ (Fig. 3(e)).

It is important to note that in the semi-classical treatment all cases $\lambda = 1, 2, 3$ correspond to regular regime as it is expected due to scaling invariance. Another situation is realized in the quantum treatment. For $\lambda = 1$ the system seems to be in regular regime. This statement is also confirmed by calculation of the Wigner function. The calculations shows that for such parameters the Wigner function has two-peak structure indicating that the regime of bistability is realized. If we increase the scaling parameter the resulted Poincaré section's shapes in a quantum trajectory in their forms seem to be close to the Poincaré sections in semiclassical treatment. So we found the parameters for which the system in classical treatment has regular dynamic while in quantum treatment its dynamic is chaotic.

The results on mean excitation numbers are presented on Fig. 3(b, d, f). As we see, for $\lambda = 1$, the oscillatory excitation numbers varies from 1 to 4 and thus the system is in deep quantum regime. The level of quantum noise is comparatively sufficient and chaos cannot exhibit itself on Poincaré section. But for scaled parameters $\lambda = 2, 3$ the excitation numbers increase and varies up to 40 (Fig. 3(d,f)). For these cases the ranges of variation are much enough to exhibit a chaotic type structures.

Note, the usefulness of the scaling procedure. It allows us to analyze classically analogy regimes for various groups of the parameters. In particular, increasing scaling parameter of the system we can investigate quantum-classical correspondence.

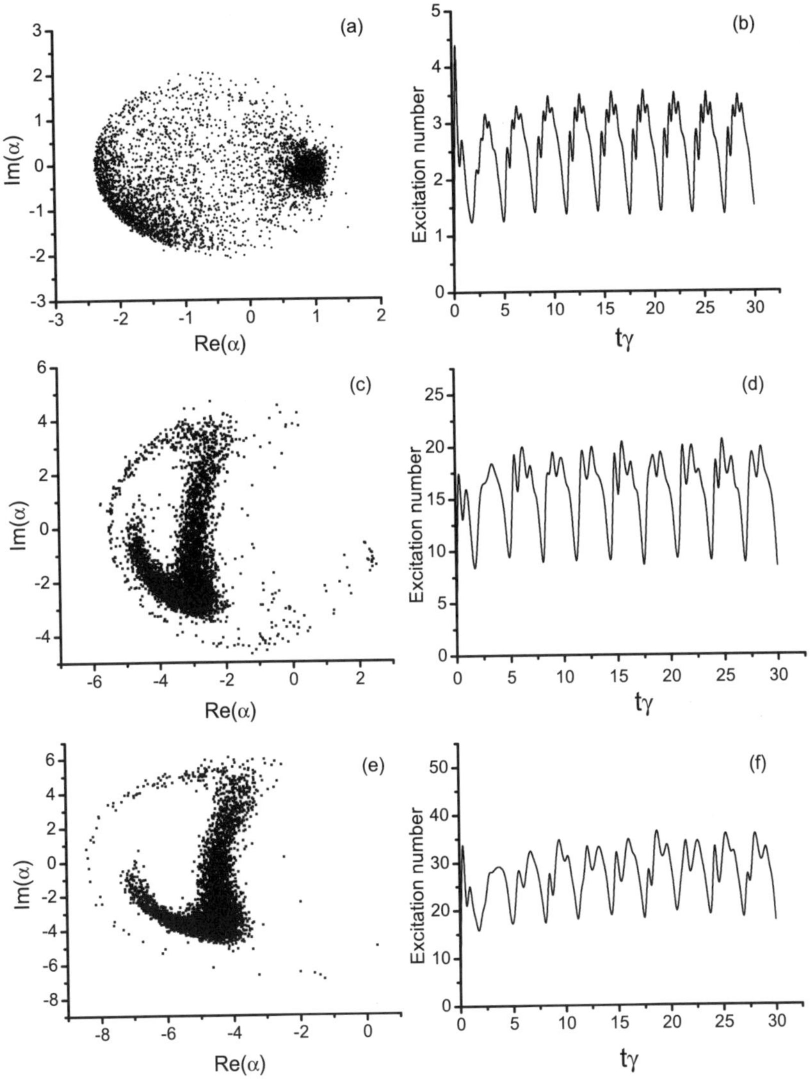

Fig. 3. The Poincaré section calculated on the base of a single trajectory for the following parameters: $\Delta/\gamma = -15$, $\chi/\gamma = 2$, $f_0/\gamma = 5.8$, $f_1/\gamma = 4.8$, $\delta/\gamma = 2$, (Fig. 3 (a)); the excitation number for the same parameters (Fig. 3(b)). Figs. 3(c) and 3(d) corresponds to scaled $\lambda = 2$ parameters. Figs. 3(e) and 3(f) corresponds to scaled $\lambda = 3$ parameters. The Poincaré sections are generated for time evaluation 5000 force periods.

5. Quantum Interference Assisted by Chaos

It is well assessed that in the case of unitary dynamics, without any losses, an anharmonic oscillator leads to sub-Poissonian statistics of oscillatory excitation number, quadratic squeezing, and superposition of macroscopically distinguishable coherent states. The dissipation and decoherence lead usually to losses of these effects. In this section, we numerically analyze the phenomena at the overlap of chaos, dissipation, and quantum effects for the time-dependent nonlinear model. It was recently shown[11–13] that physical systems based on this model have a potential for generation of high-degree sub-Poissonian light as well as for the observation of quantum-statistical effects and quantum interference that accompanied by chaotic dynamics. Here we concentrate on studies of quantum interference. We have pointed out that the time modulation of the oscillatory parameters, which are the strength of third-order nonlinearity or the amplitude of the driving force, leads to formation of the quantum-interference patterns in phase space in over transient regimes, for the definite time intervals exceeding the transient dissipation time.

It is well known that the Wigner distribution function in phase-space can simply visualize nonclassical effects including quantum-interference. For example, a signature of quantum interference is exhibited in the Wigner function by non-positive values. In this section the numerical results of the nonstationary Wigner functions in chaotic regimes of AHO are presented and discussed.

It should be noted that the most of investigations of the quantum distributions of oscillatory states, including also modes of radiations, have been made for the steady-state situations. The simplicity of Kerr nonlinearity allows to determine the Wigner function of the quantum state under time evolution due to interaction. In this sense, we note the main peculiarity of our paper in comparison with above noted important inputs. In this paper, we calculate the Wigner functions in an over transient regime, $t \gg \gamma^{-1}$, of the dissipative dynamics, however, we consider time-dependent effects which appear due to the time-modulation of the oscillatory parameters.

Below we investigate the Wigner functions for the case of time-modulated nonlinearity. Note that the dynamics of the system with time-modulated nonlinear strength AHO is chaotic in the ranges: $\delta \geq \gamma$ and $\chi_0 \simeq \chi_1$ and for negative detuning. As analysis show, controlling transition from the regular to chaotic dynamics can be realized through the intermediate ranges of bistability by varying the strength χ_1 of the modulation processes in the ranges from $\chi_1 \ll \chi_0$ to $\chi_1 \leq \chi_0$. The results of

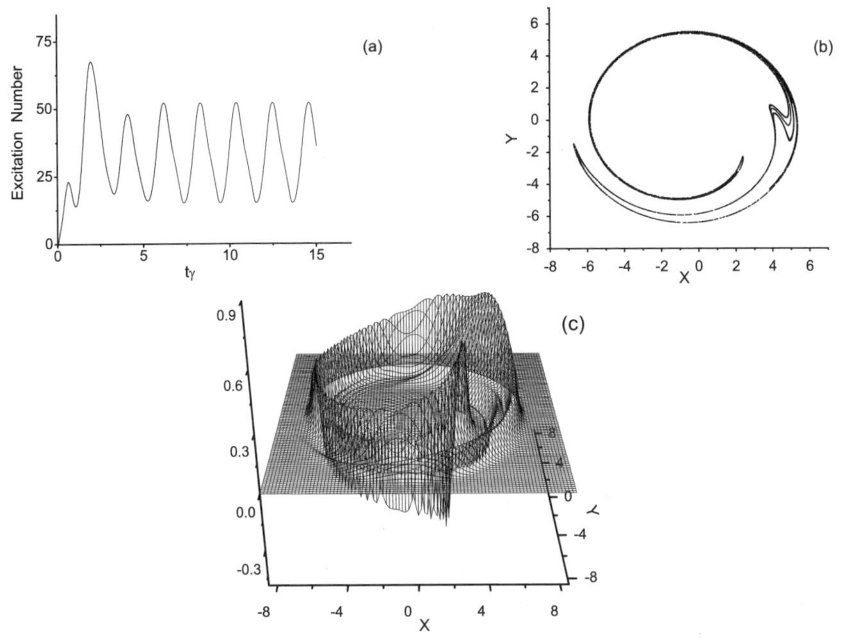

Fig. 4. The mean excitation number (a); the *Poincaré* section (b) ($\approx$ 20000 points) for the dimensionless complex amplitude α, plotted at times of the constant phase $\delta t_k = 2\pi k$ (k = 0,1,2...), when the maximal interference pattern on the Wigner function (c) is realized, for the case of time-modulated nonlinearity $\chi(t)$. The parameters are in the range of chaos: $\Delta/\gamma = 5, \chi(t)/\gamma = 0.2(1 + 0.75\sin(\Omega t)), \Omega/\gamma = 3, f/\gamma = 10$.

the ensemble averaged numerical calculation of the mean excitation number, the Poincaré section and the Wigner function are shown on the Figs. 4, respectively. The mean excitation number of the driven AHO versus dimensionless time is depicted in Fig. 4 (a). In contrast to the semiclassical result its quantum ensemble counterpart (see, Fig. 4(a)) has clear regular periodic behavior for time intervals exceeding the characteristic dissipation time, due to ensemble averaging. The Fig. 4(b) clearly indicates the classical strong attractors with fractal structure that are typical for a chaotic Poincaré section. Thus, the Wigner function (Fig. 4(c)) reflects the chaotic dynamics, its contour plots in the (x, y) plane are similar to the Poincaré section. However, the Wigner functions have regions of negative values for the definite time intervals. The example depicted on Fig. 4(c) corresponds to time intervals $\gamma t_k = 6 + \frac{2\pi k}{\delta}\gamma$ $(k = 0, 1, 2...)$, for which the mean excitation number reaches a macroscopic level, i.e., $n = 52$. The interference pattern is destroyed as the time modulation is decreased. Indeed, it is shown[33,35] that for an oscillatory mode of driven anharmonic oscillator, including dissipation, the Wigner function is positive in all phase space.

6. Conclusion

In summary, we have discussed dissipative chaos answering what is the counterpart of the semiclassical Poincaré section in quantum treatment. We have presented a type of nonstationary systems showing chaotic dynamics and intrinsically quantum properties which are modelled by a driven dissipative anharmonic oscillator with time-dependent parameters. The connection between quantum and classical treatments of chaos has been realized by means of a comparison between strange attractors on the semiclassical Poincaré section, the shapes of the Poincaré section on a single quantum trajectory and the contour plots of the Wigner functions. We have analyzed the borders of validity of scaling invariance for quantum dissipative chaos and we have demonstrated realization of long-lived quantum interference assisted by chaotic dynamics for over transient regime and for the macroscopic level of oscillatory excitation numbers.

Acknowledgments

This work was supported by NFSAT/CRDF grant UCEP-02/07 and ISTC grant A-1606.

References

1. M. C. Gutzwiller, Chaos in Classical and Quantum Systems (Springer-Verlag, Berlin, 1990); Quantum Chaos, Quantum Measurement, edited by P. Citanovic, I. Percival, and A. Wizba (Kluwer, Dordrecht, 1992); K. Nakamura, Quantum Chaos, A New Paradigm of Nonlinear Dynamics, Vol. 3 of Cambridge Nonlinear Science Series (Cambridge University Press Cambridge, 1993); Quantum Chaos, edited by G. Casati and B. Chirikov (Cambridge University Press, Cambridge, 1995); F. Haake, Quantum Signatures of Chaos (Springer-Verlag, Berlin, 2000).
2. E. Ott, M. Antonsen, Jr., and J.D. Hanson, Phys. Rev. Lett. **53**, 2187 (1984); T. Dittrich and R. Graham, Ann. Phys. (N.Y.) **200**, 363 1990).
3. A. Peres, in Quantum Chaos: Proceedings of the Adriatico Research Conference on Quantum Chaos, edited by H. A. Cerdeira et al. (World Scientific, Singapore 1991); A. Peres, Quantum Theory: Concepts and Methods (Kluwer, Dordrecht, 1993); R. Schack and C.M. Caves, Phys. Rev. Lett. **71**, 525 (1993).

4. W. H. Zurek and J. P. Paz, Phys. Rev. Lett. **72**, 2508 (1994); **75**, 351 (1995); S. Habib, K. Shizume, and W. H. Zurek, ibid. **80**, 4361 (1998); J. Gong and P. Brumer, Phys. Rev. E **60**, 1643 (1999).
5. W. H. Zurek, Nature (London) **412**, 712 (2001).
6. N. Gisin and I. C. Percival, J. Phys. A **25**, 5677 (1992); **26**, 2233 (1993); **26**, 2245 (1993); I. C. Percival, Quantum State Diffusion(Cambridge Unoversity Press, Campridge, (2000).
7. T. P. Spiller and J.F. Ralph Phys. Lett. A **194**, 235 (1994).
8. T.A. Brun, I.C. Percival, and R. Schack, J. Phys. A **29**, 2077 (1996); T. Bhattacharya, S. Habib, and K. Jacobs, Phys. Rev. Lett. **85**, 4852 (2000).
9. H. H. Adamyan, S. B. Manvelyan, and G. Yu. Kryuchkyan Phys. Rev. A **63**, 022102 (2001).
10. H. H. Adamyan, S. B. Manvelyan, and G. Yu. Kryuchkyan Phys. Rev. E**64**, 046219 (2001).
11. G. Yu. Kryuchkyan and S. B. Manvelyan, Phys. Rev. Lett. **88**, 094101 (2002);
12. G. Yu. Kryuchkyan and S. B. Manvelyan, Phys. Rev. A**68**, 013823 (2003).
13. T. G. Gevorgyan, A, R. Shahinyan, G. Yu. Kryuchkyan Phys. Rev. A. **79**, 062905, (2009).
14. I. Kozinsky, H. W. Ch. Postma, O. Kogan, A. Husain and M. L. Roukes, Phys. Rev. Lett. **99**, 207201 (2007).
15. H. W. Ch. Postma, I. Kozinsky, A. Hussian, and M. L. Roukes, Appl. Phys. Lett. **86**, 223105 (2005).
16. X. M. H. Huang, C. A. Zorman, M. Mehregany, and M. L. Roukes, Nature (London) **412**, 496 (2003); K. C. Schwab and M. L. Roukes, Phys. Today **58**, N7, 36 (2005).
17. A. N. Cleland and M. R. Geller, J. App. Phys. **92**, 2758 (2002); A. N. Cleland and M. R. Geller, Phys. Rev. Lett. **93**, 070501 (2004).
18. I. Katz, A. Retzker, R. Straub, and R. Lifshitz, Phys. Rev. Lett. **99** 040404 (2007).
19. E. Buks, E. Segev, S. Zaitsev, B. Abdon, and M. P. Blencowe, Europhys. Lett., **81** 10001 (2008).
20. R. Almog, S. Zaitsev, O. Shtempluck, and E. Buks, App. Phys. Lett. **90**, 013508 (2007); J. S. Aldridge and A. H. Cleland, Phys. Rev. Lett **94**, 156403 (2005).
21. R. Almog, S. Zaitsev, O. Shtempluck, and E. Buks, Phys. Rev. Lett. **98**, 078103 (2007).
22. E. Babourine-Brooks, A. Doherty, G. J. Milburn, quant-ph/ 0804.3618v1, (2008).
23. M. P. Blencowe, Phys. Rep. **395** 159 (2004); T. J. Kippling and K. J. Vahala, Opt. Expr. **15**,17172 (2007); M. Aspelmeyer and K. Schwab, New J. Phys. **10**, 095001 (2008).
24. A. E. Kaplan, Phys. Rev. Lett. **48**, 138 (1982).
25. D. Enzer and G. Gabrielse, Phys. Rev. Lett. **78**, 1211 (1997); G. Gabrielse, H. G. Dehmelt, and W. Kells, ibid. **54**, 537 (1985).
26. M. Rigo, G. Alber, F. Mota-Furtado, and P. F. OMahony, Phys. Rev. A**55**, 1665 (1997); A**58**, 478 (1998).

27. F. Diedrich, J. C. Berquist, W. M. Itano, and D. J. Wineland, Phys. Lett. **62**, 403 (1989); D. J. Wineland, C. Monroe, W. M. Itano, D. Liebfred, B. E. King and D. M. Mekhot, J. Res. Natn. Inst. Standt. Technol. **103**, 259 (1998).
28. S. Mancini and P. Tombesi, Phys. Rev. **A52**, 2475 (1995).
29. A. Imamoglu, H. Schmidt, G. Woods, and M. Deutsch, Phys. Rev. Lett. **79**, 1467 (1997); M. J. Werner and A. Imamoglu, Phys. Rev. A **61** 011801(R) (1999); M. Fleischhauer A. Imamoglu, and J. P. Marangos, Phys. Mod. Phys. **77**, 633 (2005).
30. P. Bermel, A. Rodriguez, J. D. Joannopoulos, and M. Soljacic, Phys. Rev. Lett. **99**, 053601 (2007).
31. F. G. S. L. Brandao, M. Hertmann, and M. B. Plenio, New J. Phys. **10**, 043010 (2008).
32. Q. A. Turchette , C. J. Hood, W. Lange, H. Mabuchi, and H. J. Kimble, Phys. Rev. Lett. **75**, 4710 (1995); K. Nemoto, W. J. Munro, Phys. Rev. Lett. **93**, 250502 (2004); W. J. Munro, K. Nemoto and T. P. Spiller, New J. Phys. **7**, 137 (2005); J. Lee, M. Paternostro, C. Ogden, Y. W. Cheong, S. Bose and M. S. Kim, New J. Phys. **8**, 23 (2006).
33. K. V. Kheruntsyan J. Opt. B: Quantum Semiclass. Opt. **1**, 225 (1999).
34. G. Yu. Kryuchkyan and K. V. Kheruntsyan Opt. Commun **120**, 132 (1996).
35. K. V. Kheruntsyan D. S. Krahmer, G. Yu. Kryuchkyan and K. G. Petrossian, Opt. Commun **139**, 157 (1997).
36. P. D. Drummond and D. F. Walls, Phys. Rev. A **23** 2563 (1981).

FREQUENCY CHIRPED LASER PULSES IN ATOMIC PHYSICS: COHERENT CONTROL OF INNER AND TRANSLATIONAL QUANTUM STATES

G. P. DJOTYAN, J. S. BAKOS, ZS. SÖRLEI, G. DEMETER, N. SÁNDOR, D. DZSOTJAN, M. Á. KEDVES, B. RÁCZKEVI, P. N. IGNÁCZ, J. SZIGETI

KFKI-Research Institute for Particle and Nuclear Physics, H-1121, Hungary
E-mail: djotjan@rmki.kfki.hu

We present our recent results on applications of the frequency modulated (chirped) laser pulses in atomic physics. These applications include coherent control of populations and coherences of multilevel atoms, as well as coherent manipulation of translational states of the atoms by frequency chirped laser pulses. Results of the theoretical analysis are compared with the results of experiments on coherent acceleration of Rb atoms cooled in magneto-optical trap.

1. Introduction

The adiabatic following method is the basis for efficient transfer of the population of an atomic (or molecular) state to a target state or for creation of a coherent superposition of quantum states.[1–12] This method is robust against small-to-medium variations of main parameters of the laser fields in contrary with the use of laser pulses having a given area of the Rabi frequency (integral of the Rabi frequency over time). The latter is sensitive already to small variations of this area and to resonance conditions between the interacting laser field and the atomic transitions.[13] The mostly used schemes of the adiabatic following include the stimulated Raman adiabatic passage (STIRAP),[1–4] Stark chirped rapid-adiabatic passage (SCRAP)[5,6] and schemes involving frequency modulated (chirped) laser radiation.[7–12] In Ref. (11), a complete population transfer between meta-stable states of the Λ-atom was demonstrated without considerable excitation of the atom using a single frequency-chirped laser pulse. A similar scheme but in a tripod-type atom was used in Ref. (12) to create a coherent superposition of three ground states of the atom. The initial coherent preparation of the medium has led to different important effects and application possibilities in the fields of quantum and nonlinear optics revealed in the last decades. They include extreme change in the group velocity of propagation of laser pulses (including even complete stopping of the pulses (see review papers (14,15)), significant enhancement of the efficiency of nonlinear optical processes with relatively low-intensity pumping,[16] applications in quantum computing,[17] optical information processing using metastable quantum states,[18–20] etc.) Variation of the energy of a quantum system is accompanied by variation of the mechanical momentum of the system. Exchange of energy and momentum between laser field and resonant atomic or molecular systems is the physical basis for the mechanical action of the laser radiation on these systems. The result of this action is greatly determined by the ability of the laser photons to transfer completely the populations between atomic quantum states (see e.g. Refs (1,21)). For some applications (e.g. in atomic interferometry, or coherent manipulation of cold atomic ensembles without heating them and preserving atomic coherency), a coherent transfer of the mechanical momentum is required. Such transfer may be completed utilizing STIRAP-based schemes with a limited number of photon momentum $\hbar k$ (with k being the wave vector of the laser radiation) transferred to the atoms.[1] The amount of the mechanical momentum transferred to an atomic ensemble may be substantially increased using sequences of pairs of short counter-propagating laser pulses,[21] which produce fast excitation and stimulated de-excitation of the atoms. As a result, the atom receives mechanical momentum equal to $2\hbar k$ after the action of a pair of the counter-propagating laser pulses. Obviously, this mechanism will work if each laser pulse is able to perform near complete transfer of the atomic populations between the states.[13,22] Short FC (frequency-chirped) laser pulses have been used for effective manipulation of the translational states of the atoms without heating the atomic ensemble in Refs. (23,24). When the counter-propagating FC pulses partially overlap, the effect of multi-photon transitions between the atomic quantum states may result in the transfer of the mechanical momentum equal to $N\hbar k$ (with integer N) instead of $2\hbar k$ in the case of the separated pulses.[25,26] The former value of the transferred momentum may substantially exceed the maximum value of the transferred momentum in the case of separated pulses in the interaction regimes when $N > 2$.

In this paper, we present our recent results on creation of coherent superposition of meta-stable (ground) states in a tripod-model atom by FC laser pulses and using them for coherent storage of optical phase and amplitude information. We also present our results on application of the FC laser pulses for efficient transfer of mechanical momentum from the laser field to an atomic ensemble. Interaction of counter-propagating and partially overlapping FC laser pulses with a cloud of Rb^{85} atoms cooled and trapped in magneto-optical trap (MOT) is modeled and the resulting acceleration of the atomic cloud is measured experimentally.

2. Coherent Control of Quantum States of a Tripod-Structured Atom

In this Section, we consider an atom with tripod-structure of the working levels in the field of three FC laser pulses. We assume that the interaction time in all cases of our study in this paper is shorter, than any relaxation time of the atomic system including decay time of the excited state and any coherence relaxation time. This allows us to deal with the Schrödinger equation for the state probability amplitudes. Each laser pulse in the proposed scheme is interacting with a corresponding allowed transition (see Inset in Fig. 1): The population transfer is only allowed between the excited state $|0\rangle$ and the meta-stable states ($|1\rangle$, $|2\rangle$ and $|3\rangle$) while single-photon transitions are forbidden between the meta-stable states. The two laser fields connecting the states $|1\rangle$ and $|2\rangle$ are in Raman resonance with the transition between these states. The third laser field has nonzero Raman detuning for the two-photon transition between the state $|3\rangle$ and each of the other two states ($|1\rangle$ and $|2\rangle$). This Raman detuning is supposed to be larger than the width of the Fourier spectrum of the pulse envelopes (without chirp) of the interacting laser fields. The Schrödinger equation for the column atom state-vector $\underline{a}$ has the form:

$$i\hbar\frac{d}{dt}\underline{a} = \hat{H}\underline{a}, \tag{1}$$

with the Hamiltonian $\hat{H}$ in the rotating-wave approximation in the form:

$$\hat{H} = \hbar \begin{pmatrix} \Delta_1 - \alpha t & \Omega_1^*(t) & \Omega_2^*(t) & \Omega_3^*(t) \\ \Omega_1(t) & 0 & 0 & 0 \\ \Omega_2(t) & 0 & -\Delta_{21} & 0 \\ \Omega_3(t) & 0 & 0 & -\Delta_{31} \end{pmatrix}, \tag{2}$$

where $2\Omega_k(t) = \frac{E_k(t) d_{0k}}{\hbar}$ is the $\mathrm{k^{th}}$ Rabi-frequency, and d_{0k} is dipole matrix element for transition between the $\mathrm{k^{th}}$ ground state and the excited state (k=1,2,3). The electric field strength of the laser field interacting with the atom has the form $E(t) = \sum_{i=1}^{3} E_i(t) \cos\left(\int_{-\infty}^{t} \omega_i(t')\, dt'\right)$. A same linear

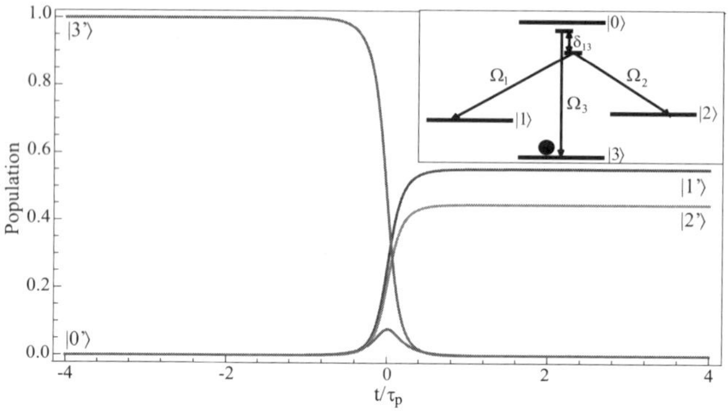

Fig. 1. Atomic states population dynamics in the field of three FC laser pulses. Inset: the scheme of the tripod-model atom with three laser pulses with same envelopes and same frequency chirp.

variation in time (linear frequency chirp) of the carrier frequencies of the interacting laser pulses is assumed in this paper: $\omega_k(t) = \omega_k^{(0)} + \alpha t$, where $\omega_k^{(0)}$ is the central carrier frequency of the kth field and α is the speed of the chirp. The central frequency detuning Δ_k from the resonance atomic frequency ω_{0k} is $\Delta_k = \omega_{0k} - \omega_k^{(0)}$, $k \in \{1,2,3\}$, and $\Delta_{j1} = \Delta_j - \Delta_1$, $j \in \{2,3\}$.

In the scheme proposed in this paper, we assume that the two laser fields $E_1(t)$ and $E_2(t)$ are in Raman resonance with the transition between the states $|1\rangle$ and $|2\rangle$, which means that $\Delta_{21} = 0$ in the Hamiltonian of Eq. (2). In contrary, the Raman detuning for the transition between the state $|3\rangle$ and each of the other two states $|1\rangle$ and $|2\rangle$ is large compared with the bandwidths $\delta\omega_K$ of the laser pulses (without chirp): $|\delta_{31}| = |\Delta_3 - \Delta_1| \gg \delta\omega_k$, (k=1,2,3).

2.1. *Coherent transfer of populations and creation of coherent superposition of metastable states*

We assume that initially the population of the atom is optically pumped into the state $|3\rangle$, (see Inset in Fig. 1). The dynamics of the atomic states populations in the field of the three FC laser pulses is shown in Fig. 1. The laser pulses have Gaussian shape: $\Omega_k(t) = |R_k| \exp\left[-\left(\frac{t}{\tau}\right)^2\right] \cdot \exp\{i\phi_k\}$ $(k = 1,2,3)$, where $R_k = |R_k| \cdot \exp\{i\phi_k\}$ is the complex amplitude and ϕ_k in the constant phase of the kth pulse Rabi frequency. $\tau = \dfrac{\tau_L}{\sqrt{2\ln 2}}$, where

τ_L is the full width at half maximum (FWHM)of the shapes of the pulses (intensity). The laser pulses are assumed to have same shapes and durations but in general, different amplitudes. A same linear chirp of the laser pulses is assumed with the chirp speed α.

As it can be seen in Fig. 1 the complete population of the initially populated state $|3\rangle$ is transferred into the initially empty states $|1\rangle$ and $|2\rangle$ with negligible population of the excited state $|0\rangle$. The underlying physics of the process may be understood in the basis of "bright" $|g_{br}\rangle$ and "dark" $|g_{dr}\rangle$ superposition states:

$$|g_{br}\rangle = \frac{R_1\,|1\rangle + R_2\,|2\rangle}{\sqrt{|R_1|^2 + |R_2|^2}} \text{ and } |g_{dr}\rangle = \frac{R_2^*\,|1\rangle - R_1^*\,|2\rangle}{\sqrt{|R_1|^2 + |R_2|^2}}. \tag{3}$$

Due to the condition of Raman resonance, the "dark" superposition state is decoupled from the interaction and only the "bright" state component interacts with the laser field. In this basis, the tripod-structured atom may be represented as a three-level Λ-structured atom with one of the ground states coinciding with the superposition state $|g_{br}\rangle$ and the other ground and excited states coinciding with the states $|3\rangle$ and $|0\rangle$ of the original tripod-atom. As it has been shown earlier (see, Refs (11,20)), a pair of FC pulses with Raman detuned carrier frequencies may transfer the complete population of the Λ-atom from one ground state into the other one. The excitation of the atom is suppressed at sufficiently high peak Rabi frequencies of the pulses exceeding the value of the Raman detuning. In the case of the tripod-atom under consideration, the complete population of the state $|3\rangle$ is transferred into the "bright" superposition state $|g_{br}\rangle$. In this way, a coherent superposition $|g_{br}\rangle$ between the states $|1\rangle$ and $|2\rangle$ is created. Because the population of this state by definition depends on the amplitude and relative phase of the two laser pulses in Raman resonance, the relative amplitude and the relative phase of these pulses may govern the composition of this state and hence, the created coherence.

For practical applications, it is particularly important that a same coherent superposition state is created in all atoms composing an optically thick medium. As it is shown above, the excitation of the tripod-atom is suppressed during the creation of the superposition state. One could anticipate a nearly lossless propagation of the interacting FC laser pulses due to the absence of the atomic excitation. The lossless propagating laser pulses will create same value of the atomic coherence throughout the optically thick medium. This process is similar to the one analyized in Ref. (27) for a medium composed of Λ-model atoms.

2.2. *Optical phase and amplitude information mapping into populations of atomic metastable states*

The populations of the states $|1\rangle$ and $|2\rangle$ in the "bright" superposition state created as a result of interaction with the three FC laser pulses depend substantially on the relative phase and amplitudes (Rabi frequencies Ω_1 and Ω_2) of the pulses in Raman resonance. The sensitivity of the resulting populations on the relative phase of these two laser pulses open a possibility for mapping of an optical phase information into the populations of the metastable atomic states. Because the information is written in the populations of these states without considerable excitation of the atom, the information storage time is scaled by the lifetime of the metastable states. It exceeds by orders of magnitude the information storage time of the EIT-based schemes that is scaled by the transverse relaxation time. It is worth noting that the proposed method is the best applicable for mapping the transverse distribution of the phase information (transverse phase images). The information has to be coded in the transverse phase distribution of one of the laser beams among the two in the Raman resonance. The second laser beam in the Raman resonance has to have homogeneous phase front. The phase information carried by the first laser pulse will be mapped in the populations of the states $|1\rangle$ and $|2\rangle$. An additional laser pulse with homogeneous transverse distribution of the intensity will be used as a reading out laser. This pulse will measure the population of a metastable state after the information writing process by transferring it to an excited state. The transverse phase information stored in the transverse distribution of the atomic state population will be transformed into the transverse distribution of the intensity of the reading pulse. A modification of the presented scheme may be utilized for amplitude information (transverse image) writing and storage in the populations of the atomic metastable states.

3. Control of Atomic Motion

As mentioned in the introduction, a series of counterpropagating frequency-chirped laser pulses may exert a sizeable force on the atoms. It has long been known, that if the first of two counterpropagating pulses excites the atoms and the second one de-excites them, a uniform momentum transfer of $2\hbar k$ can be effected over a large inhomogeneously broadened ensemble.[28–30] It has also been shown recently, that if the two pulses that arrive in succession overlap each other in the region of interaction, the result can be an adiabatic process that involves multiple photon exchanges between the two waves with the corresponding change of the atomic momentum.[25,26] The theoretical description of such transitions can be accomplished by solving Scrödinger's equation using the following Hamiltonian:

$$\hat{H} = -\frac{\hat{p}^2}{2m} + \hat{H}_a - \hat{d}E \tag{4}$$

where $\hat{p}$ is the center of mass momentum of the atom, $\hat{H}_a$ describes the internal electronic degrees of freedom (often just two levels are taken into account), $\hat{d}$ is the dipole operator and E is the classical electric field of the light composed of the two counterpropagating pulses. For the simplest case, the two-level atom, the equation for the momentum space probability amplitudes of the ground and excited states $\alpha(p,t), \beta(p,t)$ then become:

$$i\partial_t\alpha(p,t) = \frac{p^2}{2\hbar m}\alpha(p,t) - \frac{\Omega^+}{2}e^{i\phi^+}\beta(p+\hbar k,t) - \frac{\Omega^-}{2}e^{i\phi^-}\beta(p-\hbar k,t)$$
$$i\partial_t\beta(p,t) = \frac{p^2}{2\hbar m}\beta(p,t) - \frac{\Omega^+}{2}e^{-i\phi^+}\alpha(p-\hbar k,t) - \frac{\Omega^-}{2}e^{-i\phi^-}\alpha(p+\hbar k,t)$$

Here the Rabi frequencies of the forward and backward propagating waves $\Omega^\pm$ derive from the slowly varying envelope of the pulses and the dipole matrix element d_{ge} as: $\Omega^\pm(t) = d_{ge}E/\hbar$ and depend slowly on time. Likewise, the phases $\phi^\pm(t)$ that contain the detunings and chirps of the pulses also vary slowly in time with the optical frequencies removed using the rotating wave approximation. These equations can be solved conveniently using a computer to investigate the interaction between atoms and laser pulses as a function of various parameters, in particular the pulse amplitudes, chirp speeds, and the delay of the second pulse.

The result of such an investigation can be seen in Fig. 2, where two curves have been plotted as a function of pulse amplitude at a given value of the delay of the second pulse, i.e. given pulse overlap. The solid curve depicts the momentum transfered to the atoms during the interaction, while the dashed curve shows the final population of the excited state. The parameters are such, that the pulses induce adiabatic population transfer between the ground and excited states when they act separately, so a momentum transfer of $2\hbar k$ is expected in this case. The calculation with the overlapping pulses however shows, that as the pulse amplitude is increased, the momentum transfered to the atoms can be substantially greater. There are plateaus where $2N\hbar k$ momentum is obtained by the atoms ($N > 1$). Furthermore, the atomic population returns to the ground state after the interaction just as it would have after a pair of separated pulses, i.e. the process is repeatable in rapid succession. While it is difficult to explain solely

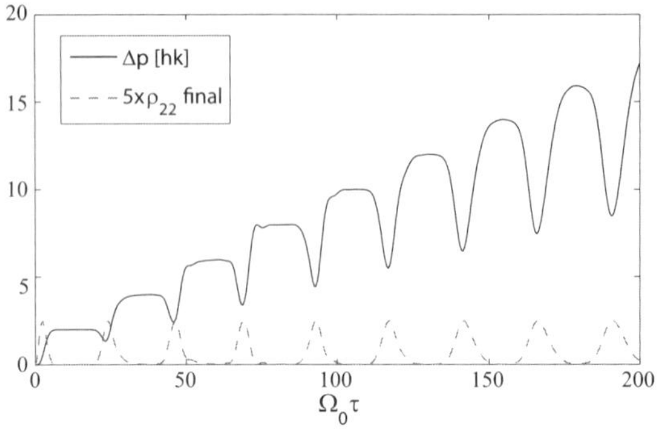

Fig. 2. Momentum transfer (in units of $\hbar k$) and final population of the excited state (multiplied by 5 to be visible) after an interaction with a pair of chirped, partially overlapping laser pulses. The pulse amplitude of the Gaussian pulses normalized by the pulse length Ω_0/τ is shown on the horizontal axis. The chirp speed and the delay of the second pulse used in this calulation were $\beta = 20/\tau^2$ and $\tau_{delay} = 1.5\tau$.

from the results of the simulations why such clean multiphoton adiabatic processes can occur, important insight can be obtained by the usage of adiabatic Floquet theory.[31–33] Starting from the Schrödinger equation for a two level atom interacting with a bichromatic electric field, one can use the Fourier transform to remove the rapid time exponential appearing in the Hamiltonian matrix that will now depend on time only through the slowly varying pulse amplitudes. The eigenvalues and eigenvectors of this so-called Floquet Hamiltonian can then be analyzed, and the possible routes that the system may take during the interaction may be determined. The "zero field" eigenstates of the Floquet Hamiltonian will be $|n, g\rangle$, (which corresponds to the atom being in the ground state and n photons exchanged between the two fields) and $|m, e\rangle$ (which corresponds to the atom being in the excited state, with m photons absorbed from one field and $m - 1$ emitted into the other). The interaction will mix these states and the analysis yields the connections that the interaction produces between them. Provided that the conditions for adiabatic following are maintained, the atom will evolve with the eigenstates. Figure 3 shows the evolution of the Floquet eigenvalues as a function of time during the action of two overlapping chirped pulses. The thick black line depicts the route that the initial $|0, g\rangle$ state takes during the interaction. As it is connected to the $|3, g\rangle$ state, the interaction will result in the atom returning to the ground state at the end with three photons exchanged between the fields, i.e. a momentum change of $6\hbar k$.

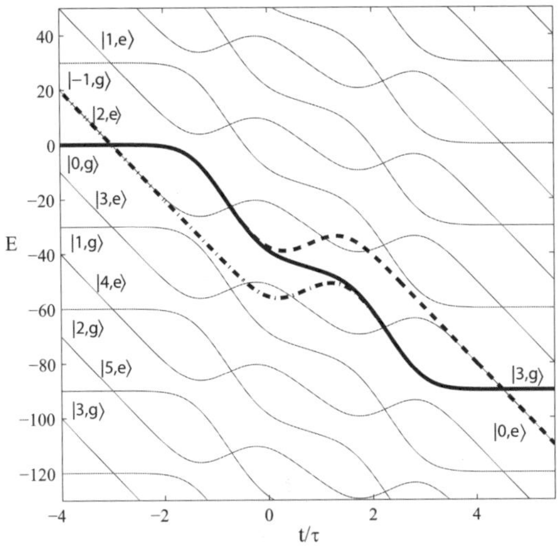

Fig. 3. Time evolution of the Floquet eigenvalues during the interaction with two overlapping chirped pulses. The solid black line shows a six photon transition from $|0, g\rangle$ to $|3, g\rangle$. The pulse amplitude $\Omega_0 = 60/\tau$, other parameters are the same as for Fig. 2.

One issue that must be considered when using chirped laser pulses to accelerate atoms is of course spontaneous emission. An exact quantification of the detrimental effects of spontaneous emission on the schemes described above can be obtained only through laborious numerical calculations (e.g. simualtion of density matrix equations, or using a Monte-Carlo wave function approach). However, for an estimate a simple model can also be constructed,[26] that yields formulas in terms of only the timing parameters of the pulses and can be shown to produce very good results. The endnote of such calculations is that exploiting the possibility of pulse overlap and multiphoton transitions is advantageous over using separate pulses in most situations.

4. Acceleration of Atoms in MOT by Counterpropagating Frequency-Chirped Laser Pulses

4.1. *The MOT arrangement*

We have carried out experiments to realize coherent transfer of mechanical momentum to a cold atomic ensemble by chirped laser pulses with high efficiency. The atomic cloud was gathered from a collimated beam of Rb atoms using the usual magneto-optical trap arrangement,[23] see Fig. 4. The resulting atomic ensemble of about 0.8 mm diameter consisted of about 10^6 atoms, and had about 100 μK temperature.

The MOT cloud was illuminated with a beam of frequency chirped laser pulses in the horizontal plane, at about 10 degrees angle with respect to the axis perpendicular to the atomic beam. The beam was retroreflected by a mirror at a distance of 45 cm from the MOT cloud, i.e. the reflected pulses returned with a delay of 3 ns after the original, forward propagating ones. The acceleration of the atomic cloud was measured with an image intensified CCD camera looking at the MOT from above at a small angle to the vertical direction, similarly to the line of sight in Fig. 4.

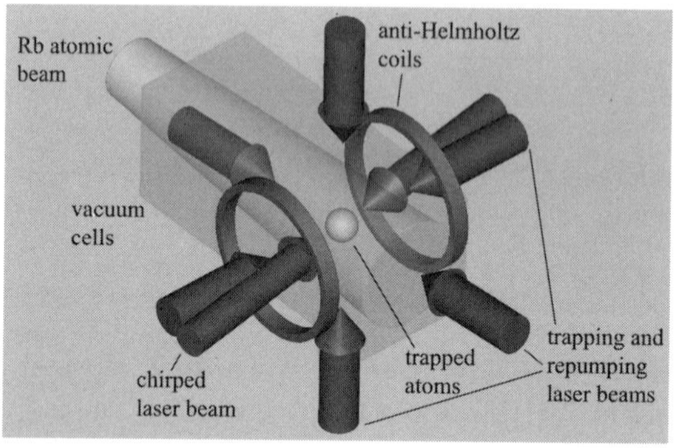

Fig. 4. Schematic diagram of the experimental arrangement for the coherent momentum transfer to Rb atoms cooled and collected in a MOT. The glass vacuum cell was connected to the large vacuum system where the Rb atomic beam was generated.

The purpose of the measurement was to determine the number of $\hbar k$ momenta that was transferred to the atoms by the counterpropagating pulse pair. Therefore, the displacement of the MOT cloud was measured both with illumination from only one direction (by blocking the retroreflector), and with to-and-fro illumination. The efficiency of the coherent momentum transfer was determined by comparing the displacement values in these two cases.

4.2. *Generation and characterization of chirped pulses*

The chirped pulses were sliced from the continuous radiation of a free-running diode laser, for which the driving current was modulated sinusoidally with a period of 60 ns around the DC current component. Therefore, besides the intensity, the frequency of the laser light oscillated with an amplitude of about 1 GHz around the mid-value determined by the DC current of the diode. This radiation passed through a Fabry-Perot interferometer (FPI) of 500 MHz bandwidth, with its transmission frequency tuned to the Rb atomic resonance. The radiation exiting the FPI consisted of pulses with frequency sweeping around the Rb resonance, and chirp speed determined by the offset frequency, which is defined as the difference between the mid-value of the laser frequency (set by the DC current) and the FPI transmission (fixed to the Rb resonance). In the course of the experiments we varied the DC current of the laser diode, and measured the displacement of the MOT cloud as a function of the offset frequency.

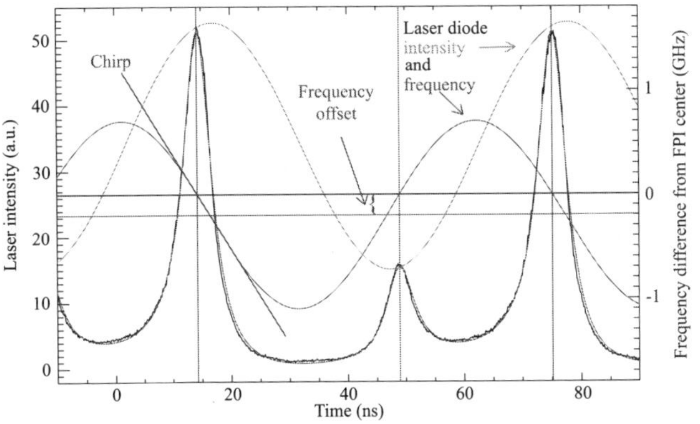

Fig. 5. Measured (noisy, black curve) and fitted (smooth, red curve) intensity, and frequency evolution of the chirped pulse train.

The characteristic parameters of the resulting pulses were measured by two methods. The spectrum of the radiation was analyzed using a combination of a monochromator and a Fabry-Perot etalon to determine the position and range of the frequency sweep. The temporal behaviour was detected by a fast photodiode and a digital oscilloscope where the shape of the pulse train was recorded. The obtained temporal function of the light intensity was then analyzed by numerical fitting the Airy function of the transmission of the FPI in order to calculate the exact frequency and intensity evolution of the radiation. The procedure is demonstrated in Fig. 5 where a recorded signal shape and the resulting fit can be seen. The obtained pulse widths were about 5-10 ns depending on the setting of the frequency offset, and so the forward and backward propagating pulses overlapped at the position of the MOT. The chirp speed value can be calculated from the derivative of the frequency evolution at the instant of the transmission maximum. This method is valid if the ringing of the FPI can be neglected at the frequency variation rates in use; the validity of this condition was carefully examined and verified as explained in detail.[24]

This method for the generation of chirped pulses of a few ns and chirp speeds of about 10-100 MHz/ns is simple and robust, however, the characteristic parameters of the resulting pulses are closely related to each other, and it is difficult to control them separately. Therefore, we have been developing more flexible techniques for this purpose, with the use of integrated electro-optical amplitude and phase modulators for controlling the intensity and frequency evolution of the laser radiation. We could produce chirped pulses with the required characteristics with the use of a single intensity modulator by exploiting the intrinsic chirp of the instrument.[34] Similar methods will be used in our future experiments on coherent population transfer with chirped pulses.

4.3. *Experimental results on coherent acceleration*

The amount of mechanical momentum transferred by the counter-propagating chirped pulses to the cold Rb atoms was determined from the displacement of the MOT cloud from the equilibrium position after the influence of a series of 5000 periods (300 μs) of the pulse train. When the atomic cloud was illuminated from only one direction, the atoms excited by the light returned to the ground state by spontaneous emission, and the maximum amount of transferred momentum per pulse was $\hbar k$ resulting in an excursion about 2 mm of the atomic ensemble. In the case of the interaction with the to-and-fro propagating pulse pair, the acceleration of the atoms was strongly dependent on the offset frequency of the modulated laser. In Fig. 6 the transferred momentum per pulse pair can be seen (scaled in units of $\hbar k$ after comparison with the one-sided illumination), as a function of the offset frequency. The dependence of the measured chirp speed on the offset frequency is also depicted in the figure.

The most important result is that the maximum momentum transfer is higher than $2\hbar k$ per pulse pair, i.e. the effect of multiphoton transitions during the overlapping pulses played an observable role in this interaction. It can be seen that momentum transfer is most efficient when the frequency offset is near its upper limit, where the chirp value is around 50 MHz/ns in the vicinity of the slowly varying interval of the modulation.

Similar conclusions can be drawn from the results of numerical simulations using a Monte-Carlo wavefunction approach (mentioned in the

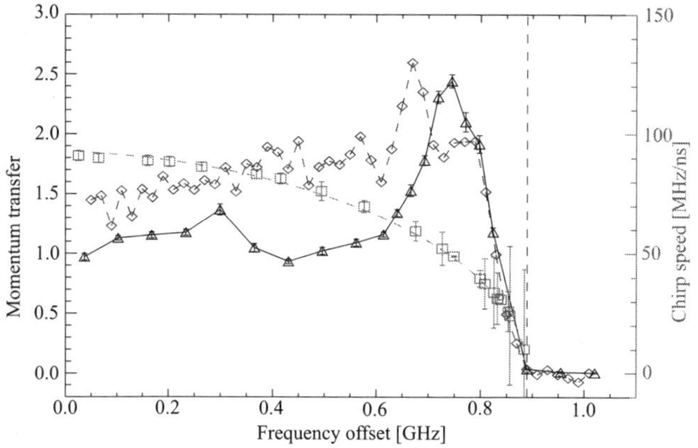

Fig. 6. Number of transferred momenta per counterpropagating pulse pair (solid curve), in units of $\hbar k$, versus offset frequency $\nu_{ofs} = \nu_{l0} - \nu_{Rb}$ (i.e. the difference between the mid-value of the laser frequency ν_{l0} and the Rb resonance ν_{Rb}) of the chirped laser. The dashed curve represents the results of the numerical simulations. The measured chirp speed data with the fit to the points assuming harmonic time dependence are depicted in the dotted curve. The dashed vertical line marks the limit of the frequency modulation, i.e. above this offset value the laser frequency is always above resonance.

previous section), assuming a two-level atom with spontaneous emission, and taking into account the actual characteristic parameters of our experiment. The qualitative agreement suggests that also the dependence on the offset frequency (chirp) can be described by the model of the multiphoton transitions during the overlapping pulses. The increasing discrepancy between the theoretical and experimental data as the offset decreases can be explained by the role of the lower lying hyperfine levels of the excited state of the Rb atom, which are not taken into account in the calculations. It is also important to note that our pulse intensities of $1 - 2\mathrm{W/cm}^2$ are relatively low for the multiphoton transitions to become significant. In our forthcoming experiments we are going to apply higher intensities in order to realize several times $2\mathrm{N}\hbar\mathrm{k}$ coherent momentum transfer by an overlapping pair of counterpropagating chirped pulses.

5. Conclusions

In this paper, we have shown that the frequency chirped laser pulses may be successfully utilized for creation of coherent superposition states involving metastable quantum states without considerable excitation of the system. This feature of the FC pulses is used for mapping and storage of the optical phase and amplitude transverse images in populations of the atomic states providing a possibility for extremely long storage times. The ability of FC laser pulses to effectively manipulate the translational degrees of freedom of atoms has been demonstrated using counter-propagated laser pulses. It has been shown that due to the multiphoton exchange of energy (and mechanical momentum), the efficiency of the momentum transfer may be increased drastically as compared with the case of separated FC laser pulses. The results of the theory are compared with the results of experiments on coherent acceleration of a cloud of Rb^{85} atoms in MOT. The presented schemes are robust against the small-to-medium variations of the parameters of the laser pulses. Due to the use of the FC pulses, the proposed schemes do not need strict resonance conditions providing high efficiency both in homogeneously and inhomogeneously broadened media including the solid state media.

Acknowledgments

The financial support of the Research Fund of the Hungarian Academy of Sciences (OTKA) under contracts K 68240, NN 78112 and F 67922 and the Janos Bolyai Research Fellowship for G. D. is gratefully acknowledged.

References

1. K. Bergmann, H. Theuer, B.W. Shore, Rev.Mod.Phys. **70**, 1003-1025 (1998).
2. R. Unanyan, M. Fleischhauer, B.W. Shore, K. Bergmann, Opt. Commun. **155**, 144-154 (1998).
3. L. P. Yatsenko, B. W. Shore, T. Halfmann, K. Bergmann, A. Vardi, Phys. Rev. A **60**, R4237 (1999).
4. F. Vewinger, M. Heinz, R. G. Fernandez, N. V. Vitanov, K. Bergmann, Phys. Rev. Lett., **91**, 213001, (2003).
5. L. Yatsenko, N. V. Vitanov, B. W. Shore, T. Rickes, K. Bergmann, Opt. Communs. **204**, 413-423 (2002).
6. A. A. Rangelov, N. V. Vitanov, L. P. Yatsenko, B. W. Shore, T. Halfmann, K. Bergmann, Phys. Rev. A **72**, 053403 (2005).
7. G. P. Djotyan, J. S. Bakos, Zs. Sörlei, Opt. Express, **4**, 113-120 (1999).
8. G. P. Djotyan, J. S. Bakos, G. Demeter, and Zs. Sörlei, J. Opt. Soc. Am. B **17**, 107-113 (2000).
9. G. Djotyan, J. Bakos, Zs. Sörlei, J. Szigeti, P. Ignacz, Z. Tóth, Laser Phys., **10**, 355-359 (2000).
10. G. P. Djotyan, J. S. Bakos, Zs. Sörlei, Phys. Rev. A, **64**, 013408 (2001).
11. G. P. Djotyan, J. S. Bakos, Zs. Sörlei, Phys. Rev. A, **70**, 063406 (2004).
12. G. P. Djotyan, J. S. Bakos, Zs. Sörlei, J. Szigeti, D. Dzsotjan, JOSA B, 25, p.166-174 (2008).
13. L. Allen and J. H. Eberly, *Optical Resonance and Two-Level Atoms* (Dover, New York, 1987).

14. M. D. Lukin, Rev. Mod. Phys., **75**, 457-472 (2003).
15. M. Fleischhauer, A. Imamoglu, J. P. Marangos, Rev. Mod. Phys., **77**, 633-673 (2005).
16. M. D. Lukin, P. R. Hemmer, M. Loeffler, M. Scully, Phys Rev. Lett.**81**, 2675-2678 (1998).
17. T. Rickes, J. P. Marangos, T. Halfmann, Opt. Commun. **227**, 133-142 (2003).
18. C. P. Williams and S.H. Clearwater, *Explorations in Quantum Quantum Computing* (Springer-Verlag, Berlin, 1997).
19. A. Mair, J. Hager, D.F. Phillips et.al., Phys. Rev. A **65**, 031802 (2002).
20. G. P. Djotyan, N. Sandor, J. S. Bakos, and Zs. Sörlei. JOSA B, **26**, 1959-1966 (2009).
21. H. Metcalf and P. van der Straten,*Laser Cooling and Trapping*, Springer, 1999.
22. G. P. Djotyan, J. S. Bakos, G. Demeter, P. N. Ignacz, M. A. Kedves, Zs. Sörlei, J. Szigeti, Z. L. Tóth, Phys Rev A **68**, 053409 (2003).
23. J. S. Bakos, G. P. Djotyan, P. N. Ignácz, M. Á. Kedves, M. Serényi, Zs. Sörlei, J. Szigeti and Z. Tóth, Eur. Phys. J. D, **39**, 59-66 (2006) .
24. J. S. Bakos, G. P. Djotyan, P. N. Ignácz, M. A. Kedves, M. Serényi, Zs. Sörlei, J. Szigeti, and Z. Tóth, Eur. Phys. J. D **44**, 141-149 (2007).
25. G. Demeter, G. P. Djotyan, Zs. Sörlei, and J. S. Bakos., Phys. Rev. A **74**, 013401 (2006).
26. G. Demeter, G.P Djotyan., J. Opt. Society of Am. B,**26**, 867-875 (2009).
27. G. Demeter, D. Dzsotjan, and G. P. Djotyan. Phys. Rev. A, **76**, 023827 (2007).
28. J. S. Bakos, G. P. Djotyan, G. Demeter, and Zs. Sörlei, Phys. Rev. A **53**, 2885-2888 (1996).
29. G. P. Djotyan, J. S. Bakos, G. Demeter, and Zs. Sörlei, J. Opt. Soc. Am. B **13**, 1697-1705 (1996).
30. G. Demeter, G. P. Djotyan and J. S. Bakos, J. Opt. Soc. Am. B **15**, 16-24 (1998).
31. J. F. Shirley, Phys. Rev. **138**, B979-987 (1965).
32. S. Guérin, L. P. Yatsenko and H. R. Jauslin, Phys. Rev. A **63**, 031403(R) (2001).
33. S. Guérin, F. Monti, J.-M. Dupont and H. R. Jauslin, J. Phys. A: Math. Gen **30**, 7193-7215 (1997).
34. J. S. Bakos, G. P. Djotyan, P. N. Ignácz, M. Á. Kedves, B. Ráczkevi, Zs. Sörlei, J. Szigeti, Opt. Las. Eng. **47** (2009).

STRONGLY CORRELATED QUANTUM DYNAMICS OF MULTIMODE LIGHT COUPLED TO A TWO-LEVEL ATOM IN A CAVITY

T. KUMAR[1], A. B. BHATTACHERJEE[2] and M. MOHAN[1]

[1] *Department of Physics and Astrophysics, University of Delhi, Delhi-110007, India*

[2] *Department of Physics, ARSD College, University of Delhi (South Campus), New Delhi-110021, India*

We study a composite multimode light-two-level atom system in a cavity. We show that coupling of the two-level atom to multiple modes of the light destroys the Mott phase of the composite system thus making the system less useful platform for developing concepts in quantum information processing.

Keywords: Multimode light, quantum phases, cavity.

1. Introduction

The search for interesting and potentially useful quantum-mechanical phenomena on a mesoscopic scale in condensed matter and atomic physics is a challenging task. The Jaynes-Cummings model describes the interaction of a single, quasi-resonant optical cavity field with a two-level atom. The optical nonlinearities generated due to the coupling between the atom and the photons leads to an effective photon-photon repulsion.[1,2] However it was shown that photon-photon repulsion degrades in the presence of many atoms.[3,4] Later, Rebic et al.[5] showed that the nonlinear interaction afforded by placing a single two-level atom inside a cavity would suffice for realizing photon blockade. Greentree et al.[6] showed that by adding photons to a two-dimensional array of coupled optical cavities each containing a single two-level atom in the photon-blockade regime, a long-lived, strongly interacting dressed states of excitations (coupled atom-photons) are formed which can undergo at zero temperature a characteristics Mott insulator to superfluid quantum phase transition. Devices based on photon blockade mechanism has a strong potential to be useful for quantum computation. In particular because of Mott's phase robustness, devices based on this effect at non-zero temperature has been suggested.[6] Motivated by such interesting developments in photon blockade schemes, we study in this work, the prospects of utilizing quantum phases of light in a single atom driven by multiple different modes of photons for possible quantum devices. In particular, we consider a two-dimensional array of photonic bandgap cavities. Each cavity contains a single two-level atom, quasi-resonant with the multiple cavity modes. In the classical limit, an array of coupled photonic bandgap cavities has been described for novel waveguide applications,[7,8] nanocavity lasers[9] and in the quantum regime a two-cavity arrangement has been proposed as a Q-switch.[10] Recently,the dynamics of linear arrays of coupled cavities containing four-level atomic systems was studied by Hartmann et al.[11] In the quantum optics context, the generalized Jaynes-Cummings model where the transition is mediated by multiple different modes of photons was first studied by Dantsker.[12]

2. The Basic Model

2.1. *Eigenvalues and Eigenfunctions*

To create an atom-multi-photon system whose dynamics we want to study, we consider a two-dimensional array of photonic band gap cavities as discussed earlier.[6] Each cavity contains a single two-level atom, quasi-resonant with the multi-modes of the cavity. To motivate the search for Hubbard-model-type interactions within the present system, we first discuss the eigenvalues, second-order coherence and the probability of finding the atom in the excited state in the presence of multi-modes.

The system we consider here is an effective two-level atom with upper and lower states denoted by $|1>$ and $|0>$, respectively. In the multi-photon processes, some intermediate states $|i>$, $i = 2, 3..$, are involved, which are assumed to be coupled to $|1>$ and $|0>$ by dipole allowed transitions. Let ω_0, ω_1, and ω_i, denote the corresponding frequency of the atomic energy level $|0>$, $|1>$ and $|i>$, respectively. We also denote ω as the transition frequency between states $|1>$ and $|0>$. The atom interacts with the n cavity fields with frequencies Ω_i, $i = 1, 2..n$, where $\Omega_1 + \Omega_2 + ...\Omega_n \cong \omega$. Also we assume that detuning between the atomic transition frequency and any one of the n modes is non-zero. Under these circumstances, the intermediate states can be adiabatically eliminated (effectively, we are neglecting the Stark effect) and the effective Hamiltonian of the two-level atom can be written in the rotating-wave approximation as

$$\hat{H}^{JC} = \hat{H}_0 + \hat{H}_I \tag{1}$$

with

$$\hat{H}_0 = \frac{\omega}{2}\hat{\sigma}_z + \sum_{i=1}^{n} \hbar\Omega_i \hat{a}_i^\dagger \hat{a}_i, \tag{2}$$

$$\hat{H}_I = \left(\prod_{i=1}^{n} g_i^{1/n} \hat{a}_i^\dagger \hat{\sigma}_- + \prod_{i=1}^{n} g_i^{1/n} \hat{a}_i \hat{\sigma}_+ \right). \tag{3}$$

Here $\hat{a}_i^\dagger$ and $\hat{a}_i$ are the creation and destruction operator of the nth cavity mode respectively. and $\hat{\sigma}_z$, $\hat{\sigma}_-$ and $\hat{\sigma}_+$ are the usual pauli spin matrices. In the bare state basis $|1, n_1, n_2, ...n_n >$ and $|0, n_1+1, n_2+1, ..n_n+1>$, the eigenvalues and the eigenvectors are calculated as

$$\lambda^{\pm} = K(n_1, n_2, ..n_n) \pm Q(n_1, n_2, ...n_n), \tag{4}$$

where

$$K(n_1, n_2, ..n_n) = \frac{\omega - \Delta}{2} + \sum_{i=1}^{n} \omega_i n_i, \tag{5}$$

$$Q(n_1, n_2, ..n_n) = \sqrt{\frac{\Delta^2}{4} + \prod_{i=1}^{n} (g_i)^{2/n}(n_i+1)}. \tag{6}$$

Here $\Delta = \omega - \sum_i \omega_i$. The corresponding eigenvectors are

$$|\pm, n_1, n_2, ..n_n >= A^{\pm}_{n_1,n_2,..n_n} |0, n_1, n_2, ..n_n > + B^{\pm}_{n_1,n_2,..n_n} |1, n_1, n_2, ..n_n >, \tag{7}$$

where,

$$A^{\pm}_{n_1,n_2,..n_n} = \frac{-\Delta/2 \mp Q}{\sqrt{2Q^2 \pm \Delta Q}} \tag{8}$$

and

$$B^{\pm}_{n_1,n_2,..n_n} = \frac{\prod_i (g_i)^{1/n} \sqrt{(n_i+1)}}{\sqrt{2Q^2 \pm \Delta Q}} \tag{9}$$

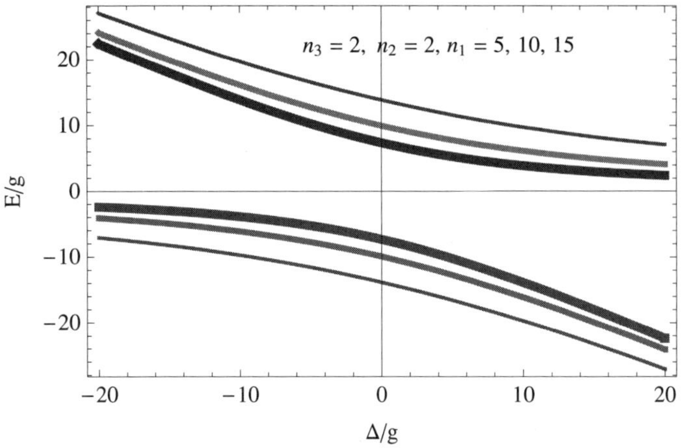

Fig. 1. Dimensionless eigenspectrum ($E = \lambda^{\pm} -$ optical energy) for a single atom interacting with 3-modes of a high Q-cavity as a function of Δ/g (dimensionless atom-cavity detuning), centered around zero ($\lambda^{\pm} = \sum_{i=1}^{3} \omega_i n_i$). The eigenspectrum splits into two branches corresponding to the dressed states, $|+, n_1, n_2, ..n_n >$ (upper branch) and $|-, n_1, n_2, ..n_n >$ (lower branch). The two branches anti-cross at $\Delta/g = 0$, with the splitting increasing with increasing excitation number $n_1 = 5, 10, 15$ (in order of decreasing thickness of the plots)

Fig. 1 shows a plot of the dimensionless eigenvalues for the 3 mode case. The photon energies have been subtracted for ease of comparison. The photon number in the first mode (n_1) is allowed to change while keeping the photon number in the other two modes (n_2, n_3) constant. The on-site photonic repulsion is evinced by the increasing energy separation with n_1. The atom-photon coupling constants g_i have been taken to same for all modes.

2.2. *Population probability and coherence function*

The results of the previous subsection leads to the following matrix representation of the time evolution operator for the given manifold $M(n_1, n_2, ..n_n)$.

$$\hat{U}(n_1, n_2, ..n_n; t) =$$
$$\begin{pmatrix} \left[Q + \frac{\Delta}{2}\right] e^{-i\lambda^+ t} + \left[Q - \frac{\Delta}{2}\right] e^{i\lambda^- t} & \prod_i (g_i)^{1/n} \sqrt{n_i + 1}(e^{-i\lambda^+ t} - e^{-i\lambda^- t}) \\ \prod_i (g_i)^{1/n} \sqrt{n_i + 1}(e^{-i\lambda^+ t} - e^{-i\lambda^- t}) & \left[Q - \frac{\Delta}{2}\right] e^{-i\lambda^+ t} + \left[Q + \frac{\Delta}{2}\right] e^{i\lambda^- t} \end{pmatrix} \tag{10}$$

The time evolution operator can now be utilized to calculate the density operator of the system at any time t with an arbitrary initial condition $\hat{\rho}(0)$ as

$$\hat{\rho}(t) = \hat{U}(t)\hat{\rho}(0)\hat{U}^{\dagger}(t) \tag{11}$$

The expectation value of any operator can then be obtained easily as

$$< \hat{O}(t) >= Tr[\hat{\rho}(t)\hat{O}(o)] \tag{12}$$

We assume that at $t = 0$ the density operator can be decomposed into its atomic and field parts, i.e.,

$$\hat{\rho}(0) = \hat{\rho}^{A}(0) \otimes \hat{\rho}^{F}(0), \tag{13}$$

where the atom is in the ith energy eigenstate $|i>$ and the fields are in a general state as

$$|f >= \sum_{n_1,n_2,..n_n} R_{n_1,n_2,..n}|n_1, n_2, ..n_n > . \tag{14}$$

Futhermore the reduced atomic density operator $\hat{\rho}^{A}(t)$ and the reduced field density operator $\hat{\rho}^{F}(t)$ are introduced by taking the trace of $\hat{\rho}(t)$ over the field states and over the atomic states respectively. Now if we denote $P_{n_1,n_2,..n_n}(0) = |R_{n_1,n_2,..n_n}|^2$ as the initial photon distribution, then the probability of finding the atom in the excited state $P_1(t)$ for $|i >= |0 >$ is calculated as

$$P_1(t) = \rho^{A}(t) = \sum_{n_1,n_2,..n_n} P_{n_1,n_2,..n_n}(0)\frac{\prod_i g_i^{2/n}(n_i+1)\sin^2 Qt}{Q^2} \tag{15}$$

The initial photon distribution is assumed to be coherent which implies that $R_{n_1,n_2,..n_n}$ is written as

$$R_{n_1,n_2,..n_n} = \frac{exp\left[-\frac{\sum_i \bar{n}_i}{2}\right]\prod_i(\bar{n}_i)^{n_i/2}}{\sqrt{\prod_i n_i!}} \tag{16}$$

Here $\bar{n}_i$ is the mean photon number in the ith mode. The degree of nth order coherence with zero-time delay is defined as

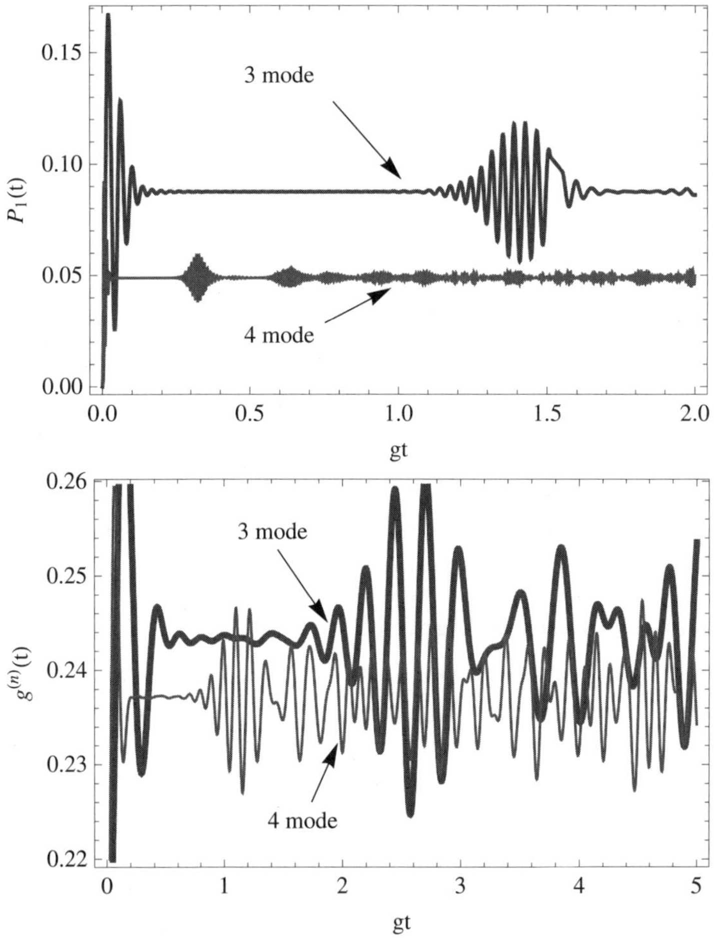

Fig. 2. Probability $P_1(t)$ (top plot) of finding the atom in the excited state and coherence function (bottom plot) as a function of gt. The initial field states are prepared in 3-modes and 4-modes of coherent states with mean photon number $\bar{n}_i = 20$ in each mode.

$$g^{(n)} = \frac{< \prod_{i=1}^{n} \hat{n}_i(t) >}{\prod_{i=1}^{n} < \hat{n}_i(t) >}, \tag{17}$$

whose magnitude controls the n-photon transition rate.

If we assume the atom to be initially in the excited state then mean photon number of the ith mode and the expectation value of the correlation of photon numbers for the n modes that is related to the degree of second order coherence is

$$< \hat{n}_i(t) >= Tr[\hat{\rho}(t)\hat{n}_i], \tag{18}$$

$$< \hat{n}_1\hat{n}_2...\hat{n}_n >= Tr[\hat{\rho}(t)\hat{n}_1\hat{n}_2...\hat{n}_n]. \tag{19}$$

The time evolution of the probability of finding the atom in the excited state ($P_1(t)$) and the nth order coherence function (g^n) in the presence of 3-modes and 4-modes of the coherent state are plotted in Fig. 2. As seen from the figure the quantum revival and collapse of the Rabi oscillations are more compact for the 4-mode case than the 3-mode case. The more interesting observation is the fact that P_1 is less for the 4-mode case than the 3-mode case. Similarly the coherence function $g^4 < g^3$. This result has some interesting consequence for the quantum phase diagram of the composite photon-atom system to be calculated in the next section.

3. Quantum Phases of Light

The Hamiltonian of our system is given by a combination of the Jaynes-Cummings Hamiltonian with photon hopping between cavities and the chemical potential term. Here we assume that only one mode (k^{th}) mode is able to hope between the cavities. The hopping is achieved by evanescent coupling between the cavities. The tunneling frequency is approximately given by $\kappa = \Omega_k/Q$ (Ω_k is the frequency of the kth mode and Q is the quality factor of the cavities). If Ω_k is kept large as compared to the frequencies of the other mode then one can achieve tunneling of only the kth mode.

$$H = \sum_l H_l^{JC} - \kappa \sum_l \hat{a}_{k,l}^\dagger \hat{a}_{k,l\pm 1} - \mu \sum_l \left(\sum_i \hat{a}_{i,l}^\dagger \hat{a}_{i,l} + \sigma_l^+ \sigma_l^- \right) \tag{20}$$

Here l is the site index. We are assuming nearest neighbour hopping and the hopping frequency κ is assumed to be same. The chemical potential is same for all site if we assume zero disorder. Here $\left(\sum_i \hat{a}_{i,l}^\dagger \hat{a}_{i,l} + \sigma_l^+ \sigma_l^-\right)$ is the total number of atomic and photonic excitations (the conserved quantity in our system). We introduce a superfluid order parameter $\psi =< \hat{a}_l >$, which we take to be real and use the decoupling approximation, $\hat{a}_l^\dagger \hat{a}_m =< \hat{a}_l^\dagger > \hat{a}_m + < \hat{a}_l > \hat{a}_m^\dagger - < \hat{a}_l^\dagger >< \hat{a}_m >$. The resulting effective mean-field Hamiltonian can be written as a sum over single sites

$$H^{MF} = \sum_l H_l^{JC} - z\kappa\psi \sum_l (\hat{a}_{k,l}^\dagger + \hat{a}_{k,l}) + z\kappa|\psi|^2$$
$$- \mu \sum_l \left(\sum_i \hat{a}_{i,l}^\dagger \hat{a}_{i,l} + \sigma_l^+ \sigma_l^- \right) \tag{21}$$

Here $z = 3$ is the number of nearest neighbours. To obtain the system's zero-temperature phase diagram, we use the procedure of ref.[13] All energies are now scaled with respect to $z\kappa$. The unperturbed ground state energy of the state with exactly $\left(\sum_i \hat{a}_{i,l}^\dagger \hat{a}_{i,l} + \sigma_l^+ \sigma_l^-\right)$ particles is $\lambda^0 = \bar{\lambda}^- - \bar{\mu}\left(\sum_i \hat{a}_{i,l}^\dagger \hat{a}_{i,l} + \sigma_l^+ \sigma_l^-\right)$, where $\bar{\lambda}^- = \lambda^-/z\kappa$. We only need to consider the negative branch for the purpose of determining the ground state since $\lambda^- < \lambda^+$. A change in the total number of quasi-excitations per site will occur when $\bar{\lambda}_{n_k+1}^- - \bar{\mu}(n_k+1) = \bar{\lambda}_{n_k}^- - \bar{\mu} n_k$. We can determine the critical chemical potential, $\bar{\mu}_c(n_k)$, where the system will change from n_k to $n_k + 1$ quasi-excitations per site as

$$\bar{\mu}_c(n) = \bar{\lambda}_{n_k+1}^- - \bar{\lambda}_{n_k}^-, \tag{22}$$

The second order correction to the energy with the well known expression

$$E_m^{(2)} = \sum_{n_i \neq m_i} \frac{| < -, m_1, m_2, ..m_n|V|-, n_1, n_2, ..n_n > |^2}{\lambda_{m_k}^{(0)} - \lambda_{n_k}^{(0)}}, \tag{23}$$

where

$$V = -\left(\hat{a}_k^\dagger + \hat{a}_k\right) \tag{24}$$

Here we take the unperturbed eigenvector $|-, n_1, n_2, ..n_n >$ corresponding to $\sum_i n_i$ particles. Since the interaction V couples only to states with one more or less excitations than in the ground state, we find

$$E_m^{(2)} = \frac{\left(A_{m_k}^- A_{m_k-1}^- \sqrt{m_k} + B_{m_k}^- B_{m_k-1}^- \sqrt{m_k - 1}\right)^2}{\bar{\lambda}_{m_k}^- - \bar{\lambda}_{m_k-1}^- - \bar{\mu}}$$
$$+ \frac{\left(A_{m_k}^- A_{m_k+1}^- \sqrt{m_k + 1} + B_{m_k}^- B_{m_k+1}^- \sqrt{m_k}\right)^2}{\bar{\lambda}_{m_k}^- - \bar{\lambda}_{m_k+1}^- + \bar{\mu}} \tag{25}$$

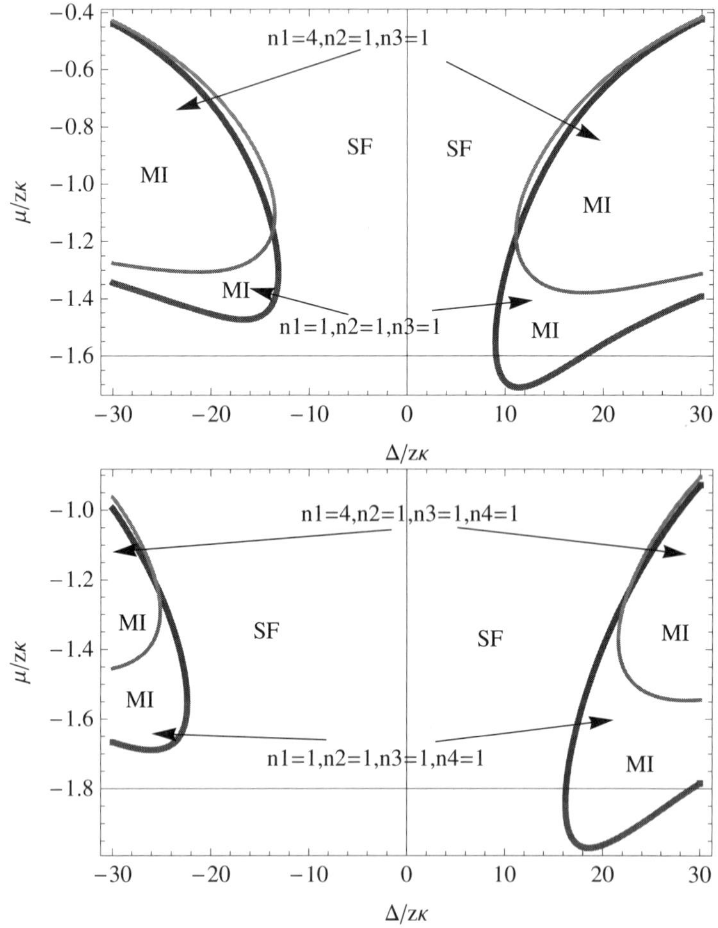

Fig. 3. Phase diagram of the coupled multimode light and the two-level atom syatem in a cavity. The top plot is for 3-mode case and the bottom plot is for 4-mode case. The extreme right-hand and the extreme left-hand edge (where photonic repulsion dominates at higher detuning) is the Mott-insulator (denoted by MI) and the superfluid phase is found in between the MI regions (denoted by SF).The size of the Mott region is found to decrease with the quasi-excitations for both the 3-mode and 4-mode case.The size of the Mott lobes is found to be smaller for the 4-mode case as compared to the 3-mode case.

According to the Landau procedure for second-order phase transitions, we write the ground state as an expansion in ψ

$$E_m(\psi) = a_0(m, \bar{\mu}, \bar{\Delta}) + a_2(m, \bar{\mu}, \bar{\Delta})\psi^2 + O(\psi^4), \tag{26}$$

$E_m(\psi)$ is minimized as a function of the superfluid order parameter ψ. We find that $\psi = 0$ when $a_2 > 0$ and that $\psi \neq 0$ when $a_2 < 0$. This means that $a_2 = 0$ signifies the boundary between the superfluid and insulator phases of light. This yields

$$\mu_{\pm} = \frac{1}{2}\left(f_1^2 - f_2^2 + \delta\lambda_1 - \delta\lambda_2\right) \pm \frac{1}{2}\sqrt{(f_1^2 - f_2^2)^2 + (\delta\lambda_1 + \delta\lambda_2)^2 + 2(f_1^2 + f_2^2)(\delta\lambda_1 + \delta\lambda_2)} \quad (27)$$

where

$$f_1 = \left(A^-_{m_k} A^-_{m_k-1}\sqrt{m_k} + B^-_{m_k} B^-_{m_k-1}\sqrt{m_k - 1}\right) \quad (28)$$

$$f_2 = \left(A^-_{m_k} A^-_{m_k+1}\sqrt{m_k + 1} + B^-_{m_k} B^-_{m_k+1}\sqrt{m_k}\right) \quad (29)$$

$$\delta\lambda_1 = \lambda^-_{m_k} - \lambda^-_{m_k-1} \quad (30)$$

$$\delta\lambda_2 = \lambda^-_{m_k} - \lambda^-_{m_k+1} \quad (31)$$

The subscript $\pm$ in Eqn. (27) denotes the upper and lower halves of the Mott insulating regions of phase space. Fig. 3 shows the plot of Eqn. (27) for the 3 mode case (left plot) and 4 mode case (right plot). By equating $\bar{\mu}_-$ and $\bar{\mu}_+$ we can find the point of largest Δ_{max} for each Mott region (MI). The dynamics illustrated in Fig. 3 is extremely rich. The extreme right-hand and the extreme left-hand edge (where photonic repulsion dominates at higher detuning) is the Mott-insulator (denoted by MI) and the superfluid phase is found in between the MI regions (denoted by SF). The size of the Mott region is found to decrease with the quasi-excitations for both the 3-mode and 4-mode case. The transition from the SF to the MI phase of the quasi-excitations occurs at Δ_{max}. The size of the Mott lobes is found to be smaller for the 4-mode case as compared to the 3-mode case. This observation can be correlated with the observations noted in Fig. 2. For the 4-mode case, the probability of transition to the excited state is less as compared to the 3-mode case. The photonic repulsion is a nonlinear process and is directly dependent on the probability of transition to the excited state. A lower transition probability implies a decreased photonic repulsion and hence a smaller Mott region. To find the eigenenergies and to experimentally identify the various phases of the coupled atom-photon system, one can perform a transmission spectroscopy with the scattered light by direct readout of the number of photons coming out of the cavity. Photon loss can be minimized by using high Q cavities and thus ensuring that the light field remains quantum-mechanical for the duration of the experiment.

4. Conclusions

In summary, we have studied the strongly correlated dynamics of a two-level atom coupled to multi-mode cavity photons in the photon blockade regime. We have shown that coupling of the two-level atom to multiple modes of the light destroys the Mott phase of the composite system thus making the system less useful platform for developing concepts in quantum information processing.

Acknowledgements

One of the authors Tarun Kumar acknowledges the Council of University Grants Commission, New Delhi for the financial support under the Junior Research Fellowship scheme Sch/JRF/AA/30/2008-2009.

References

1. K .M. Birnbaum et al., Nature, **436**, 87 (2005).
2. A. Imamoglu, A. Schmidt, G. Woods and M. Deutsch, Phys. Rev. Letts. **79**, 1467 (1997).
3. P. Grangier et al., Phys. Rev. Lett., **81**, 2833 (1998).
4. A. D. Greentree et al., J. Opt. B: Quantum Semiclass. Opt., **2**, 252 (2000).
5. S. Rebic, A. S. Parkins and S. M. Tan, Phys. Rev. A, **65**, 063804 (2002).
6. A. D. Greentree et al., Nature Physics, **2**, 856 (2006).
7. R.J. Lang and A. Yariv, IEEE J. Quantum Elecronics, **24**, 66 (1988).
8. E. Ozbay et al., IEEE J. Quantum Electronics, **38**, 837 (2002).
9. H. Altug et al., Nature Phys., **2**, 484 (2006)
10. A. D. Greentree et al., Phys. Rev A., **73**, 013818 (2006).
11. M. H. Hartmann et al., Nature Physics, **2**, 849 (2006).
12. A. R-Dantsker, J. Mod. Opt., **39**, 1471 (1992).
13. D. van Oosten, P. van der Straten and H.T.C. Stoof, Phys. Rev. A, **63**, 053601 (2001).

FEEDBACK-DRIVEN ADIABATIC QUANTUM DYNAMICS

A. E. ALLAHVERDYAN

Yerevan Physics Institute, Alikhanian Brothers Street 2, Yerevan 375036, Armenia

G. MAHLER

Institute of Theoretical Physics I, University of Stuttgart, Pfaffenwaldring 57, 70550 Stuttgart, Germany

We study quantum adiabatic dynamics, where the external field moves slowly in time and is influenced by system's state (feedback). The information for the feedback is gained from non-disturbating measurements done on an ensemble of identical non-interacting systems. This situation without feedback is governed by the adiabatic theorem: adiabatic energy level populations stay constant, while the adiabatic eigenvectors get a specific phase contribution (Berry phase). Under feedback the adiabatic theorem does not hold: the adiabatic populations satisfy a closed equation of motion that coincides with the replicator dynamics well-known by its applications in evolutionary game theory. The feedback generates a new gauge-invariant adiabatic phase, which is free of the constraints on the Berry phase (e.g., the new phase is non-zero even for real adiabatic eigenfunctions).

Keywords: Quantum mechanics, Berry's phase, Quantum ensemble theory.

1. Introduction

The adiabatic theorem of quantum mechanics governs the evolution of a quantum system subject to slowly varying external fields. This is one of the oldest and most widely used general tools.[1] It emerged in early days of quantum mechanics as an heuristic method of quantization.[2] Its modern applications span a vast array of fields, such as two-level systems (nuclei undergoing magnetic resonance or atoms interacting with a laser field),[1] quantum field theory (where a low-energy effective theory is derived by integrating out fast high-energy degrees of freedom), Berry's geometric phase,[3,4] inflationary cosmology and cosmic microwave background radiation.[5] The Berry phases and the adiabatic theorem also found applications in quantum information processing.[6,7]

A control-theoretic perspective on the quantum adiabatic physics is that it studies a system subject to a slow, *open loop* (i.e., no feedback) control, where the evolution of the external fields is given *a priori* via the specific time-dependent parameters of the system Hamiltonian. In view of numerous application of this setup, it is natural to wonder about the quantum adiabatic *closed-loop* control, where the external controlling fields evolve under feedback from the controlled quantum system. Any feedback needs information on the evolution of the system. This information is to be gained via measurements, which in the quantum situation are normally related with unpredictable disturbances and irreversibility. This is the main difference of the quantum feedback as compared to its classical counterpart. In this context people studied quantum control using open-system dynamics.[8]

However, quantum measurements need not be disturbing. This is the case for N non-interacting quantum particles (spins, *etc*) gently coupled to the proper measuring apparatus. For $N \gg 1$ [analog of the classical limit] one can measure single-particle observables (almost) without disturbing the single-particle density matrix. The knowledge of these observables allows to implement feedback.[9] Non-disturbing measurements on ensembles of few-level systems are routinely employed in NMR physics and quantum optics;[10] their general theory is given in Ref. 11. The feedback generated by such measurements was implemented experimentally in NMR physics.[12]

Following Refs. 13, 14 we review here the adiabatic theory under feedback obtained via such non-disturbing measurements. Next section describes the general equations of the approach. Section 3 studies the linear feedback, which is perhaps the simplest feedback mechanism to consider. Section 4 presents the main physical results of the paper. Several concrete examples are studied in section 5. We summarize in the last section.

2. Basic Equations

Consider a d-level quantum system described by a pure state $|\psi\rangle$ (generalization to mixed states is indicated below). The system evolves according to the Schroedinger equation with Hamiltonian $H[R(t)]$, where $R(t)$ is a classical controllling parameter ($\hbar = 1$):

$$i|\dot{\psi}(t)\rangle = H[R(t)]\,|\psi(t)\rangle. \tag{1}$$

By means of a continuous non-disturbing measurement performed on an ensemble of identical, non-interacting systems (each one described by $|\psi(t)\rangle$) one finds the average $\langle\psi(t)|A|\psi(t)\rangle$ of a monitoring observable A (in NMR physics A typically corresponds to the magnetization). This average enters the feedback dynamics of R

$$\dot{R} = \varepsilon F\left(R, \langle\psi(t)|A|\psi(t)\rangle\right), \tag{2}$$

where $\varepsilon \ll 1$ is a small dimensionless parameter. We assume that $F(.,.)$ is bounded from above, which means that R is a slow variable: its derivative is bounded by a small number. For $F = F(R)$ (no feedback) we recover the standard adiabatic setup. The dynamics (1) conserves the purity of $|\psi(t)\rangle$, but the overlap $\langle\phi(t)|\psi(t)\rangle$ between two different wave-functions is not conserved in time, since H depends on $|\psi(t)\rangle\langle\psi(t)|$ via (2).

For developing an adiabatic approach, we introduce the adiabatic eigenresolution of the Hamiltonian for a fixed value of the parameter R ($n = 1, .., d$):

$$H[R]\,|n[R]\rangle = E_n[R]\,|n[R]\rangle, \qquad \langle n[R]\,|\,m[R]\rangle = \delta_{mn}. \tag{3}$$

For simplicity we assume that the adiabatic energy levels are not degenerate. The representation (3) has a gauge-freedom: $|n[R]\rangle \to e^{i\alpha_n[R]}\,|n[R]\rangle$, where $\alpha_n[R]$ is an arbitrary single-valued function depending on n and R. All observables should be gauge-invariant. Expand $|\psi(t)\rangle$ as

$$|\psi(t)\rangle = \sum\nolimits_n c_n(t)\, e^{i\gamma_n(t)}\, |n[R(t)]\rangle, \tag{4}$$

$$\gamma_n(t) \equiv -\int_0^t \mathrm{d}t'\, E_n[R(t')],\;\; c_n(t) = \langle n[R(t)]\,|\psi\rangle\, e^{-i\gamma_n(t)},$$

where $\gamma_n(t)$ are the dynamical phases, while c_n are the adiabatic amplitudes. One gets from (1, 2):

$$\dot{c}_n = -\sum\nolimits_k c_k \langle n|\dot{k}\rangle\, e^{i(\gamma_k(t)-\gamma_n(t))} \tag{5}$$

$$= -\varepsilon \sum\nolimits_k c_k \langle n|k'\rangle\, F(R, c, e^{i\Delta\gamma(t)})\, e^{i(\gamma_k(t)-\gamma_n(t))}, \tag{6}$$

where $|k'\rangle = \partial_R |k[R]\rangle$. The amplitudes c_n and R are slow variables, since, e.g., $|\dot{c}_n|$ is bounded from above by the small ε in (6). However, the contribution from the dynamical phases γ_n changes fast, since on the slow time $\tau = \varepsilon t$ it behaves as $\sim e^{i\tau/\varepsilon}$; see (4). If the spacings between the adiabatic energy levels $E_n[R]$ remain large enough, the existence of some intermediate time τ_f is guaranteed, over which the dynamical phase contribution performs many oscillations, but c_n and R do not change appreciably. The adiabatic approximation divides c_n into the time-averaged (over τ_f) part $\bar{c}_n$ and the small (at least as $\mathcal{O}(\varepsilon)$) oscillating part: $c_n = \bar{c}_n + \delta c_n$. To the leading order one neglects in the RHS of (6) all the oscillating factors and substitute $c \to \bar{c}$ and $R \to \bar{R}$:[15]

$$\overset{\bullet}{\bar{c}}_n \equiv \frac{\mathrm{d}\bar{c}_n}{\mathrm{d}\tau} = -\sum\nolimits_k \bar{c}_k \langle n|k'\rangle\, \overline{F(\bar{R}, \bar{c}, e^{i\Delta\gamma})\, e^{i(\gamma_k-\gamma_n)}}, \tag{7}$$

where $\tau = \varepsilon t$, $\bar{X} \equiv \int_0^{\tau_f} \frac{ds}{\tau_f} X(s)$. Likewise, (2) produces

$$\bar{R}^{\bullet} = \overline{F(\bar{R}, \bar{c}, e^{i\Delta\gamma})}. \tag{8}$$

3. Linear Feedback

The simplest situation is

$$F = \langle\psi|A|\psi\rangle = \sum\nolimits_{n,m} c_n^* \, c_m e^{i(\gamma_m - \gamma_n)} \langle n|A|m\rangle, \tag{9}$$

where the operator A may still depend on $\bar{R}$. Eq. (9) can be regarded as the first term of the Taylor expansion assuming that $F(\langle\psi|A|\psi\rangle)$ depends weakly (small ε) on $\langle\psi|A|\psi\rangle$. Eq. (7) leads to

$$\bar{c}_l^{\bullet} = -\sum_{k,n,m} \bar{c}_k \, \langle l|k'\rangle \, \bar{c}_n^* \, \bar{c}_m \, \langle n|A|m\rangle \, \overline{e^{i(\gamma_m - \gamma_n + \gamma_k - \gamma_l)}}. \tag{10}$$

In working out (10) we shall assume that the time-integrated energy-level differences are distinct:

$$\gamma_m(t) - \gamma_n(t) \neq \gamma_l(t) - \gamma_k(t), \ \text{ if } m \neq n \text{ and } m \neq l. \tag{11}$$

This condition is generic for few-level systems and was employed in Ref. 18 when discussing the quantum ergodicity. Eq. (11) does not hold for cases like harmonic oscillator, which should be separately worked out from (10). Now in the RHS of (10) the non-zero terms are those with $m = n$ and $l = k$, and those with $m = l$ and $k = n$ (but $n \neq l$, not to count twice the term $m = n = k = l$):

$$\bar{c}_l^{\bullet} = -\bar{c}_l \, \langle l|l'\rangle \, \bar{R}^{\bullet} - \bar{c}_l \sum\nolimits_{n(\neq l)} |\bar{c}_n|^2 \, \langle l|n'\rangle \, \langle n|A|l\rangle. \tag{12}$$

Note that $\langle l|l'\rangle$ is imaginary due to $\partial_R \langle l|l\rangle = \partial_R 1 = 0$. Introducing the phase and module of $\bar{c}_n$,

$$\bar{c}_n = \sqrt{p_n} \, e^{i\phi_n}, \qquad \sum\nolimits_n p_n = 1, \tag{13}$$

we get from (12)

$$p_l^{\bullet} = -2p_l \sum\nolimits_{n(\neq l)} p_n \, \Re\{ \, \langle l|n'\rangle \, \langle n|A|l\rangle \, \}, \tag{14}$$

$$\phi_l^{\bullet} = i\langle l|l'\rangle \, R^{\bullet} - \sum\nolimits_{n(\neq l)} p_n \, \Im\{ \, \langle l|n'\rangle \, \langle n|A|l\rangle \, \}, \tag{15}$$

while analogous manipulations with (8) lead to

$$\bar{R}^{\bullet} = \sum\nolimits_n p_n \langle n|A|n\rangle. \tag{16}$$

Eqs. (14)-(16) are our central results. Before exploring them in more detail let us discuss the standard (open-loop, i.e., no feedback) adiabatics, where $A = A(R)$ is a c-number: R moves in a prescribed way according to $R^\bullet = A(R)$. Eq. (14) leads to the conservation of the probabilities $p_l^\bullet = 0$ (adiabatic theorem): the system does not get enough energy to move out of the given energy level. The RHS of (15) reduces to Berry's factor $\phi_{\mathrm{B},l}^\bullet = i\langle l|l'\rangle\, R^\bullet$. As seen from (4), though $\phi_{\mathrm{B},l}$ is by itself not gauge-invariant, it does bring an observable (Berry phase) contribution in a non-diagonal average over the state $|\,\psi(t)\,\rangle = \sum_n c_n(0)e^{i\phi_{\mathrm{B},n}(\tau)+i\gamma_n(t)}$. The Berry phase was observed in numerous experiments; see Refs. 3,4 for review. It is constrained by the following conditions.

1. The Berry phase nullifies, $\langle l|l'\rangle = 0$, if the adiabatic eigenvectors $|l\rangle$ can be made real via a suitable gauge transformation. This happens for several important situations, e.g., a spinless particle without magnetic field,[1] or a system with spin $\mathbf{S}$ in a two component magnetic field $\mathbf{h}$, since in the corresponding Hamiltonian $\propto h_x S_x + h_z S_z$, S_x and S_z can be chosen to be real matrices. In these cases the Berry phase is zero.[19]

2. $\phi_{\mathrm{B},l} = 0$ for a cyclic motion of a single slow parameter R, where R is switched on at the initial time and then switched off at the final time, thereby leaving intact the indetity of the system. The Berry phase may be different from zero if there is more than one slow parameter $\mathbf{R} = (R_1, R_2, ...)$ on a closed curve $\mathcal{C}$: $\mathbf{R}(0) = \mathbf{R}(\tau)$.[3] Then one gets a gauge-invariant expression $\phi_{\mathrm{B},l} = i\oint_{\mathcal{C}} \mathrm{d}\mathbf{R}\,\langle l|\partial_{\mathbf{R}}\, l\rangle$.[3,4]

4. Closed-Loop Adiabatics

Let us now return to (14). This closed equation for p_n arises out of the averaging over the fast dynamic phases under condition (11). Eq. (14) is non-linear over p_n due to the feedback. The probabilities p_n are no longer conserved [due to the resonance between the oscillations of c_n and those of R, see (10)], and if p_n's are known, the phases ϕ_l are obtained directly from (15). The matrix

$$a_{ln} \equiv -2\Re\{\,\langle l|\partial_R n\rangle\,\langle n|A|l\rangle\,\}, \qquad a_{ln} = -a_{ln}, \tag{17}$$

in (14) is antisymmetric; in particular, $a_{ll} = 0$, which means that the probability conservation holds with (14): $\sum_l p_l(\tau) = 1$. The edges of the probability simplex, e.g. $p_l = \delta_{l1}$, are (possibly unstable) stationary solutions of (14), so that $p_l(\tau)$ remains non-negative.

It is noteworthy that (14) coincides with the replicator equation for a zero-sum population game.[20,21] Recall that a two-player game consists of strategies $l = 1, ..., d$. When the first (second) player applies strategy l (n), he recieves payoff a_{ln} (a_{nl}). The condition $a_{nl} = -a_{ln}$ means a zero-sum game (e.g., poker): the gain of one player equals to the loss of the other. In the evolution game theory one considers a population of agents, who adopt the strategy n with probability p_n.[21] p_n change in time, dependent on success relative to other strategies. The success is defined via the pay off matrix, while actual mechanism of this change depends on the concrete implementation of the model (inheritance, learning, imitation, infection, etc).[21] The main approach of this theory is based on the replicator equation, where the frequency change $p_l^\bullet$ is proportional to p_l itself (the *autocatalyst principle* of the population biology), while the proportionality coefficient is the average payoff of the strategy l: $p_l^\bullet = p_l \sum_n a_{ln} p_n$.[20,21] We thus return to (14), where the population game, with (in general) τ-dependent payoffs a_{ln}, is now played by the energy levels. Interesting features of the replicator equation can be found without solving it; see (18)-(21).

For the open-loop control changing of R on the slow time-scale is mandatory, otherwise no adiabatic motion occurs at all. The closed-loop situation is different, since now for $\langle n|A|n\rangle = 0$ the slow motion of R is absent, $\bar{R}^\bullet = 0$ [see (16)], with still non-trivial adiabatic dynamics. R does move on the fast time, but this motion averages out on the slow time. Let us focus on this situation, since we cannot study (14)-(16) in full generality.

Eqs. (14), (17), now with τ-independent a_{ln}, is conveniently studied via the time-averages:[21]

$$\frac{1}{T} \ln \frac{p_l(T)}{p_l(0)} = \sum_n a_{ln} \bar{p}_n(T), \quad \bar{p}_n(T) = \int_0^T \frac{d\tau}{T} p_n(\tau). \tag{18}$$

There are now two different dynamic scenarios depending on the concrete form of τ-independent a_{lp} in (14), (17).

1. If all $p_l(t)$ (which were non-zero at the initial time $\tau = 0$) stay non-zero for all times, $\ln p_l(T)$ in the LHS of (18) is limited, which means that this LHS can be neglected for $T \to \infty$. We then get from (18)[20,21]

$$\sum\nolimits_n a_{ln} \bar{p}_n(\infty) = 0. \tag{19}$$

Thus all $p_l(t)$ may remain non-zero for all times provided that there is a probability vector $\bar{p}(\infty)$ that satisfies (19). Clearly, $\bar{p}(\infty)$ is a stationary state of (14), (17). Recall that the [non-negative] relative entropy is defined as

$$S[\bar{p}(\infty)|p(\tau)] = \sum\nolimits_l \bar{p}_l(\infty) \ln \left[\bar{p}_l(\infty)/p_l(\tau) \right], \tag{20}$$

where $p(t)$ is a time-dependent solution of (14). $S[\bar{p}_n(\infty)|p(\tau)]$ is equal to zero if and only if $\bar{p}(\infty) = p(\tau)$. Due to (19), $S[\bar{p}_n(\infty)|p(t)]$ is a constant of motion [thus an adiabatic invariant], since

$$S^{\bullet}[\bar{p}(\infty)|p(\tau)] = \sum\nolimits_{ln} p_l(\tau) a_{ln} \bar{p}_n(\infty). \tag{21}$$

Ref. 20 shows that this constant can be associated with Hamiltonian, and then Eqs. (14) can be recast into a Hamiltonian form. Note that the non-linearity of this dynamics is essential, since in some closely related Lotka-Volterra and bi-matrix replicator dynamics the chaotic behavior was seen explicitly in Refs. 22 and 23, respectively.

2. If the matrix a_{ln} is such that (19) does not have any probability vector solution, $\frac{1}{T} \ln \frac{p_l(T)}{p_l(0)}$ in (18) is necessarily finite for at least one l. The corresponding probability $p_l(T)$ goes to zero (for a large T): $p_l(T) \to p_l(\infty) = 0$, so that for all k one has

$$\sum\nolimits_{n} a_{kn} \bar{p}_n(\infty) \leq 0. \tag{22}$$

This inequality is strict at least for $k = l$. Eqs. (21), (22) show that $S[\bar{p}(\infty)|p(\tau)]$ now decays to zero meaning that $p(\tau)$ *relaxes* to $\bar{p}(\infty)$. This relaxation is due to the non-linearity of (14); it is impossible without feed-back.

Eq. (15) for the phases integrates as

$$\phi_l(\tau) = -\tau \sum\nolimits_{n(\neq l)} \bar{p}_n(\tau) b_{ln}, \; b_{ln} \equiv \Im\{\langle l|n'\rangle \langle n|A|l\rangle\}, \tag{23}$$

where $\bar{p}_n(\tau)$ satisfies to the algebraic equation (18). Note that b_{ln} is symmetric: $b_{ln} = b_{nl}$. Eq. (23) gives the phases of the adiabatic (linear) feedback control. It is clear that $\phi_l(\tau)$ is free of the constraints for the open-loop (Berry) phase $\phi_{\mathrm{B},l}$:

i) it is explicitly gauge-invariant together with b_{ln};

ii) its existence does not require complex adiabatic eigenvectors $|l\rangle$, provided that the monitoring observable A has at least some complex elements $\langle n|A|l\rangle$;

iii) it does not require several control parameters for cyclic processes;

iv) even if a_{nl}, defined via (17), is zero, i.e., if the probabilities p_n are conserved, the feedback-driven phases ϕ_l in (23) can be non-zero.

Note that $\phi_l = 0$ if the evolution starts from a strictly one adiabatic eigenvector $p_n(0) = \delta_{nk}$ (however this stationary state of (14) need not be stable, as we above). In contrast, the Berry phase may be non-zero even for this case, though its observation does require interference with another eigenstate.

5. Examples

We now apply our findings to two simple examples: For a two-level system (14), (17) reduce to

$$\dot{p}_1 = a_{12}(1-p_1)p_1, \qquad p_1(\tau) = \frac{e^{a_{12}\tau}}{\frac{1-p_1(0)}{p_1(0)} + e^{a_{12}\tau}}, \tag{24}$$

which means that independent of the initial value $p_1(0)$, $p_1 \to 1$ ($p_1 \to 0$) if $a_{12} > 0$ ($a_{12} < 0$). Properly choosing the time τ and a_{12}, and knowing $p_1(0)$, we can reach any value $0 \le p_1(\tau) \le 1$. Eq. (23) leads to

$$\phi_{1,2}(\tau) = \pm\frac{b_{12}}{a_{12}} \ln\left[p_1(0)e^{\mp a_{12}\tau} + 1 - p_1(0)\right]. \tag{25}$$

Two basic examples of two-level systems are the spin-$\frac{1}{2}$ and the polarization states of light. The standard Berry phase was observed in both these cases.[3]

For the three-level situation the internal stationary vector is obtained from (19) (up to normalization)

$$\bar{p}_1(\infty) \propto a_{23}/a_{12}, \quad \bar{p}_2 \propto -a_{13}/a_{12}, \quad \bar{p}_3 \propto 1, \tag{26}$$

provided these probabilities are positive, i.e., (using game-theoretic terms for the levels as strategies) the strategy 1 beats 2, 2 beats 3, but 3 beats 2 (rock-swissor-paper game). Now the τ-dependent solution $p(\tau)$ of (14) oscillates around (26).

If $\frac{a_{23}}{a_{12}} < 0$ or $\frac{a_{13}}{a_{12}} > 0$, the internal stable state is absent and we get the following attractor

$$\bar{p}^{[1]}(\infty) = (1,0,0) \quad \text{for} \quad a_{12} > 0,\, a_{13} > 0, \tag{27}$$
$$\bar{p}^{[2]}(\infty) = (0,1,0) \quad \text{for} \quad a_{12} < 0,\, a_{23} > 0, \tag{28}$$
$$\bar{p}^{[3]}(\infty) = (0,0,1) \quad \text{for} \quad a_{23} < 0,\, a_{13} < 0. \tag{29}$$

The game-theoretical meaning of (27–29) should be obvious. Recall that the existence of probability vectors (26) and of (27)-(29) is mutually exclusive. Thus from an arbitrary initial state $p(0)$ we may drive the system toward one of $\bar{p}^{[k]}(\infty)$, provided that a_{ln} is chosen properly.

5.1. *Slow motion of the controlling parameter*

R is allowed for $\langle n|A|n\rangle \neq 0$. Now (14) becomes a driven replicator equation, since a_{ln} are τ-dependent. For a general theory of driven replicator equation see Ref. 24. Here are, however, two cases, where the above results suffice for analyzing the driven situation. In the above three-level example assume that conditions $\frac{a_{23}(\tau)}{a_{12}(\tau)} < 0$ or $\frac{a_{13}(\tau)}{a_{12}(\tau)} > 0$, are satisfied for all τ. Then the same argument on the convergence to $\bar{p}^{[k]}$ defined by (27)-(29) applies. On the other hand, assume that the opposite conditions $\frac{a_{23}(\tau)}{a_{12}(\tau)} > 0$ and $\frac{a_{13}(\tau)}{a_{12}(\tau)} < 0$ hold for all τ. Now all $p_l(\tau)$ are non-zero for all times.

5.2. *Mixed states*

Above we assumed that the state of the quantum system is pure, see (1). However, the typical quantum states in NMR and quantum optics (two fields where the collective, non-disturbing measurements are especially feasible) are normally mixed. Since the closed-loop dynamics (1), (2) is not linear, the mixed-state dynamics (in general) does not reduce to the pure-case situation; its analysis along the lines (3)-(8) is presented elsewhere.[13] Nevertheless, there is one important situation, where the presented formalism can be applied directly. Mixed states in NMR are frequently of the form:

$$\rho = (1-\eta)\frac{\hat{1}}{d} + \eta|\psi\rangle\langle\psi|, \tag{30}$$

where $\hat{1}$ is the unit matrix, and where $0 < \eta < 1$ is a parameter.[10] Now the feedback goes via $\mathrm{tr}(A\rho)$ and we define c_n analogously to (4). Since $\hat{1}$ is an invariant of the dynamics (1, 2), we get the same adiabatic equations (14)-(16) with $\langle n|A|l\rangle \to \eta\langle n|A|l\rangle$.

6. Conclusion

We studied how the feedback generated by non-disturbing (ensemble) measurements effects the adiabatic (i.e., slowly driven) quantum dynamics. Without feedback this dynamics is well-known and characterized by two basic results: conservation of the adiabatic energy populations and the Berry phase. For the simplest linear feedback we have found that

- The populations are no longer constant. Instead, they satisfy the canonical [replicator] equation of the population game theory, allowing us to visualize the corresponding dynamics as a zero-sum game played by the adiabatic energy levels. This relation may be empoyed for an accurate experimental realization of the replicator dynamics, since all its biological applications are only indirect. Many features of the [non-linear] replicator equation are obtained without solving it.

- In addition to the Berry phase, the feedback generates a new, explicitly gauge-invariant phase, which [as compared to the Berry phase] exists under a wider range of conditions. In particular, there are scenarios of feedback, where the probabilities are constant (resembling the ordinary situation), but the new phases are still non-trivial.

The work was supported by Volkswagenstiftung grant "Quantum Thermodynamics: Energy and information flow at nanoscale".

References

1. A. Messiah, *Quantum mechanics* (North-Holland, Amsterdam, 1962), Vol. 2.
2. P. Ehrenfest, Naturwissenschaften, **11**, 543 (1923).
3. A. Shapere and F. Wilczek, *Geometric Phases in Physics* (World Scientific, Singapore, 1989).
   Y. Ben-Aryeh, arXiv:quant-ph/0402003.
4. G.G. de Polavieja and E. Sjoqvist, Am J Phys **66**, 431 (1998).
5. L. Parker, Phys. Rev. D **3**, 346 (1971).
   U. H. Danielsson, Class. Quant. Grav. **22**, S1 (2005).
6. J. Pachos *et al.*, Phys. Rev. A **61**, 010305 (2000).
   J.A. Jones *et al.*, Nature **403**, 869 (2000).
7. E. Farhi *et al.*,Science **292**, 472 (2001).
8. V.P. Belavkin, Rep. Math. Phys. **45**, 353 (1999). M. A. Armen, *et al.*, Phys. Rev. Lett. **89**, 133602 (2002). J. Geremia *et al.*, Science **304**, 270 (2004). P. Bushev *et al.*, Phys. Rev. Lett. **96**, 043003 (2006). A. C. Doherty *et al.*, Phys. Rev. A **62**, 012105 (2000). K. Jacobs, Phys. Rev. A **67**, 030301(R) (2003). Special issue on quantum control, J. Opt B, **7**, no. 10 (2005).
9. C. Altafini, Quantum Information Processing, **6**, 9 (2007).
10. O.R. Ernst, G. Bodenhausen and A. Wokaun, *Principles of Nuclear Magnetic Resonance in One and Two Dimensions* (Oxford University Press, Oxford, 1987).
11. S. Lloyd and J.-J. Slotine, Phys. Rev. A, **62** 012307 (2000). D. Poullin, Phys. Rev. A, **62** 012307 (2005).
12. P. Broekart and J. Jeener, J. Magn. Reson. A **113**, 60 (1995).
13. A. E. Allahverdyan and G. Mahler, EPL **84**, 40007 (2008).
14. A. E. Allahverdyan and D. B. Saakian, EPL **81**, 30003 (2008).
15. This way of applying the adiabatic approximation coincides with the general method proposed in Ref. 16. The method also allows to find the higher-order corrections and determine the applicability region of the adiabatic approach. For a recent discussion on the validity of the ordinary (open-loop) adiabatic approach see Ref. 17.
16. N.N. Bogoliubov and Yu.A. Mitropolskii, *Asymptotic Methods in the Theory of Nonlinear Oscillations* (Gordon & Breach, New York, 1961); see chapter V.

17. S. Duki et al., Phys. Rev. Lett. **97**, 128901 (2006). J. Ma et al., Phys. Rev. Lett. **97**, 128902 (2006). K.-P. Marzlin and B.C. Sanders, Phys. Rev. Lett **97**, 128903 (2006). D.M. Tong et al., Phys. Rev. Lett. **98**, 150402 (2007). R.MacKenzie et al., Phys. Rev. A **73**, 042104 (2006).
18. J. von Neumann, Z. Phys., **57**, 30 (1929).
19. This statement does not hold if there are level-crossings.
20. E. Akin and V. Losert, J. Math. Biology, **20**, 231 (1984).
21. J. Hofbauer and K. Sigmund, *Evolutionary Games and Population Dynamics* (Cambridge Univ. Press, Cambridge, 1998).
22. G. Picard and T.W. Johnston, Phys. Rev. Lett. **48**, 1610 (1982).
23. Y. Sato, E. Akiyama and J.D. Farmer, PNAS **99**, 4748 (2002).
24. A. E. Allahverdyan and Chin-Kun Hu, Phys. Rev. Lett. **102**, 058102 (2009).

LANDAU-ZENER TRANSITION IN NONLINEAR QUANTUM SYSTEMS

A. M. ISHKHANYAN
Institute for Physical Research of NAS of Armenia,
Ashtarak, 0203, Armenia
E-mail: aishkhanyan@gmail.com

A comprehensive theory of the generalized Landau-Zener problem for quadratic- and cubic-nonlinear quantum two-state models is developed. Combining analytical and numerical methods, a simple analytic formula involving elementary functions only is constructed for the final transition probability for the *quadratic*-nonlinear Landau-Zener problem. The formula provides a highly accurate approximation for the whole rage of the variation of the Landau-Zener parameter. Further, a rigorous analysis of the *cubic*-nonlinear Landau-Zener problem applicable, e.g., to the interacting Bose-Einstein condensates is presented. It is shown that for any set of involved parameters the time evolution of the system is accurately described by a two-term variational ansatz. Applying an exact third order nonlinear differential equation a fifth order polynomial equation is constructed for the final transition probability. A root of this equation can be viewed as a generalized Landau-Zener formula for cubic-nonlinear quantum systems.

Keywords: Landau-Zener transition, Term crossing, Nonlinear two-state model.

1. Introduction

The linear-in-time term-crossing two-state model by Landau,[1] Zener,[2] Stuckelberg,[3] and Majorana[4] is a long-standing quantum paradigm dating back to 1930s. It describes the transition dynamics in a two-level quantum-mechanical system with a time-dependent Hamiltonian when the energy separation between the two states is linearly swept in time. This is a well-appreciated approximation that have played a prominent role in studying of a number of important physical phenomena in many branches of contemporary physics ranging from radiation-matter interactions to collision physics. Since the realization of the Bose-Einstein condensates in dilute atomic gases[5] different nonlinear generalizations of this problem have been suggested and explored both theoretically and experimentally. It should be noted that there are two basic modifications of the original Landau-Zener model which differ by the basic type of the involved nonlinearity. In some cases, e.g., when discussing the interparticle interactions in Bose-condensates the nonlinearity is accounted for by adding *cubic terms* to the basic *linear* equations describing the original *linear* two-level problem (see, e.g., Refs. 6–10). In contrast, in photo- and magneto-association of ultra-cold atoms an effective two-state problem is encountered where the motion equations do not involve linear terms so that the basic physical process is principally of nonlinear nature.[11,12] In the present paper we discuss both cases. We first present a comprehensive analysis of a model described by a *quadratic-nonlinear* set of two-level equations that have been a subject of considerable interest during the last years.[11–22] This is a basic version of the nonlinear two-state problem arising in all the nonlinear field theories involving a Hamiltonian with a Fermi 2:1 resonance.[23] Applying a two-term ansatz proposed earlier to describe the time-dynamics of the transition probability,[22] we construct a highly accurate formula for the final Landau-Zener transition probability. The formula is compact, involves elementary functions only and is applicable for the whole variation range of the input Landau-Zener parameter. Further, we discuss the cubic-nonlinear model. Here, we suggest an analogous two-term variational ansatz (written as a sum of a solution to a limit first-order nonlinear ordinary differential equation and a scaled solution to the basic linear Landau-Zener problem) that accurately, in detail, describes the time evolution of the system in the whole time domain. We show that the role of cubic terms is effectively described by a sole combined parameter. A remarkable final result of the present development is a highly accurate generalized formula for the final Landau-Zener transition probability which is applicable for the whole variation range of the involved parameters, both the Landau-Zener parameter and the mentioned combined cubic-interaction parameter.

2. Landau-Zener Transition in Quadratic-Nonlinear Two-State Systems

We consider the following semiclassical nonlinear two-state model treating the atomic and molecular modes as classical fields[11,12]

$$
\begin{aligned}
i\frac{da_1}{dt} &= U(t)e^{-i\delta(t)}\bar{a}_1 a_2, \\
i\frac{da_2}{dt} &= \frac{U(t)}{2}e^{i\delta(t)}a_1 a_1,
\end{aligned}
\tag{1}
$$

where t is the time, a_1, a_2 are the atomic and molecular state probability amplitudes, $\bar{a}_1$ is the complex conjugate of a_1, $U(t)$ is the Rabi frequency, and the detuning modulation function $\delta(t)$ is the integral of the detuning ε ($\varepsilon = d\delta/dt$) of the associating field frequency from that of the transition from the atomic state to the molecular one. This model is encountered, e.g., in the theories of cold atom production in atomic Bose-Einstein condensates via laser Raman photoassociation or magnetic Feschbach resonance,[24] and in the second harmonic generation in non-linear optics,[25] etc.

For the Landau-Zener model under consideration the Rabi frequency is constant, $U = U_0 = \text{const}$, the detuning is assumed to linearly-in-time cross the resonance at time point $t = 0$: $d\delta/dt = 2\delta_0 t$, and the conventional Landau-Zener parameter λ is defined as $\lambda = U_0^2/\delta_0$. System (1) describes a lossless process so that the total number of particles is conserved: $|a_1|^2 + 2|a_2|^2 = const = 1$. Finally, we treat the basic case when the system starts from the all-atomic state: $a_1(-\infty) = 1$, $a_2(-\infty) = 0$.

In order to describe the *time evolution* of the transition probability $p(t) = |a_2(t)|^2$ (note that because of the convention applied here $p(t) \in [0,\, 1/2]$), we have previously introduced[22] a two-term ansatz involving three variational parameters, A, C_1, and λ_1:

$$p = p_0(A,t) + C_1 \frac{p_{LZ}(\lambda_1, t)}{p_{LZ}(\lambda_1, \infty)}, \tag{2}$$

where, $p_{LZ}(\lambda_1, t)$ is the solution of the *linear* Landau-Zener problem for an effective λ_1 and $p_0(A,t)$ is a root of the following *quartic* polynomial equation:

$$\frac{\lambda}{4t^2} = \frac{p_0(p_0-\beta_1)(p_0-\beta_2)}{9(p_0-\alpha_1)^2(p_0-\alpha_2)^2}, \tag{3}$$

where the involved parameters are defined as

$$\alpha_{1,2} = \frac{1}{3} \mp \frac{1}{6}\sqrt{1+\frac{6A}{\lambda}}, \beta_{1,2} = \frac{1}{2} \mp \sqrt{\frac{A}{2\lambda}}. \tag{4}$$

Note that both $p_0(A,t)$ and $p_{LZ}(\lambda_1, t)$ satisfy the initial condition $p(t = -\infty) = 0$.

It has been demonstrated numerically that this ansatz produces highly accurate results. Namely, the numerical simulations have shown that for any given value of the input Landau-Zener parameter, $\lambda \in [0, \infty)$, one can always find A, C_1, and λ_1 so that function (2) accurately fits the numerical solution to the exact equation for the molecular state probability in the whole time domain – the graphs produced by the formula are practically indistinguishable from the numerical solution. For arbitrary time points, the absolute error is commonly of the order of 10^{-4} (a slightly less accurate result, $\sim 10^{-3}$, is observed for points in a small region embracing the first local maximums and minimums of $p(t)$ after the resonance crossing point has been passed) and for the final transition probability $p(+\infty)$ the proposed approximation assures an absolute error of the order of 10^{-5}.[22]

Further, it has been shown that the variational constant λ_1, i.e., the effective Landau-Zener parameter involved in the linear solution $p_{LZ}(\lambda_1, t)$, is expressed in terms of the final transition probability $p(t = +\infty)$:

$$\lambda_1 = \lambda\,(1 - 3p(+\infty)). \tag{5}$$

Then, to construct analytic approximations for the remaining two variational parameters $A(\lambda)$, $C_1(\lambda)$, and, hence, to eventually determine the final transition probability written in terms of these parameters as

$$p(+\infty) = \left(\frac{1}{2} - \sqrt{\frac{A}{2\lambda}}\right) + C_1, \tag{6}$$

a two-parametric fit involving the Gauss hypergeometric functions ${}_2F_1$[25] has been applied (Eqs. (11), (12), Ref. 22).

The principal development we report in the present paper is that we show that the scaling parameter $C_1(\lambda)$ can be chosen as

$$C_1 = \frac{P_{LZ}}{4}\sqrt{\frac{2}{\lambda}A}, \tag{7}$$

where P_{LZ} is the linear Landau-Zener transition probability $P_{LZ} = 1 - e^{-\pi\lambda}$. This particular choice that is justified by examining the next approximation term turns out to be rather productive since it suggests a simpler, *one-parametric* fit. Note that the proposed choice causes only a slight decrease in the accuracy of the fit for the function $p(t)$. A slight increase in the deviation from the numerical solution is observed only for the points of the small region embracing the first local maximums and minimums of the function $p(t)$ after the resonance crossing point has been passed. However, we note that the accuracy of the approximation remains of the same order as before, when the more general two-parametric fit was applied. Importantly, this approach does not affect the accuracy of the approximation for the final transition probability $p(+\infty)$. In the meantime, the proposed choice leads to simple analytic formulas involving elementary functions only for both $A(\lambda)$ and $p(+\infty)$.

The appropriate form of the Gauss hypergeometric function ${}_2F_1$ to be applied to fit the variational function $A(\lambda)$ if the parameters $\lambda_1(\lambda)$ and $C_1(\lambda)$ are fixed by Eqs. (5) and (7) is given as follows

$$A = \frac{\lambda}{2} \cdot {}_2F_1\left(1\,,\,1+a\,;\,2;\,-\frac{\lambda^2}{2}\right) \tag{8}$$

(compare with Eq. (11), Ref. 22). We note that now the acting variational *constant*, a (this constant should not depend on λ), is one of the two upper parameters of the hypergeometric function, not the lower one, as applied in Ref. 22. The advantage of varying the upper parameter of the hypergeometric function of the given form [i.e., with the parameters as indicated in Eq. (8)] is that in this case one derives analytic expressions in terms of elementary functions. Indeed, for an arbitrary a the function $A(\lambda)$ is reduced to

$$A = \frac{1}{a\lambda}\left(1 - \left(1 + \frac{\lambda^2}{2}\right)^{-a}\right). \tag{9}$$

Accordingly, the final transition probability is written as

$$p(+\infty) = \frac{1}{2} - \frac{a_0(2 - P_{LZ})}{\lambda}\sqrt{1 - \left(1 + \frac{\lambda^2}{2}\right)^{-a}}, \tag{10}$$

where $a_0 = 1/\sqrt{8a}$. This formula is the main result of the present paper.

Since in the limit $\lambda \to +\infty$ holds $P_{LZ} = 1$ and for a positive a the term under the square root tends to unity, the asymptote of $p(+\infty)$ is given as

$$p(+\infty) = \frac{1}{2} - \frac{a_0}{\lambda}. \tag{11}$$

This power-law dependence has been indicated by several authors (see, e.g., Refs. 14–22). We have for the first time estimated the value of the constant a_0 as $a_0 = (2/3)/\pi \approx 0.2122$.[14] Further studies improved the estimate to 0.2214.[15] Given that the *exact* value of this constant is established as $(\ln 2)/\pi$[21] we finally obtain

$$a_0 = \frac{\ln 2}{\pi} \approx 0.22063560, \tag{12}$$

$$a = \frac{1}{8a_0^2} \approx 2.56778606. \tag{13}$$

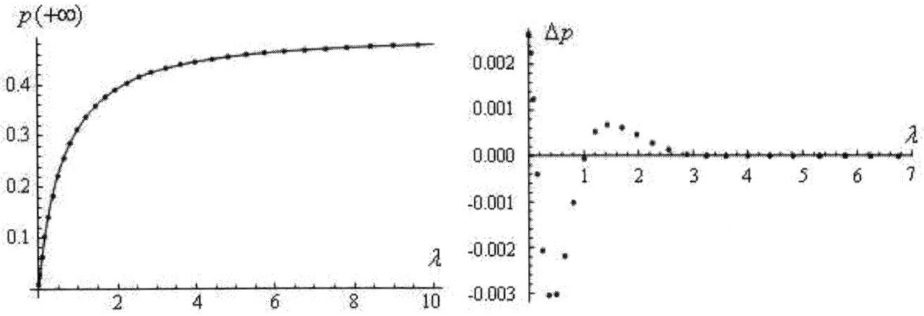

Fig. 1. a) Final transition probability to the molecular state as a function of λ. Circles show the numerical result, the solid line is calculated using formula (10). b) Deviation of the approximation defined by formulas (10), (12), (13) from the numerical solution. Starting from $\lambda = 3$ Eq. (10) provides at least 5-digit accuracy.

The derived solution defines a fairly good approximation. Comparison of the final transition probability given by the analytic formula (10) with the numerical result is made in Fig. 1. It is seen that for the whole variation range of the input Landau-Zener parameter λ, deviation of the formulas from the numerical result does not exceed $3 \cdot 10^{-3}$. We note that starting already from $\lambda = 3$ the formula provides 5-digit accuracy.

Since all the parameters involved in the general ansatz (2) describing the *whole time dynamics* of the system are eventually written in terms of the principal variational parameter A, in order to fulfill the development we examine the behavior of this parameter. The dependence $A(\lambda)$ is shown in Fig. 2. We note that two clearly marked interaction regimes are observed: the weak coupling regime corresponds to $\lambda < 1$, and the strong interaction occurs at $\lambda > 1$ (A reaches its maximum 0.252396 at $\lambda = 0.947445$). The asymptotic behavior of the parameter A in these two regions is: $A \sim \lambda/2$ at $\lambda \to 0$ and $A \sim 1/(a\lambda)$ when $\lambda \to +\infty$.

3. Landau-Zener Transition in Cubic-Nonlinear Systems

The time-dependent mean-field two-level problem involving third-order nonlinearities we consider is defined by the following set of coupled first-order nonlinear equations[6–10]:

$$
\begin{aligned}
i\frac{da_1}{dt} &= U(t)e^{-i\delta(t)}a_2 + (\Lambda_{11}|a_1|^2 + \Lambda_{12}|a_2|^2)a_1, \\
i\frac{da_2}{dt} &= U(t)e^{i\delta(t)}a_1 + (\Lambda_{21}|a_1|^2 + \Lambda_{22}|a_2|^2)a_2,
\end{aligned}
\tag{14}
$$

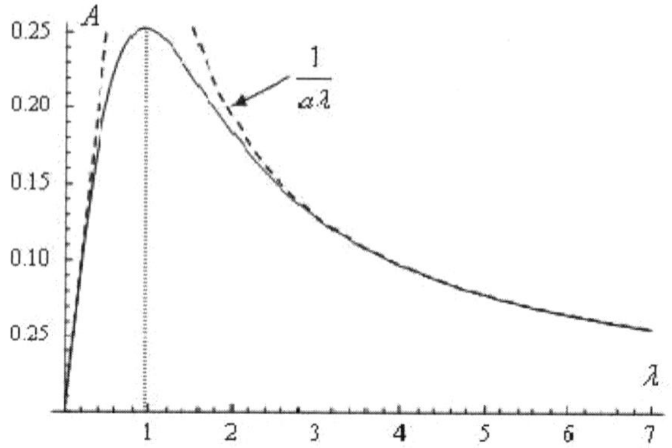

Fig. 2. The principal variational parameter A as a function of the Landau-Zener parameter λ. The function reaches maximum at $\lambda = 0.947445$. The weak and strong interaction asymptotes are given as $A \sim \lambda/2$ at $\lambda \to 0$ and $A \sim 1/(a\lambda)$ when $\lambda \to +\infty$.

where t is the time, a_1,a_2 are the fist and second states' probability amplitudes, respectively, $U(t)$ is referred to as the Rabi frequency of the laser field, $\delta(t)$ is the corresponding frequency detuning modulation function, and the cubic nonlinearities describing the inter-particle elastic interactions are characterized by (real) coefficients Λ_{jk}. The Landau-Zener field configuration $\{U(t), \delta(t)\}$ we discuss is the one for which the detuning is assumed to be a linear function of time: $\delta_t(t) = 2\delta_0 t$, and the Rabi frequency is supposed to be constant: $U(t) = U_0 = \text{const}$. We treat the basic case when the evolution of the system starts from the first level, i.e., the initial conditions imposed are $a_1(-\infty) = 1$ and $a_2(-\infty) = 0$.

We start by noting that the transition probability $p = |a_2|^2$ satisfies the following nonlinear ordinary differential equation of the third order:

$$\left(\frac{d}{dt} - \frac{G_t}{G}\right)\left(\frac{d^2p}{dt^2} - \frac{U_0^2}{2}(4 - 8p)\right) + G^2\frac{dp}{dt} = 0, \qquad (15)$$

where $G_t = dG/dt$, $G = \delta_t - \Lambda_a + 2\Lambda_s p$,

$$\Lambda_a = \Lambda_{11} - \Lambda_{12}, \Lambda_s = \frac{\Lambda_{11} - 2\Lambda_{12} + \Lambda_{22}}{2}, \qquad (16)$$

$(\Lambda_{12} = \Lambda_{21})$ and λ is the conventional Landau-Zener parameter: $\lambda = U_0^2/\delta_0$.

This equation suggests a few immediate observations. First, it is seen that if the cubic nonlinearities are not taken into account the function $G(t)$ coincides with the Landau-Zener detuning $\delta_t = 2\delta_0 t$. Hence, this function plays the role of an effective (nonlinear) detuning and the point $t = t_{res}$ defined from the condition $G(t_{res}) = 0$ is the point of the effective resonance crossing. Thus, we conclude that the introduction of the cubic nonlinearities results in a *nonlinear shift of the resonance position*. Next, the structure of the effective detuning G suggests that at sufficiently large absolute values of the time variable t, when the condition $|2\delta_0 t| >> |\Lambda_a - 2\Lambda_s p|$ holds, the role of the nonlinear terms proportional to the interaction parameters Λ_{jk} becomes negligible if the time dynamics at this region is considered.

Further, we notice that the parameter Λ_a merely leads to a constant shift in the detuning which can be eliminated by the change of the time variable $t' = t - \Lambda_a/(2\delta_0)$. This change does not affect the initial conditions (since they are imposed at infinity $t = -\infty$) as well as the final transition probability at $t \to +\infty$. For simplicity of notation, below we omit the prime (this is formally equivalent to putting $\Lambda_a = 0$). Thus, we put $\Lambda_a = 0$ and the interparticle interaction is now described by a sole combined parameter Λ_s. As it is seen from Eq. (16), there exist nonzero parameters Λ_{jk} for which the interparticle interactions result only in the shift of the detuning by a constant which is eliminated by the mentioned change of the time variable. Of course, this occurs when the parameter Λ_s is equal to zero.

Based upon our previous experience, to describe the time behavior $p = p(t)$, we introduce the following *two-term ansatz*

$$p = p_0(t) + C_1 \frac{p_{LZ}(\lambda_1, t - t_1)}{p_{LZ}(\lambda_1, \infty)}. \tag{17}$$

Here, C_1 is a scaling parameter, t_1 is a time-shift parameter, $p_{LZ}(\lambda_1, t)$ is the solution of the *linear* Landau-Zener problem for an effective Landau-Zener parameter λ_1:

$$\left(\frac{d}{dt} - \frac{1}{t}\right)\left(\frac{d^2 p_{LZ}}{dt^2} + 4\lambda_1 p_{LZ} - 2\lambda_1\right) + 4t^2 \frac{dp_{LZ}}{dt} = 0, \tag{18}$$

and $p_0(t)$ is the solution of a nonlinear first order *augmented limit equation*[22] (both $p_{LZ}(\lambda_1, t)$ and $p_0(t)$ are supposed to satisfy the initial condition $p(-\infty) = 0$). The latter equation is constructed trying to replace in Eq. (15) the second derivative d^2p/dt^2 by a function that resembles the essential features of this term but for which the solution to the equation can be derived. It has previously been shown that in the case of the quadratic-nonlinear Landau-Zener problem the simplest choice of this replacement by a constant term A leads to highly accurate results.[22] The idea here is to justify the introduced constant in a way to cancel the mean action of the terms controlled by the logarithmic derivative G_t/G. To proceed with this idea, one should look at the most effective time point. It is readily understood from physical point of view and it is immediately seen mathematically that this point is the resonance crossing point t_{res} because here the mentioned derivative diverges. It turns out that in the case considered in the present paper more productive is the replacement $A_0 + A_1 p$. Indeed, the numerical simulations reveal that one can always find specific values of the involved variational parameters A_0, A_1 such that the function (17) accurately fits the numerical solution to the exact equation (15) in the whole time domain.

The resultant limit first order differential equation is readily integrated to produce a *quartic* polynomial equation for limit function $p_0(t)$ with variational parameters α_1 and β_1 (originating from A_0 and A_1):

$$\frac{U_0^2}{G^2(p_0)} = -\frac{1}{4}\frac{p_0(p_0 - \beta_1) + C_0}{(p_0 - \alpha_1)^2}. \tag{19}$$

Note that for the initial condition $p(-\infty) = 0$ the integration constant C_0 vanishes: $C_0 = 0$. Further, the meaning of involved variational parameters α_1 and β_1 is readily understood when examining the limits $t \to t_{res}$ and $t \to +\infty$. Since in the first case the left-hand side of Eq. (19) diverges, it follows that at this point p_0 should adopt value α_1, i.e., $p_0(t_{res}) = \alpha_1$, so that α_1 is the transition probability at resonance crossing point $t = t_{res}$ given by the limit function p_0. In the second limit, at $t \to +\infty$, the left-hand-side of the equation vanishes, hence, should be $p_0(+\infty) = \beta_1$, thus, β_1 defines the final transition probability achieved by the limit function $p_0(t)$ (Fig. 3). It is readily understood that in the linear limit $\Lambda_s \to 0$ the function p_0 should vanish, hence, at this limit both α_1 and β_1 tend to zero.

The ansatz (17) with Eq. (19) provides a highly accurate approximation if $|\Lambda_s| < U_0$. An example of the fit produced by this formula is shown in Fig. 4. It is seen that the suggested solution is practically indistinguishable from the exact numerical result almost in the whole time domain. A slight deviation is observed only in a small region embracing a few first oscillation periods after the system passes the resonance, though even here the deviation is only on the level of a few percents. In order to better examine the substantial progress achieved by the presented development we note the difference of the corresponding nonlinear and linear solutions (Fig. 3).

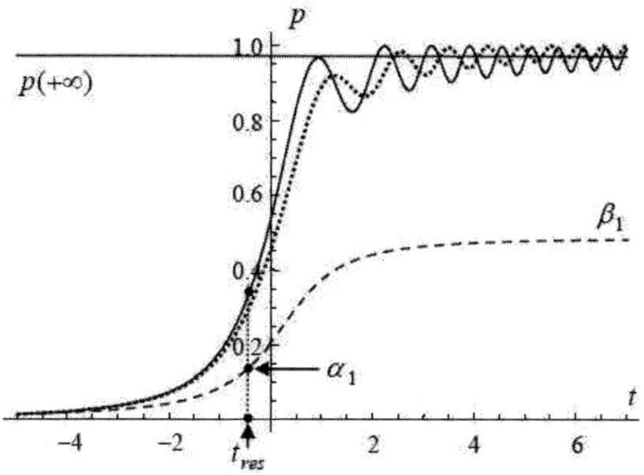

Fig. 3. Nonlinear (solid line) and linear (dotted line) transition probabilities as a function of time at $\lambda = 1.5$ and $\Lambda_s = 0.5$ (horizontal solid line is the final transition probability $p(+\infty)$). Dashed line shows the limit solution p_0 defined by Eq. (19). The position of the resonance crossing $t = t_{res}$ is marked by filled circles.

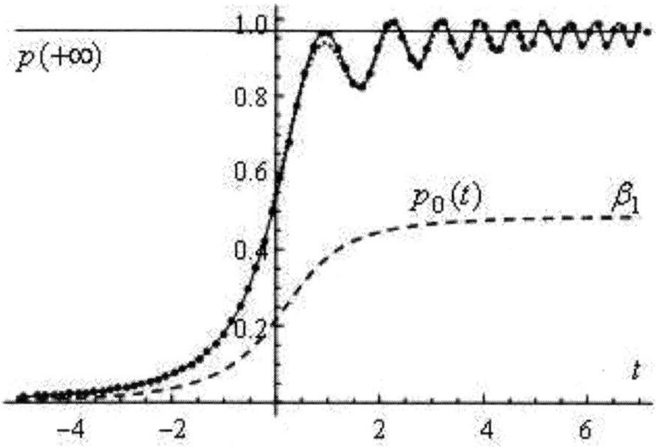

Fig. 4. Nonlinear transition probability p as a function of time at $\lambda = 1.5$ and $\Lambda_s = 0.5$. Dotted line presents the approximation by ansatz (17) with limit Eq. (19), dashed line is the limit solution p_0, horizontal solid line is the final transition probability $p(+\infty)$, the filled circles show the time points used for fit. A slight deviation from the exact result (solid line) is observed only in a small region embracing a few first oscillation periods.

Discussing now the final transition probability p_{inf} at $t \to +\infty$ we note that a much better fit is achieved when approximating the second derivative term d^2p/dt^2 in Eq. (15) by a *quadratic* polynomial in p. In this case we derive a *fifth-order* limit equation of the form

$$\frac{\lambda}{G^2(p_0)} = \frac{k^2}{9} \frac{(p_0 - \beta_0)(p_0 - \beta_1)(p_0 - \beta_2)}{(p_0 - \alpha_1)^2 (p_0 - \alpha_2)^2}, \tag{20}$$

where for definiteness, without loss of generality, we adopt the numeration of the involved parameters as $\alpha_1 < \alpha_2$ and $\beta_0 < \beta_1 < \beta_2$. This equation suggests several developments. A notable result is achieved when applying a limiting procedure involving simultaneous limits $t \to +\infty$ and $\delta_0 \to 0$ in a way that the function $G(t)$ tends to $2\Lambda_s p_{\mathrm{inf}}$ so that the left-hand side of Eq. (20) becomes $\lambda/(2\Lambda_s p_{\mathrm{inf}})^2$. Of particular interest is now the linear limit $\Lambda_s = 0$. Since in this case the left-hand side of the equation diverges it is clear that at this point the parameter α_1 adopts the ordinary linear Landau-Zener transition probability $P_{LZ} = 1 - e^{-\pi\lambda}$, i.e., $\alpha_1(\Lambda_s = 0) = P_{LZ}$. In the same way, considering the limits $\Lambda_s \to \pm\infty$ we conclude that $\beta_0 = 0$ and $\beta_1 = 1$. The resulting equation for p_{inf} thus reads

$$\frac{9\lambda}{(2k\,\Lambda_s p_{\mathrm{inf}})^2} = \frac{p_{\mathrm{inf}}(p_{\mathrm{inf}} - 1)(p_{\mathrm{inf}} - \beta_2)}{(p_{\mathrm{inf}} - P_{LZ})^2 (p_{\mathrm{inf}} - \alpha_2)^2}. \tag{21}$$

Numerical testing shows that this equation accurately describes the final transition probability for all the variation range of input parameters of the problem under consideration, U_0, δ_0 and Λ_{jk}. The accuracy of the suggested approximation is demonstrated in Figs. 5 and 6 for several values of the Landau-Zener parameter λ ($\lambda = 10,\ 1,\ 0.2,\ 0.05$).

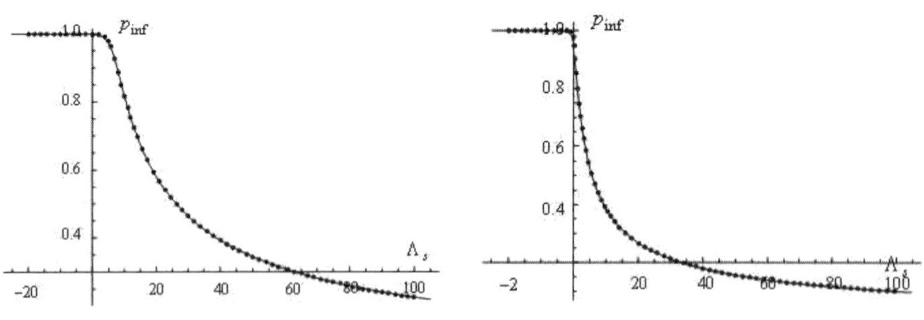

Fig. 5. Final transition probability p_{inf} as a function of Λ_s for $\lambda = 10$ (left figure) and $\lambda = 1$ (right figure). Solid line - Eq. (21), filled circles – numerical result.

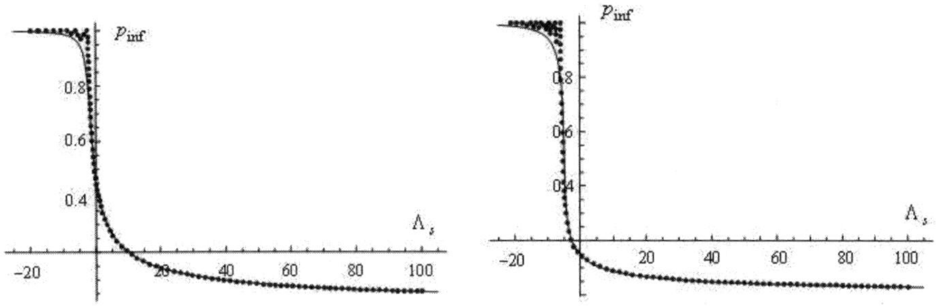

Fig. 6. Final transition probability p_{inf} as a function of Λ_s for $\lambda = 0.2$ (left figure) and $\lambda = 0.05$ (right figure). Solid line - Eq. (21), filled circles – numerical result.

We note that the description is very good for $\lambda \geq 1$ (Fig. 5). Furthermore, it is seen that for $\lambda < 1$ the coincidence is good almost everywhere. Except a small region of negative Λ_s (Fig. 6), the absolute error is of the order of 10^{-4}. Note that for this region of negative Λ_s and relatively small λ the deviation is also rather small. A notable observation is that for small $\lambda << 1$ and negative Λ_s the final transition probability undergoes abrupt, nearly vertical jump to $p_{\text{inf}} = 1$ followed by expressed oscillations which dump to the line $p_{\text{inf}} = 1$ (Fig. 6). This feature is not described by the suggested equation (21). However, we have checked that a modification of the proposed approach is potent to describe this peculiarity as well. We will address this point in a separate publication because these oscillations suggest a different physical behavior that deserves a separate treatment.

Finally, we numerically find that the variational parameters k, α_2 and β_2 are approximated as

$$k = 1 - e^{-\lambda}, \alpha_2 = 2.5, \beta_2 = 1 + \frac{12}{(1+\lambda)^3} \tag{22}$$

(the accuracy of the last formula is shown on Fig. 7). These relations together with Eq. (21) fulfill the development. The derived equation (21) can be viewed as a generalized Landau-Zener formula for the two-level interacting Bose-Einstein condensates. For completeness, on Fig. 8 we show the family of curves $p_{\text{inf}}(\Lambda_s)$ defined by this formula for several λ.

Analyzing the curves shown on Fig. 8 we note that one may distinguish two interaction regimes depending on whether the second derivative $d^2p_{\text{inf}}/d\Lambda_s^2$ is negative or positive. For the positive case we note that for $\lambda >> 1$ we have $k \approx 1$ and $P_{LZ} \approx \beta_2 \approx 1$, hence, Eq. (21) is considerably simplified reducing to a cubic one from which we readily obtain the

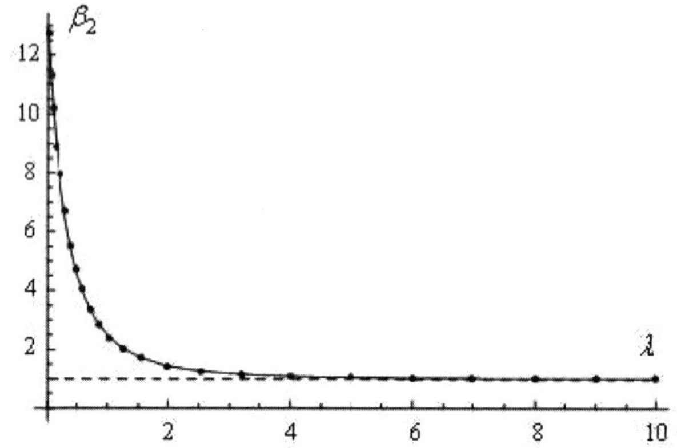

Fig. 7. The parameter β_2 as a function of λ.

asymptotic behavior of $p_{\rm inf}$ for $\Lambda_s \to +\infty$:

$$p_{\rm inf} \sim \left(\frac{3U_0}{2\Lambda_s}\right)^{2/3} \left[\frac{5}{2} - \left(\frac{15U_0}{4\Lambda_s}\right)^{2/3}\right]^{2/3} . \tag{23}$$

This formula provides highly accurate approximation starting already from $\lambda = 3$.

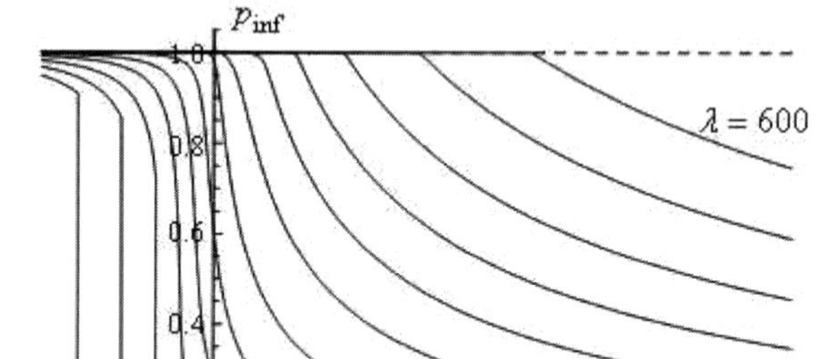

Fig. 8. From left to right: the dependence $p_{\rm inf}(\Lambda_s)$ defined by Eq. (21) for λ = 0.01, 0.015, 0.025, 0.05, 0.1, 0.3, 1.25, 5, 15, 40, 100, 250, and 600.

4. Conclusion

To summarize, we have developed a comprehensive theory of the Landau-Zener transition in quadratic- and cubic-nonlinear two-state systems. We have derived compact analytic formulas for the final Landau-Zener transition probability that present highly accurate approximations for the whole rage of the variation of input parameters for both physical situations.

Acknowledgments

This research has been conducted in the scope of the International Associated Laboratory IRMAS. The work was supported by the Armenian National Science and Education Fund (ANSEF Grants No. 2009-PS-1692 and No. 2010-MP-2186).

References

1. L.D. Landau, Phys. Z. Sowjetunion, **2**, 46 (1932).
2. C. Zener, Proc. R. Soc. London, Ser. A, **137**, 696 (1932).
3. E.C.G. Stuckelberg, Helv. Phys. Acta., **5**, 369 (1932).
4. E. Majorana, Nuovo Cimento, **9**, 45 (1932).
5. K.B. Davis, M.-O. Mewes, M.R. Andrews, N.J. van Druten, D.S. Durfee, D.M. Kurn, and W. Ketterle, Phys. Rev. Lett., **75**, 3969 (1995); M.H. Anderson, J.R. Ensher, M.R. Matthews, C.E. Wieman, and E.A. Cornell, Science, **269**, 198 (1995).
6. O. Zobay and B.M. Garraway, Phys. Rev. A, **61**, 033603 (2000).
7. V.A. Yurovsky, A. Ben-Reuven, P.S. Julienne, and C.J. Williams, Phys. Rev. A **62**, 043605 (2000); V.A. Yurovsky, A. Ben-Reuven, and P.S. Julienne, Phys. Rev. A, **65**, 043607 (2002).
8. J. Liu, L. Fu, B.-Y. Ou, S.-G. Chen, D.-Il Choi, B. Wu, and Q. Niu, Phys. Rev. A **66**, 023404 (2002); B. Wu and Q. Niu, Phys. Rev. A **61**, 023402 (2000); J. Liu, B. Wu, and Q. Niu, Phys. Rev. Lett. **90**, 170404 (2003); J. Liu, B. Liu, and L.-B. Fu, Phys. Rev. A **78**, 013618 (2008).
9. D. Witthaut, E.M. Graefe, and H.J. Korsch, Phys. Rev. A, **73**, 063609 (2006).
10. M. Jona-Lasinio, O. Morsch, M. Cristiani, N. Malossi, J.H. Muller, E. Courtade, M. Anderlini, and E. Arimondo, Laser Phys. Lett. **1**, 147 (2004); M. Jona-Lasinio, O. Morsch, M. Cristiani, E. Arimondo, and C. Menotti, Laser Physics **15**, 1180 (2005).
11. M. Kostrun, M. Mackie, R. Cote, and J. Javanainen, Phys. Rev. A, **62**, 063616 (2000).
12. D.J. Heinzen, R. Wynar, P.D. Drummond, and K.V. Kheruntsyan, Phys. Rev. Lett., **84**, 5029 (2000).
13. F.A. van Abeelen and B.J. Verhaar, Phys. Rev. Lett., **83**, 1550 (1999).
14. A. Ishkhanyan, M. Mackie, A. Carmichael, P.L. Gould, and J. Javanainen, Phys. Rev. A, **69**, 043612 (2004).

15. A. Ishkhanyan, J. Javanainen, and H. Nakamura, J. Phys. A, **39**, 14887 (2006); A. Ishkhanyan, J. Javanainen, and H. Nakamura, J. Phys. A, **38**, 3505 (2005).
16. E. Pazy, I. Tikhonenkov, Y.B. Band, M. Fleischhauer, and A. Vardi, Phys. Rev. Lett., **95**, 170403 (2005).
17. R.A. Barankov and L.S. Levitov, arXiv:cond-mat/0506323v1 (2005).
18. B.E. Dobrescua and V.L. Pokrovsky, Physics Letters A, **350**, 15 (2006).
19. J. Liu, B. Liu, and L.-B. Fu, Phys. Rev. A, **78**, 013618 (2008).
20. A. Altland and V. Gurarie, Phys. Rev. Lett., **100**, 063602 (2008); A. Altland, V. Gurarie, T. Kriecherbauer, and A. Polkovnikov, Phys. Rev. A, **79**, 042703 (2009).
21. A.P. Itin and P. Trm, Phys. Rev. A, **79**, 055602 (2009); A.P. Itin and S. Watanabe, Phys. Rev. E, **76**, 026218 (2007).
22. A. Ishkhanyan, B. Joulakian, and K.-A. Suominen, J. Phys. B, **42**, 221002 (2009).
23. E. Fermi, Z. Phys., **71**, 250 (1931); E. Fermi and F. Rasetti, Z. Phys., **71**, 689 (1931).
24. T. Kohler, K. Goral, and P.S. Julienne, Rev. Mod. Phys., **78**, 1311 (2006).
25. Y.R. Shen, The Principles of Nonlinear Optics (Wiley, New York, 2002); R.W. Boyd, Nonlinear Optics (Academic Press, Boston, 1992).
26. M. Abramovitz, I.A. Stegun, Handbook of mathematical functions, (Dover , New York, 1972).

PART B

Photonic and Electronic Phenomena in Structured Media

MULTIPLE INTERACTIONS IN MULTILAYERED STRUCTURES OF NONLINEAR MATERIALS

D. A. ANTONOSYAN* and G. Yu. KRYUCHKYAN**

Yerevan State University, Alex Manoogian 1, 0025, Yerevan, Armenia

Institute for Physical Researches, National Academy of Sciences,
Ashtarak-2, 0203, Ashtarak, Armenia
**E-mail: antonosyand@ysu.am*
***E-mail: kryuchkyan@ysu.am*
www.ysu.am

We study multiple quasi-phase interactions in periodically poled grating configurations which are designed as one-dimensional multilayered media consisting of layers with different $\chi^{(2)}$ susceptibilities and refractive indexes. We analyze the spectral properties of photon pairs produced by means of pulsed parametric down-conversion in a modified periodically poled structure that additionally involve dispersive segments. The method of superlattices in multilayered media is used for designing of a phase-reversed configuration that simultaneously phase-match two interactions.

Keywords: Multiple optical interactions, quasi phase matching, down-conversion, superlattices.

1. Introduction

Multiple optical interactions in $\chi^{(2)}$ nonlinear media, leading to the simultaneous quasi-phase matching (QPM) of several parametric processes, result in many phenomena of optics and quantum physics. Traditionally, the technique of QPM between interacting waves is realized in periodically poled crystals by applying a periodic variation of the sign of the second order nonlinear coefficient at the period of the oscillating mismatched phase or in photonic crystals (see the reviews[1,2]). In order to simultaneously phase match two interactions, aperiodic modulation of the nonlinear coefficient is required that allows each nonlinear process to take place in its own grating section.

Applications of QPM cascaded processes include the generation of large nonlinear phase shift, cascaded generation of the third and fourth harmonics, multi-photon down conversion and others (see, for a review[3] and papers[4–11]). Recently, it has been shown that QPM multiple interactions are also extremely promising for generation of nonclassical states of light and open interesting perspectives for quantum information technologies. Among these, we note the generation of squeezed light in cascaded frequency doubler,[12] cascaded down-conversion and generation of "three-photon" light,[13] and generation on nonclassical light beams in periodically poled crystals,[14] the production of multipartite entanglement between the resonator's modes,[15] realization of continuous-variable cluster states for universal computation.[16] These important developments initiate the further investigations in area of QPM structures and parametric generation of quantum states with the specific modal character required for different applications. In the paper[17] the spectral properties of joint states of photon pairs produced by pulsed parametric down-conversion in a multilayered structures (superlattice structure) of second-order nonlinear and linear materials have been investigated. The key to this approach is the idea of manipulating overall group delay mismatches between the various fields in structured materials for synthesis of twin photon states. Recently, this approach has been developed in the series of papers (see, Ref. 18 and references therein).

In this paper, we analyze several multiple interactions in finite multilayered nonlinear structures pumped by laser pulses in application to cascaded parametric processes. Such structures consist of layers with different susceptibilities of nonlinearity and with different refractive indexes. The main advantage of this approach is that the design of multiple QPM interactions can be done in more controllable form by detailed description of elementary interactions in each layers including dispersions of interacting waves. Moreover, new schemes leading to sufficiently strong $\chi^{(2)}$ interactions due to the corrected phase-matching in each of layers can be elaborated in this way. This approach is differ in some points from the standard treatment of cascaded parametric processes in which only the periodic or aperiodic properties of $\chi^{(2)}$ nonlinearity are exploited (see, for example, Refs. 1,2,9,10). On the whole, the approach discussed leads to formation of new spectral effects in the multiple parametric interactions.

In this paper we concentrate on discission of the structures involving finite number of alternating positive and negative susceptibilities that are usually realized in periodically poled nonlinear crystal (PPNC). The question turns into how to construct a structure so that multiple high efficiency QPM parametric processes can be made simultaneously. For this goal the method of nonlinear multilayered structures will be used for designing of so-called phase-reversed configuration that allows simultaneously phase-match two parametric processes. In the other part of the paper we analyze a technique in which optical linear elements are included in PPNC configuration for both compensation of dispersions exhibited by the fields in nonlinear medium.

The cascaded three-wave process can be completely described by high-order terms on $\chi^{(2)}$ interaction. This approach is usually based on Dyson expansion of time-evolution matrix with the interaction Hamiltonian (2). In this paper, we discuss the other alternative approach based on the Magnus expansion.

The present article is structured as follows. In section II we discuss three-wave type-II interaction in one-dimensional multilayered $\chi^{(2)}$ materials and spectral amplitude of two-photon radiation. In section III we discuss high-order terms on $\chi^{(2)}$ interaction. In section IV we consider the superlattice configuration that involves alternating positive and negative second-order susceptibilities, $\chi_+ = \chi^{(2)} > 0$ and $\chi_- = \chi^{(2)} < 0$ which are interchanged by the dispersive spacer. This configuration involves both the grating vector and the dispersive parameters, and thus seems to be more flexible for engineering of various two-photon Gaussian wave packets in the process of down-conversion, as it is demonstrated here. Section IV contains application of the superlattice approach in investigation of the double phase-matching. The concrete calculations center on the phase-reversed configuration, on calculation of the assemble quasi-pase matched function that allows simultaneously phase-match two parametric processes. We summarize our results in Section VI.

2. Parametric Three-Wave Interaction in Composite Nonlinear Media

In this section, we briefly discuss three-wave type-II interaction in one-dimensional nonlinear composite media considering two important for applications configurations. In one of this media consists of alternating layers with different coefficients of nonlinearity and three fields involved are normal to each of the interfaces between layers. This approach also involves layered structure of $\chi^{(2)}$ crystals and dispersive linear optical elements having various refractive indexes, that is illustrated on Fig. 1. The interaction Hamiltonian reads as

$$H(t) = H^-(t) + H^+(t), \tag{1}$$

$$H^{-}(t) = \int_0^L dz\chi^{(2)}(z)E_p(z,t)E_1^{(-)}(z,t)E_2^{(-)}(z,t), \tag{2}$$

where $\chi^{(2)}(z)$ is the second-order nonlinearity, L is the total length of a medium and $E_p, E_1^{(-)}, E_2^{(-)}$ represent the electric field operators associated with each of the interacting field. We assume that propagation of involved three fields is normal to each of the interfaces between different layers and restrict ourselves to a scalar modes. In the general case, the dynamics of the electromagnetic waves in this medium also includes the reflectivity of propagating waves at boundaries of neighboring segments. Thus, inside each layer the fields can be expressed as superposition of forward and backward propagating plane waves with wave vectors $|k_0| = \frac{\omega}{c}n_p$, $|k_i| = \frac{\omega_i}{c}n_i$, $(i = 1, 2)$.

In this case, the interaction Hamiltonian can be expressed as the sum of interactions in each layer in terms of the electric fields in nth layer, $E_{pn}(z,t)$, $E_{in}^{(-)}(z,t)$

$$H^{-}(t) = \sum_n \int_{z_n}^{z_{n+1}} dz\chi^{(2)}(z)E_{pn}^{*}(z,t)E_{1n}^{(-)}(z,t)E_{2n}^{(-)}(z,t). \tag{3}$$

We assume the pump field E_{pn} is classical, it can be expressed as

$$E_{pn}(z,t) = \int d\omega e^{i\omega t}(\alpha_{pn}(\omega)\exp\{-ik_{0n}(\omega)(z-z_n)\} + \beta_{pn}(\omega)\exp\{ik_{0n}(\omega)(z-z_n)\}), \tag{4}$$

through the spectral amplitudes of the forward and backward waves. The both other fields are quantized and their positive frequency part can be expressed as

$$E_{jn}^{(-)}(z,t) = \int d\omega e^{i\omega t}(\alpha_{jn}^{*}(\omega)\exp\{-ik_n(\omega)(z-z_n)\}a_j^{+}(\omega) + \beta_{jn}^{*}(\omega)\exp\{ik_n(\omega)(z-z_n)\}b_j^{+}(\omega)), \tag{5}$$

where $j = 1, 2$, the annihilation operators $a_j(\omega)$ and $b_j(\omega)$ correspond to the forward and backward plan waves, respectively, i.e. $a_i(k) \to a_i(\omega)$, and $a_i(-k) \to b_i(\omega)$.

Following a standard perturbation approach, the amplitude of emitting a subharmonic photon at frequency ω_1 together with photon at frequency ω_2 is determined as

$$\Phi(\omega_1, \omega_2) = \langle 0|a_1(\omega_1)a_2(\omega_2)\left(-\frac{i}{\hbar}\right)\int_{\infty}^{t} H(t^{'})dt^{'}|0\rangle. \tag{6}$$

In the collinear type-II, non-degenerate down-conversion process the amplitude is calculated as

$$\begin{aligned}\Phi(\omega_1, \omega_2) &= \frac{-2\pi i}{\hbar}\sum_n \int_{z_n}^{z_{n+1}} \alpha_{pn}(\omega_0)\alpha_{1n}^*(\omega_1)\alpha_{2n}^*(\omega_2)\chi_n e^{i\Delta k_n(z-z_n)}dz \\ &= \sum_n \alpha_{pn}(\omega_0)\alpha_{1n}^*(\omega_1)\alpha_{2n}^*(\omega_2)\chi_n l_n e^{i\frac{\Delta k_n l_n}{2}} sinc\left(\frac{\Delta k_n l_n}{2}\right). \end{aligned} \tag{7}$$

Here: $\Delta k_n = k_{0n}(\omega_0) - k_{1n}(\omega_1) - k_{2n}(\omega_2)$ is the phase-matching, χ_n is the nonlinear susceptibilities in the segments, $\omega_0 = \omega_1 + \omega_2$ and $l_n = z_{n+1} - z_n$. The amplitudes $\alpha_{jn}(\omega)$ and $\beta_{jn}(\omega)$ describe the electric fields propagating in the forward and backward directions, correspondingly, in one-dimensional composite materials, have been studied in Ref. 24,25. Thus, the result (7) is a generalization of the well known analogues results presented in Refs. 17,18 on the cases when the refractivities of the interacting fields are taking into account.

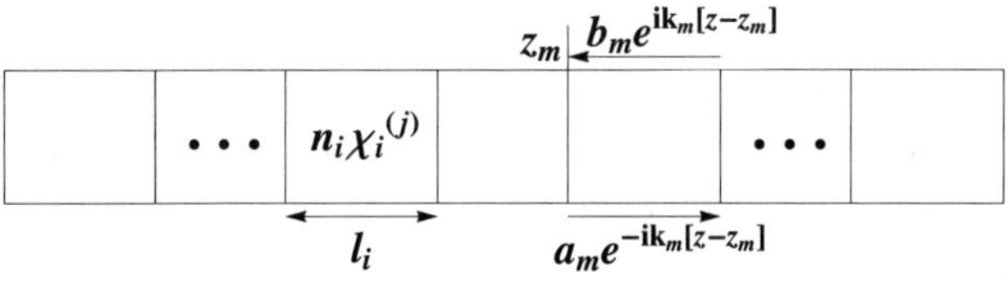

Fig. 1. Scheme of layer structure, $\chi_i^{(j)}$ is the j-order nonlinearity in the i-th layer, l_i is the length and $n_i$is the refractive index of i-th layer

In approximation that the effects of refraction from each of the layers are small, $\beta_{jn}(\omega) \simeq 0$ and then $\alpha_{jn}(\omega)$ describes the phase incursion of forward propagating wave in the nth layer. For the scheme presented in Fig. 1, we have

$$\alpha_{jn}(\omega_j) = \alpha_{j1} exp\{-i\sum_m^{n-1} k_{jm} l_m\}, \tag{8}$$

$$\alpha_{pn}(\omega_0) = E_0(\omega_0) exp\{-i\sum_m^{n-1} k_{0m} l_m\}. \tag{9}$$

In this approximation, we get for the amplitude of two-photon state

$$\Phi(\omega_1,\omega_2) = \frac{-2\pi i}{\hbar} E_0(\omega_1+\omega_2)\alpha_{11}\alpha_{21}G(\Delta k), \tag{10}$$

where

$$G(\Delta k) = \sum_m l_m \chi_m e^{-i\left(\varphi_m + \frac{\Delta k_m l_m}{2}\right)} sinc\left(\frac{\Delta k_m l_m}{2}\right),$$

$$\varphi_m = \sum_n^{m-1} l_n \Delta k_n, \qquad \varphi_1 = 0. \tag{11}$$

Note, that the quantity $G(\Delta k)$ can be interpreted as a Fourier transforms of the effective nonlinear coefficient of three-wave interaction in one-dimensional structured media. Really, it is easy to check that

$$\begin{aligned} G(\Delta k) &= \int_0^L \chi^{(2)}(z) e^{i\Delta k(z)} dz \\ &= \sum_n \int_{z_n}^{z_{n+1}} \chi_n e^{i\Delta k_n z} dz, \end{aligned} \tag{12}$$

leading to the expression (11).

We consider below two applications of these general expressions in area of PPNC assuming that $\chi^{(2)}(z)$ in the segments consists of nonlinear coefficients $\chi_m = \pm\chi$ with sign changes occurring at the position z_m.

2.1. *Interaction in TE and TM polarization basis*

Below we present results for the configuration involving a classical pump field that propagates under the angle with respect to z axis.[19,26] The wave vector $\mathbf{k}$ propagates in the yz plane (with polarization $\mathbf{e}$, which is perpendicular to $\mathbf{k}$) at an angle Θ with respect to z axis, k_y is the component of the wave vector, which stays constant by propagation through the media. The $\mathbf{e}_{pF,TE(TM)}$ denotes an incident $TE(TM)$-wave polarization direction. Electric field inside each homogeneous layer could be presented as a sum of forward and backward propagating wave amplitudes. The positive-frequency pump-field amplitude $\mathbf{E}^{(+)}_{np,\alpha}(z,t)$ with polarization $\alpha(\alpha = TE, TM)$ in the nth layer can be written as

$$\mathbf{E}_{pn}(z,t) = \mathbf{E}_{pn,TE}(z,t) + \mathbf{E}_{pn,TM}(z,t), \tag{13}$$

$$\mathbf{E}_{np,\alpha}(z,t) = e^{-ik_y y}\int d\omega e^{i\omega t}\left(A^{(n)}_{pF,\alpha}\mathbf{e}^n_{pF,\alpha}\exp\{ik^{(n)}_{p,z}(z-z_n)\}\right.$$
$$\left. + A^{(n)}_{pB,\alpha}\mathbf{e}^n_{pB,\alpha}\exp\{ik^{(n)}_{p,z}(z-z_n)\}\right), \tag{14}$$

where $A^{(n)}_{pF,\alpha}$ and $A^{(n)}_{pB,\alpha}$ describe the fields of polarization $\alpha = TE, TM$ in the nth layer propagating in the forward and backward directions, correspondingly.

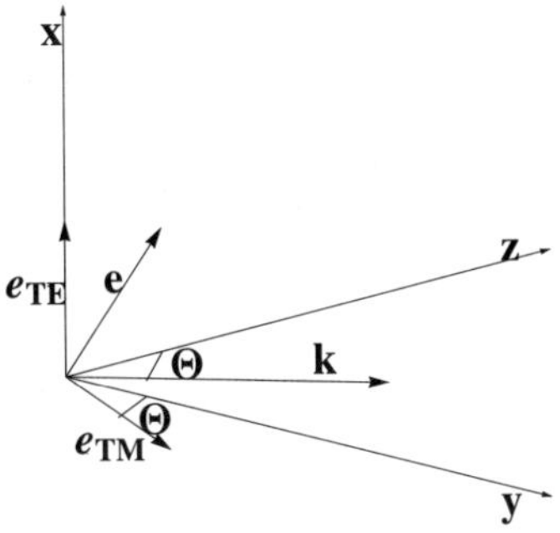

Fig. 2. A field with wave-vector **k** (lying in yz plane) and polarization vector **e** (perpendicular to **k**) propagates at an angle Θ with respect to z axis.The field is linearly polarized. The polarization vector $\mathbf{e}_{TE}$ of TE wave is parallel to x axis whereas the polarization vector $\mathbf{e}_{TM}$ wave lyes in yz plane.

The positive-frequency electric-field operators $E^{(+)}_1(z,t)$ and $E^{(+)}_2(z,t)$ for the two subharmonic signal and idler fields in the nth layer can be decomposed into TE and TM-wave contributions

$$\mathbf{E}^{(+)}_{jn}(z,t) = \mathbf{E}^{(+)}_{jn,TE}(z,t) + \mathbf{E}^{(+)}_{jn,TM}(z,t), \tag{15}$$

$$\mathbf{E}^{(+)}_{jn,\alpha}(z,t) = e^{-ik_y y}\int d\omega e^{i\omega t}\left(\gamma_{jnF,\alpha}\mathbf{e}^{(n)}_{jF,\alpha}(\omega)a_{jF,\alpha}\exp\{ik^{(n)}_z(z-z_n)\}\right.$$
$$\left. + \gamma_{jnB,\alpha}\mathbf{e}^{(n)}_{jB,\alpha}(\omega)a_{jB,\alpha}\exp\{-ik^{(n)}_z(z-z_n)\}\right), \tag{16}$$

$k_z = \sqrt{\left(\frac{n\omega}{c} - k_y^2\right)}$, where $j = 1, 2$, $\alpha = TE, TM$, n is refractive index of the medium, a_{jF} and a_{jB} are annihilation operators of forward and backward waves, correspondingly. The interaction Hamiltonian is written as

$$H(t) = \sum_n \int_{z_n}^{z_{n+1}} dz\chi(z) : \sum_{\alpha,\beta,\gamma=TE,TM} \times \left[\mathbf{E}_{np,\alpha}(z,t)\mathbf{E}^{(-)}_{1n,\beta}(z,t)\mathbf{E}^{(-)}_{2n,\gamma}(z,t) + h.c.\right], \tag{17}$$

where χ denote a third-order tensor of nonlinear coefficients, the symbol : is shortened of the tensor χ with respect to its three indices. If we consider one dimensional photonic crystal consisting of a series of alternative layers of materials a and b we will have two equations like Eq. (16) for a and b layers

$$\gamma_{jnF(B),\alpha}(\omega) = \begin{cases} \alpha^{F(B)}_{jn,\alpha}(\omega) \text{ for material } a; \\ \beta^{F(B)}_{jn,\alpha}(\omega) \text{ for material } b. \end{cases}$$

Matching the fields at the boundaries we obtain relations between the amplitudes in neighboring slabs, which can be reduced to a relation in the same material in the neighboring cells:

$$\begin{pmatrix} \alpha^F_{n-1} \\ \alpha^B_{n-1} \end{pmatrix} = \begin{pmatrix} A & B \\ C & D \end{pmatrix} \begin{pmatrix} \alpha^F_n \\ \alpha^B_n \end{pmatrix}. \tag{18}$$

The coefficients for TE and TM waves are

$$A_{TE} = \exp^{ik^a a}\left[\cos(k^b b) + \frac{i}{2}\left(\frac{k^b}{k^a} + \frac{k^a}{k^b}\right)\sin(k^b b)\right],$$

$$B_{TE} = \exp^{-ik^a a}\left[-\frac{i}{2}\left(\frac{k^b}{k^a} - \frac{k^a}{k^b}\right)\sin(k^b b)\right],$$

$$A_{TM} = \exp^{ik^a a}\left[\cos(k^b b) + \frac{i}{2}\left(\frac{n_b^2 k^a}{n_a^2 k^b} + \frac{n_a^2 k^b}{n_b^2 k^a}\right)\sin(k^b b)\right],$$

$$B_{TM} = \exp^{-ik^a a}\left[-\frac{i}{2}\left(\frac{n_b^2 k^a}{n_a^2 k^b} - \frac{n_a^2 k^b}{n_b^2 k^a}\right)\sin(k^b b)\right], \tag{19}$$

with $C_{TE/TM} = B^*_{TE/TM}$, $D_{TE/TM} = A^*_{TE/TM}$.

The $\beta_n^{F(B)}$ coefficients are related to the $\alpha_n^{F(B)}$ via $(\beta^F_n, \beta^B_n) = M(\alpha^F_n, \alpha^B_n)$, where M is given by:

$$M_{TE} = \begin{pmatrix} \frac{(k^a + k^b)}{2k^b} e^{ia(k^a - k^b)} & \frac{(k^b - k^a)}{2k^b} e^{-ia(k^a + k^b)} \\ \frac{(k^a - k^b)}{2k^b} e^{ia(k^a + k^b)} & \frac{(k^a + k^b)}{2k^b} e^{-ia(k^a - k^b)} \end{pmatrix}, \tag{20}$$

$$M_{TM} = \begin{pmatrix} \frac{(n_b^2 k^a + n_{ba}^2 k^b)}{2n_a n_b k^b} e^{ia(k^a - k^b)} & \frac{(n_b^2 k^a - n_a^2 k^b)}{2n_a n_b k^b} e^{-ia(k^a + k^b)} \\ \frac{(n_b^2 k^a - n_a^2 k^b)}{2n_a n_b k^b} e^{ia(k^a + k^b)} & \frac{(n_a^2 k^b + n_b^2 k^a)}{2n_a n_b k^b} e^{-ia(k^a - k^b)} \end{pmatrix}. \quad (21)$$

3. High-Order $\chi^{(2)}$ Processes in Superlattice Structures

Following a standard perturbation approach the time evaluation of quantum states is described in the Dyson expansion form as

$$|\psi(t)\rangle = U(t, t_0)|\psi_{\text{in}}\rangle = T \exp\left[-\frac{i}{\hbar}\int_{t_0}^{t} H(t^{'})dt^{'}\right]|\psi_{\text{in}}\rangle, \quad (22)$$

where T is the time ordering operator and $|\psi_{\text{in}}\rangle$ is an initial state of the system. We consider below high-order terms of the unitary Dyson expansion with the interaction Hamiltonian (2). In general, this Hamiltonian does not commute at different times due to Dyson-type time dependent integrations are appeared in the perturbation expansion. This circumstance usually makes the analysis of high-order processes complicated. Note, that the unitary evaluation operator $U(t, t_0)$ can be also presented in the other form: so-called Magnus expansion (see, for example Ref. 27), which seems to be more preferable for some specific systems than the standard Dyson expansion. In this representation the time-evaluation operator is a true exponential operator

$$U(t, t_0) = \exp\{\sum_k \Omega_k(t)\}, \quad (23)$$

given by the Magnus expansion. The tree terms of the series are given as

$$\Omega_1(t) = -\frac{i}{\hbar}\int_{t_0}^{t} dt_1 H(t_1), \quad (24)$$

$$\Omega_2(t) = -\frac{1}{2\hbar^2}\int_{t_0}^{t} dt_1 \int_{t_0}^{t_1} dt_2 [H(t_1), H(t_2)], \quad (25)$$

$$\Omega_3(t) = \frac{i}{6\hbar^2}\int_{t_0}^{t} dt_1 \int_{t_0}^{t_1} dt_2 \int_{t_0}^{t_2} dt_3 \Big([H(t_1), [H(t_2), H(t_3)]] + [H(t_3), [H(t_2), H(t_1)]] \Big). \quad (26)$$

For instance, it is easy to check that the second-order Dyson term corresponds to the second-order Magnus expansion. It follows from the following equality

$$\int_{t_0}^{t} dt_1 \int_{t_0}^{t} dt_2 T\left(H(t_1)H(t_2)\right) = \left(\int_{t_0}^{t} dt_1 H(t_1)\right)^2 + \int_{t_0}^{t} dt_1 \int_{t_0}^{t_1} dt_2 [H(t_1), H(t_2)]. \quad (27)$$

The main advantage of such representation is that the time ordering operator has no effect in the leading terms resulting in the usual Teylor series for the $\chi^{(2)}$ expansion. Below we consider the process of down-conversion in the second-order on $\chi^{(2)}$ interaction.

3.1. *Vector state in the second-order on* $\chi^{(2)}$

Using the formula (1) and $H^{(+)}(t)|0\rangle = 0$, the vector state describing three-photon process in the second-order is calculated as

$$|\psi^{(2)}\rangle = \left(1 - \frac{1}{\hbar^2}\int_{-\infty}^{t} dt_1 \int_{-\infty}^{t_1} dt_2 [H^+(t_1), H^-(t_2)]\right)|0\rangle - \frac{1}{2\hbar^2}\left(\int_{\infty}^{t} dt_1 H^-(t_1)\right)^2 |0\rangle. \quad (28)$$

Then, we calculate the commutator of two Hamiltonians at different times.Using the formulas(2) and (28) we obtain

$$[H^+(t_1), H^-(t_2)] = \int d\omega_0 d\omega_0^{'} d\omega_1 \omega_2 F(\omega_1, \omega_2, \omega_0) F^*(\omega_1, \omega_2, \omega_0^{'}) e^{i(\omega_1+\omega_2-\omega_0)t_2} e^{-i(\omega_1+\omega_2-\omega_0^{'})t_1}, \quad (29)$$

where

$$F(\omega_1, \omega_2, \omega_0) = \sum_n \int_{z_n}^{z_{n+1}} dz \chi_n \alpha_{1n}(\omega_1)\alpha_{2n}(\omega_2)\left(\alpha_{pn}(\omega_0)\right)^* e^{i\Delta k_n (z-z_n)}. \quad (30)$$

Thus, the interaction Hamiltonian does not commute at different times and therefore the Dyson time ordering applies, instead of the Teylor series, when calculating higher order terms in U operator. Nevertheless,the difference between Taylor and Dyson series at second-order terms is the second-order correction to vacuum state. Really, the second-order vector state is calculated as

$$|\psi^{(2)}\rangle = (1+R)|0\rangle - \frac{1}{2\hbar^2}\left(\int_{\infty}^{t} dt_1 H^-(t_1)\right)^2 |0\rangle, \tag{31}$$

where

$$R = i\frac{2\pi}{\hbar^2}\int \frac{d\omega_1 d\omega_2 d\omega_0}{\omega_1+\omega_2-\omega_0-i\lambda}|F(\omega_1,\omega_2,\omega_0)|^2, \tag{32}$$

and $\lambda \to 0$, is the radiative correction to the vacuum state $|0\rangle$. We consider the physical meaning of R considering the case of a monochromatic pump field at the frequency ω_0. In this case the quantity (32) is calculated as the following

$$R = T(\Delta - i\Gamma), \tag{33}$$

where T is a time interval. The quantity

$$\Delta = \frac{1}{\hbar^2}\int \frac{d\omega_1 d\omega_2}{\omega_1+\omega_2-\omega_0}|F(\omega_1,\omega_2,\omega_0)|^2 \tag{34}$$

is interpreted as the radiative shift of the frequency $\omega_0 + \Delta$ due to virtual two-photon process, while

$$\Gamma = \frac{\pi}{\hbar^2}\int d\omega_1 |F(\omega_1,\omega_0-\omega_1,\omega_0)|^2 \tag{35}$$

is interpreted as the decay rate of the initial pump state due to two-photon process. On the whole the unitary operator is expressed as a Teylor series

$$U = 1 + R - \frac{i}{\hbar}\int_{-\infty}^{t} H^-(t^{'})dt^{'} + \left(\frac{i}{\hbar}\int_{-\infty}^{t} H^-(t^{'})dt^{'}\right)^2 + \cdots \tag{36}$$

4. Down-Conversion in a Periodically-Poled Structure of Nonlinear and Linear Segments

The phase-matching function describing the optical properties of the nonlinear crystal crucially depends on the wave fields dispersion. In this section we analyze QPM in PPNC configuration that also includes linear optical elements. This structure consists of $\frac{N}{2}$ segments of length l_1 with positive χ and negative $-\chi$ susceptibilities, which are seperated by $\frac{N}{2}$ linear optical spacers of length l_2.

We assume two mismatch functions Δk and $\Delta\kappa$ corresponding to nonlinear $n = 1, 3, 5, ...$ and $n = 2, 4, ...$ segments. According to the formulas (11) we get

$$G(\Delta k, \Delta\kappa) = l_1 \chi e^{-i\phi} sinc\left(\frac{l_1}{2}\Delta k\right) \frac{\sin\left(\frac{N(l_1+l_2)\Delta K}{4}\right)}{\sin\left(\frac{(l_1+l_2)\Delta K}{4}\right)}. \quad (37)$$

Here $\Delta K = \bar{l_1}\Delta k + \bar{l_2}\Delta\kappa - q_m$, $\phi = \frac{1}{2}l_1\Delta k + \frac{1}{2}\Delta K(l_1 + l_2)(\frac{N}{2} + 1)$. $L = \frac{N}{2}(l_1 + l_2)$ - the total length of the medium, $\bar{l_i} = \frac{l_i}{l_1+l_2}$, $(i = 1, 2)$, $\bar{l_1} + \bar{l_2} = 1$ and $q_m = \frac{2\pi m}{d}$ is the harmonic grating wave vector, m is an arbitrary odd number, which is specified for a concrete process, $d = 2(l_1 + l_2)$. Thus, the total effective interaction coefficient is presented as the product of two separate functions. One describes each individual nonlinear crystal segment and the other function describes the modifications of the QPM function in the superlattice. Such function has been obtained in the paper,[18] however, for the other configuration of the superlattice. The difference is that in the configuration considered in our paper the assembly QPM function also involves mth harmonic grating wave vector.

4.1. *Compensation of dispersion*

We illustrate now Eq. (37) for a parametric down-conversion in the pulsed laser light beam assuming a Gaussian pulse of duration τ_p and at the central frequency ω_0 , $E(\omega) = E_0 \exp\left(-\frac{\tau_p^2}{2}(\omega - \omega_0)^2\right)$. In this case, the amplitude of photon pairs generated at frequencies ω_1 and ω_2 is given by the product of the pump envelope function $E(\omega_1 + \omega_2)$ and the $G(\Delta k, \Delta\kappa)$ function, see the formula (10).

The phase-mismatch in each of the crystal is given by

$$\Delta k = k_0(\omega_0) - k_1(\omega_1) - k_2(\omega_2), \quad (38)$$

where $k_i = |\overrightarrow{k_i}|$ represents the wave-number of each of the three fields $(i = 0, 1, 2)$, pump field and fields of the subharmonics, $i = 1, 2$, taking into account dispersion in the nonlinear segments. The phase mismatch introduced by each of the linear element is equivalently given by

$$\Delta\kappa = \kappa_0(\omega_0) - \kappa_1(\omega_1) - \kappa_2(\omega_2). \quad (39)$$

In the papers[17,18] authors demonstrated that it is possible to control the mode structure of twin photons by the superlattice of $\chi^{(2)}$ martials with linear, dispersive spacer. We believe that in the configuration considered here, which additionally involves grating wave vector q_m this technique become more flexible. We consider, for simplicity the type-I interaction (for which $k = k_1 = k_2$) by using the Taylor expansions of two phase mismatches up to the second derivation from the central frequency $\omega_1 = \omega_2 = \frac{\omega_0}{2}$, $\Delta k = \Delta k^{(0)} + \delta k$, $\Delta\kappa = \Delta\kappa^{(0)} + \delta\kappa$. We also assume that each of the $\chi^{(2)}$ crystals is aligned such that the constant term $\Delta k^{(0)} = 0$, i.e. the perfect phase matching takes place at the central down-conversion frequency $\omega_0/2$. Thus,

$$\Delta k = \delta k = \left(\beta_0^{(1)} - \beta^{(1)}\right)(\omega_1 + \omega_2 - \omega_0) - \frac{1}{2}\beta^{(2)}\left[\left(\omega_1 - \frac{\omega_0}{2}\right)^2 + \left(\omega_2 - \frac{\omega_0}{2}\right)^2\right] + \frac{1}{2}\beta_0^{(2)}(\omega_1 + \omega_2 - \omega_0)^2, \tag{40}$$

where $\beta^{(j)} = \frac{d^j k_1(\omega)}{d\omega^j}|_{\omega=\frac{\omega_0}{2}} = \frac{d^j k_2(\omega)}{d\omega^j}|_{\omega=\frac{\omega_0}{2}}$ and $\beta_0^{(j)} = \frac{d^j k_0(\omega)}{d\omega^j}|_{\omega=\omega_0}$ are the dispersion coefficients for nonlinear segments. By analogous

$$\delta\kappa = \left(\alpha_0^{(1)} - \alpha^{(1)}\right)(\omega_1 + \omega_2 - \omega_0) - \frac{1}{2}\alpha^{(2)}\left[\left(\omega_1 - \frac{\omega_0}{2}\right)^2 + \left(\omega_2 - \frac{\omega_0}{2}\right)^2\right] + \frac{1}{2}\alpha_0^{(2)}(\omega_1 + \omega_2 - \omega_0)^2, \tag{41}$$

where $\alpha^{(j)} = \frac{d^j \kappa_1(\omega)}{d\omega^j}|_{\omega=\frac{\omega_0}{2}} = \frac{d^j \kappa_2(\omega)}{d\omega^j}|_{\omega=\frac{\omega_0}{2}}$ and $\alpha_0^{(j)} = \frac{d^j \kappa_0(\omega)}{d\omega^j}|_{\omega=\omega_0}$ are the dispersion coefficients for linear medium. In this case the mismatch ΔK of the assembly is denoted by the constant phase term $\Delta\kappa^{(0)}$ in linear segments and the dispersion coefficients of the expansions

$$\Delta K = \bar{l_1}\delta k + \bar{l_2}(\Delta\kappa^{(0)} + \delta\kappa) - q_m. \tag{42}$$

The condition $\Delta K = 0$ requires that the quantities $\delta\kappa$ and δk exhibit opposite signs and hence the group velocity mismatch in the $\chi^{(2)}$ segments and linear spacer is vanished i.e. $\bar{l_1}\delta k = -\bar{l_2}\delta\kappa$. In this case the second requirement formulated for the constant term of the spacer phase-mismatch as $\bar{l_2}\Delta\kappa^{(0)} = q_m$. According to this scenarios the spectral distribution of twin photons is simulated only by one of the $\chi^{(2)}$ segments, however, the amplitude of spontaneous parametric generation is the coherent sum of all partial amplitudes. Thus, the corresponding probability is increased on N^2 times, so in the approximation $N \gg 1$

$$|\Phi(\omega_1, \omega_2)|^2 \sim N^2 l_1^2 sinc^2\left(\frac{l_1}{2}\delta k\right). \tag{43}$$

Thus, in this scheme the efficient generation of twin photon states with the specific modal character can be realized in a short crystal regime.

4.2. *Gaussian spectral distribution for twin photons amplitude*

In this section, we consider the case $\Delta k^{(0)} = 0$, $\bar{l_2}\Delta\kappa^{(0)} = q_m$, however, for more general situation without cancelation of the dispersions. From the formulas (40) and (42) we conclude that the two-photon amplitude can be written in the Gaussian form as follows

$$\Phi(\omega_1, \omega_2) = \sqrt{\frac{2N_{ph}}{\pi\Omega\sigma}} \exp\{-\frac{(\omega_1 + \omega_2 - \omega_0)^2}{2\sigma^2} - \frac{(\omega_1 - \omega_2)^2}{2\Omega^2}\}, \tag{44}$$

where the spectral widths are

$$\frac{1}{\sigma^2} = \frac{a^2 + \rho^2}{a^2\rho^2} + \tau_p^2, \quad \frac{1}{\Omega^2} = \frac{b^2 + P^2}{b^2 P^2},$$

and

$$\frac{1}{\rho^2} = \frac{1}{40}\left[l_1\left(\beta^{(1)} - \beta_0^{(1)}\right) + l_2\left(\alpha^{(1)} - \alpha_0^{(1)}\right)\right]^2 N^2,$$

$$\frac{1}{P^2} = \frac{1}{12}N\left[l_1\beta^{(2)} + l_2\alpha^{(2)}\right],$$

$$\frac{1}{a^2} = \frac{l_1^2}{10}\left[\beta^{(1)} - \beta_0^{(1)}\right]^2, \quad \frac{1}{b^2} = \frac{l_1}{6}\beta^{(2)}. \tag{45}$$

The quantity N_{ph} is the mean number of photon pairs expressed as

$$N_{ph} = \int d\omega_1 d\omega_2 |\Phi(\omega_1, \omega_2|^2 = \frac{E_0^2\chi^2 l_1 N^2 \pi\sigma\Omega}{2(1 + \sigma^2\tau_p^2)}. \tag{46}$$

Note, that for the case of pulsed down-conversion in a bulk $\chi^{(2)}$ crystal and without linear segments, (case $N = 1$, $l_2 = 0$, $q_m = 0$) these results are coincides with the results of the paper.[28]

Two parameters Ω and σ of the two-photon Gaussian wave packet determine the bandwidths of twin-photon light and the spectral correlation between photons in pair, perspectively. For an ordinary down-conversion, typically $\Omega \gg \sigma$ and therefore the bandwidth of the process is approximately proportional to Ω. For the crystal assembly involved both nonlinear and linear segments such situation can be changed. At first, we note that for $N \gg 1$ or $L \gg l_1$ the bandwidths take the form

$$\frac{1}{\sigma^2} = \frac{1}{\rho^2} + \tau_p^2, \quad \Omega = P. \tag{47}$$

Thus, the bandwidth Ω is proportional to $1/\sqrt{N}$, while for long assembly σ is proportional $1/N$. Let us now consider a specific case fulfilling the compensation requirement: the nonlinear and linear materials should be chosen such that the group velocity mismatch for the $\chi^{(2)}$ crystal and linear material have opposite signs. We assume that for the crystal assembly, the first order dispersion coefficients as well as the length l_1 and l_2 should all satisfy the condition

$$l_1\left(\beta^{(1)} - \beta_0^{(1)}\right) + l_2\left(\alpha^{(1)} - \alpha_0^{(1)}\right) = 0. \tag{48}$$

If this condition is fulfilled, for short-length nonlinear segments, the bandwidth of spectral correlation can be in the scale of the bandwidth σ.

5. Phase-Reversed Superlattice Structure

A method of the double phase-matching, studied extensively both theoretically and experimentally so-called the phase-reversed QPM structure was employed for the wavelength conversion,[29] and for the cascaded single-crystal third harmonic generation process.[30,31] This method is based on the use of the phase-reversed-sequence of many equivalent uniform short QPM domains in which neighboring nonlinear blocks have the opposite signs of the second-order susceptibility. The phase-reversed structures arranged in such a way that at the place of the joint two ends of the domains segments have the same sign of the $\chi^{(2)}$ coefficient (see Fig. 2). On the whole, this structure is characterized by two grating wave vectors.

We explain phase-reversed QPM structure as the definite assembly of nonlinear segments on the base of the formula (11). In the result, the effective nonlinear coefficient $G(\Delta k)$ is calculated in the following form

$$G(\Delta k) = L\chi_0 e^{-i\alpha(\Delta k)} Y_{M,N}(\Delta k), \tag{49}$$

where χ_0 is the absolute value of $\chi^{(2)}$ nonlinearity coefficient in each layer.

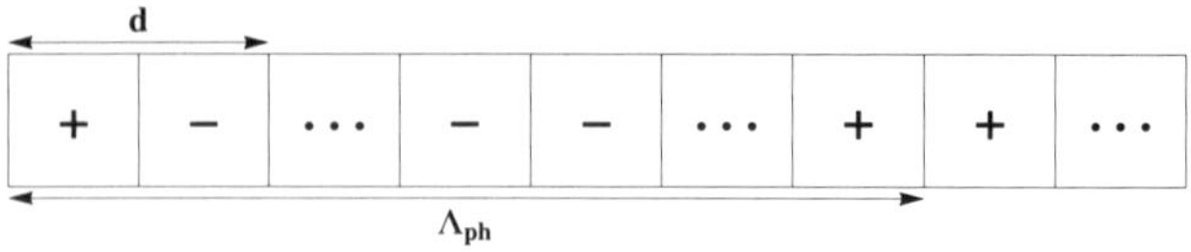

Fig. 3. Scheme of PPNC which involves nonlinear domains of length l with $\chi_+ > 0$ and $\chi_- < 0$ with phase-reversal grating (with period Λ_{ph}) upon a uniform QPM grating (with period d).

We introduce the QPM function $Y_{M,N}(\Delta k)$ which is the product of the standard phase-matching spectral function for a segment, i.e. $sinc\left(\frac{l\Delta k}{2}\right)$, and the assemble function

$$Y_{M,N}(\Delta k) = sinc\left(\frac{l\Delta k}{2}\right)\frac{1}{MN}\frac{\sin\left[\frac{Nl}{2}(\Delta k - G\right]}{\sin\left[\frac{l}{2}(\Delta k - G\right]}\frac{\sin\left[\frac{MNl}{2}(\Delta k - F)\right]}{\sin\left[\frac{Nl}{2}(\Delta k - F)\right]}. \quad (50)$$

Here: N-is the number of domains in each block, M-is the number of blocks, $\alpha(\Delta k) = \frac{Nl}{2}[(\Delta k - G) + (M-1)Nl(\Delta k - F)]$, l-is the length of each domain, $G = 2\pi/d$, $F = 2\pi/\Lambda_{ph}$, $d = 2l$, $\Lambda_{ph} = 2Nl$, $L = MNl$. It is easy to check that the function $Y_{M,N}(\Delta k)$ has another representation as the product of expansions on the number of domains and blocks

$$Y_{M,N}(\Delta k) = sinc\left(\frac{l\Delta k}{2}\right)\frac{1}{MN}e^{i\varphi}\sum_{n=1}^{N} e^{-i(n\Delta k - G_n)}\sum_{m=1}^{M} e^{-i(mN\Delta k - F_m)}, \quad (51)$$

where $\varphi = \frac{1}{2}[(N-1)l(\Delta k - G) + (M-1)Nl(\Delta k - F)]$. The above expressions (49), (51) derived on the base of the method of superlatices elaborated on the multilayered media differs from the formulas which are obtained in the standard, phenomenological approach. The results (50), (51) depend on the number of the domains and blocks and also describe the case of small numbers of the segments. It is remarkable, that in this case the QPM function displays specific interference effects. Nevertheless, as shows our analysis the formula (49) qualitatively coincide with standard results, for $N \gg 1$, $M \gg 1$.

This QPM function $Y_{M,N}(\Delta k)$ allows simultaneously phase-match two parametric processes that can be achieved by the use of the two grating vectors. In order to illustrate the possibility of the double phase-matching in such superlattice structure we present a detailed analysis of the function $Y_{M,N}(x)$ in dependence on the parameter $x = \frac{l\Delta k}{2}$ in the graphical form. Unlike to the assembly function discussed for one-matching case,[18] our function describing double matching has complex structure, particularly, each peak consists of thin structure of twin-narrow peaks. The $sinc\left(\frac{l\Delta k}{2}\right)$ multiplier make to decrease the amplitude of the consecutive peaks. The behavior of $Y_{M,N}(x)$ function is observed for different values of M and N and the result for defined values of the M and N is depicted in Fig. 4. The existence of twin peaks and non-vanishing of all amplitudes is playing important role in simultaneously phase-matching of two parametric processes.

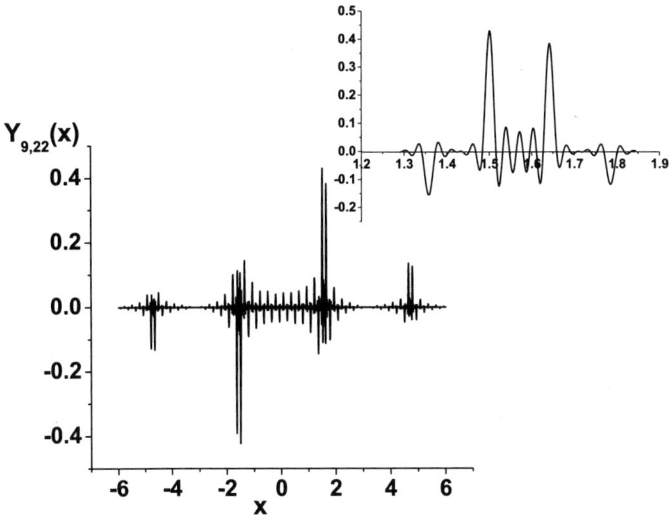

Fig. 4. Plot of the function $Y_{M,N}(x)$ for even numbers of $N = 22$ and odd numbers $M = 9$. Each of the narrow peaks consist of components of amplitudes of the same sign.

6. Conclusion

In this paper, we considered multiple QPM interactions in one-dimensional finite multilayered nonlinear media and discussed how to achieve phase-matching for several parametric processes. Such structures consist of layers with different $\chi^{(2)}$ susceptibilities and with different refractive indexes. This consideration has definite advantages in comparison with the standard treatment of multiple QPM interactions that may typically be described only by an aperiodic modulation of the $\chi^{(2)}$ susceptibility within the assumption of the translation invariance along z-axis. Note, that real applications employ crystals of finite length and the assumption of translation invariance is an approximation. In this area, we have continued the recent investigations of twin-photon wave packets considering their controllable production in periodically poled structures that involve additionally $\chi^{(1)}$ dispersive spacer. We examine how the waves dispersion in these configurations affect both the profiles of spectrally Gaussian input photons and the affectivity of down-conversion. On the other side, we have examined the problem of double phase-matching for a realistic phase-reversed configuration. We have derived the QPM function of two parametric interactions as the product of the standard phase-matching spectral function for a $\chi^{(2)}$ segment and the assembly function. We have demonstrated that assembly function strongly depends on the numbers of the layers and the phase-reversed blocks and displays specific interference effects in dependence of the mismatch parameter $x = \frac{l\Delta k}{2}$ in contrast to what happens in the case of ordinary theory.

Acknowledgments

The research was partly founded by the CRDF/NFSAT/SCS, grants UCEP 02/07 and ECSP-09-53, A-03, by the ISTS, grant A-1606. We gratefully acknowledge these supports.

References

1. R.L. Byer, J. Nonlin. Opt. Phys. Mater. **6**, 549 (1997).
2. D.S. Hum and M.M. Fejer, C.R. Physique **8**, 180 (2007).
3. S.M. Saltiel, A.A. Sukhorukov and Y.S. Kivshar, Prog. in. Opt. **44**, 1 (2004).
4. L.E. Myers, et. all., Opt. Lett. **20**, 52 (1995); J. Opt. Soc. Am. **B12**, 2102 (1995) .
5. K. Karlson, and F. Laurell, Apl. Phys. Lett. **71**, 3474 (1997).
6. S.N. Zhu, Y.Y. Zhu, and N.B. Ming, Science, **278**, 843 (1997).
7. Y. Du, et. all., Apl. Phys. Lett. **81**, 1573-1575 (2002).
8. K.S. Zhang, T. Coudreau, M. Martimelli, A. Maitre, and C. Fabre, Phys. Rev. **A64**, 033815 (2001).
9. K. Fradkin-Kashi, A. Arie, P. Urenski, and G. Rosenman, Phys. Rev. Lett. **88**, 023903 (2001).
10. R. Lifshitz, A.Arie, and A. Bahabad, Phys. Rev. Lett. **95**, 133901 (2005).
11. P.V. Gorelik, F.N.C. Wong, D. Kolker, and J.J. Zondy, Opt. Lett. **31**, 2039 (2006).
12. K.V. Kheruntsyan, G.Yu. Kryuchyan, N.T. Muradyan, K.G. Petrosyan, Phys. Rev. **A57**, 535 (1998).
13. G.Yu. Kryuchkyan, N.T. Muradyan, Phys. Lett. **A286**, 113 (2001).
14. S. Suzuki, H. Yanezawa, F. Kannari, M. Sosaki, A. Furusawa, Appl. Phys. Lett. **89**, 061116 (2006); K. Yoshino, T. Aoki, A. Furusawa, Appl. Phys. Lett. **90**, 041111 (2007).

15. O. Pfister, S. Feng, G. Jennings, R. Pooser and D. Xie, Phys. Rev. **A70**, 020302(R) (2004).
16. N.C. Menicucci, S.T. Flammia, H. Zaidi and O. Pfister, Phys. Rev. **A76**, 010302(R) (2007);
N.C. Menicucci, S.T.Flammia and O. Pfister, Phys. Rev.Lett. **101**, 130501 (2008).
17. D.N. Klyshko, JETF **77**, 222 (1993).
18. A.B. U'Ren, C. Silberhorn, K. Banaszek, I.A. Walmsley, R.K. Erdmann, W.P. Grice, and M.G. Raymer, Quant.Opt. **15**, 146-161 (2005); A.B. U'Ren, R.K. Erdmann, M. de la Cruz-Gutierrez and I.A. Walmsley, Phys. Rev. Lett. 97, 223602 (2006).
19. William T.M. Irvine, Michiel J.A. Dood, and Dirk Boumeester, Phys. Rev. **A72**, 043815 (2005).
20. A.N. Vamivakas, B.E.A. Saleh, A.V. Sergienko, and M.C. Teich, Phys. Rev. **A70**, 043810 (2004).
21. Leung P.M. Irvine, W.J. Munro, K. Nemoto, and T.C. Ralph, arXiv:quant-phys/0811.2038 v2 (2008).
22. G. D'Aguauno,M. Centini, et. all., Phys. Rev. **A 56**, 3166-3174 (1997).
23. M. Soljacic, and J.D. Joannopoulos, Nature (London), **3**, 211-219 (2004).
24. M.F. Saleh, L.D. Negro, and B.E.A. Saleh, Opt. Express **16**, 5261 (2008).
25. A. Yariv and P. Yeh, Optical Waves in Crystals (John Willey and Sons, New York, 1984).
26. Jan Perina, Marco Centini, Concita Sibilia, Mario Berlotti, and Michael Scalora, Phys. Rev. **A73**, 033823 (2006) .
27. S. Blanes, F. Casas, J.A. Oteo,and J. Ros, Phys. Reports **470**, 5-6,151 (2009).
28. W.Wasilewski, A.I. Lvovsky, K. Banaszek, and C. Radzewicz, Phys. Rev. **A73**, 063819 (2006).
29. M.H. Chou, K.R. Parameswaran and M.M. Fejer, Opt.Lett. **24**, 1157 (1999).
30. Z.W Liu, S.N. Zhu, Y.Y. Zhu, Y.Q. Qin, J.L. He, C. Zhang, H.T. Wang, N.B. Ming, X.Y. Liang, and Z.Y. Xu, Jpn. J. Appl. Phys. **40**, 6841-6844 (2001).
31. Z.W Liu, Y. Du, J.Liao, S.N. Zhu, Y.Y. Zhu, Y.Q. Qin, H.T. Wang, J.L. He, C. Zhang, and N.B. Ming, J. Opt. Soc. Am. **B 19**, 1676-1684 (2002).

INTEGRATED PHOTONIC DEVICE STRUCTURES WITH NANO-SCALE FEATURES: FOR SENSITIVE APPLICATIONS

R. M. DE LA RUE*

DOptoelectronics Research Group, Department of Electronics and Electrical Engineering, University of Glasgow,
Glasgow G12 8QQ, Scotland, U.K.
**E-mail: r.delarue@elec.gla.ac.uk*

Photonics/optoelectronics research at Glasgow University is partially reviewed. The research activity reviewed is also notably connected to research groups furth of Glasgow. Silicon waveguide photonics at the sub-micrometre/nanometre scale is emphasized, including photonic wires, strong Bragg gratings and micro-/nano-cavity resonators and slow light in photonic crystal channel waveguides. High-volume processes for the fabrication of photonic quasi-crystal structures that enhance and shape the light extracted from GaN blue LEDs will also be described. Finally split-ring resonator (SRR) array based metamaterials are featured - together with their application for organic sensing.

Keywords: Silicon photonics, optical waveguide sensors, organic sensors, nanophotonics, metamaterials, nano-imprint lithography, photonic (quasi-) crystals, split-ring resonators.

1. Introduction

This review of activity in the Optoelectronics Research Group at the University of Glasgow will only provide a partial view, with an emphasis on waveguide silicon photonics. Some consideration is also given to device structures based on epitaxial III-V semiconductor heterostructures. A substantial part of the research has involved collaboration with research workers elsewhere in the world, as will emerge in the course of this chapter. Several of the device structures that will be described use either photonic wire or photonic crystal concepts, while others can be categorized as metamaterials. The scale of the device structures and the detailed features that they involve imply that they could legitimately be described as having a micro-photonic character, but also as falling within the domain of nanophotonics. Photonic-wire optical waveguides realised, in many cases, on a silicon-on-insulator (SOI) base are now routinely called 'photonic nanowires' by those carrying out this research. Metamaterial structures in the form of arrays of carefully shaped metallic patterns such as split-ring resonators (SRRs) may involve sub-micrometre characteristic dimensions and controlled gaps with dimensions as small as a few tens of nanometres, but their fundamental response can still occur at optical wavelengths in the infra-red part of the spectrum.

2. Waveguide Silicon (Nano-)Photonics

Waveguide silicon photonics begins with a basic planar optical waveguide consisting of a sub-micrometre thick layer of lightly doped single-crystal silicon bonded on to and supported by a lower cladding of thermally grown silicon dioxide - both layers being, in turn, supported by a large-diameter wafer of single-crystal silicon. Such SOI wafers are available commercially at moderate cost. The silicon layer may then be patterned into sub-micrometre wide stripes that are effectively single-mode for each of the two basic polarisations, (quasi-)TE and (quasi)TM of the infra-red light that is typically launched into the light at a cleaved edge of a silicon wafer section. Such narrow stripe waveguides are often called photonic wires or photonic nanowires. An alternative way of launching light into and extracting light from silicon photonic wire waveguides is to use a grating coupler arrangement. The simple 'straight' photonic wire - whether fabricated in silicon or, using a somewhat different geometry, in a III V semiconductor epitaxial system such as the AlGaAs/GaAs system - is already an interesting medium for investigating non-linear propagation effects such as solitonic propagation and four-wave mixing.[1,2] The photonic wire may also be folded into a meander line or a double-spiral to provide moderately low-loss delay-line type optical storage - or bent round upon itself to form a ring-resonator. Photonic-wire ring-resonators can be brought together in a chain of rings that may be called coupled-resonator optical waveguides (CROWs) - or coupled individually or separately to one or more photonic-wire 'bus-bars'.

2.1. *Silicon photonic wire Bragg gratings*

The photonic wire geometry, in particular its cross-section, can be modified in various ways to add useful functionality. One interesting photonic-wire based structure is the Bragg-grating[3] formed by periodically alternating regions of wider waveguide and narrower waveguide. An electron micrograph of such a Bragg-grating is shown in Fig. 1(a), while Fig. 1(b) is an expanded view of the section identified by a square in the first micrograph. The width of the photonic wire as it arrives at the Bragg grating is 500 nm. With careful attention to the fabrication process details, Bragg grating structures with smooth (sub-1 nm average roughness) sidewalls and precise (few nanometre error) dimensional specifications are realisable - and these factors justify the structure being considered as a nano-photonic entity. Nanometre-scale dimensional reproducibility issues related to HSQ resist storage lifetime are also considered in Ref. 3.

Fig. 1. (a) Photonic-wire Bragg-grating realised in silicon-on-insulator. (b) Detail from Fig. 1(a).

2.2. *Stop-band closure in photonic wire Bragg-gratings*

As might be expected, Bragg-gratings with strong stop-bands that build on the large refractive index contrast between silicon and silica or air are possible.[4] An interesting and somewhat surprising phenomenon can appear when the grating has a sufficiently 'square' structure. That phenomenon is the 'disappearing stop-band', which occurs under particular conditions of strong grating feature indentation. It is a simultaneous manifestation of collinear Bragg interaction and the polarisation dependent Brewster's angle phenomenon, so it can be described as doubly British. The disappearing stop-band behaviour only occurs under quite restricted conditions and a narrow range of wavelengths and/or indentation dimensions, suggesting possible applicability in sensing. Additional insight into the disappearing stop-band phenomenon arises from detailed electromagnetic modelling, as illustrated by the series of intensity plots shown in Fig. 3, which are accompanied by the corresponding dispersion curves. The stop-band disappears at the transition point between situations where most of the Bragg-mode optical power is in the narrow rib-sections, for the upper side-band (at smaller indentation), and situations where most of the optical power is in the wider fin-sections (at larger indentation) for the upper side-band - and vice versa for the lower side-band.

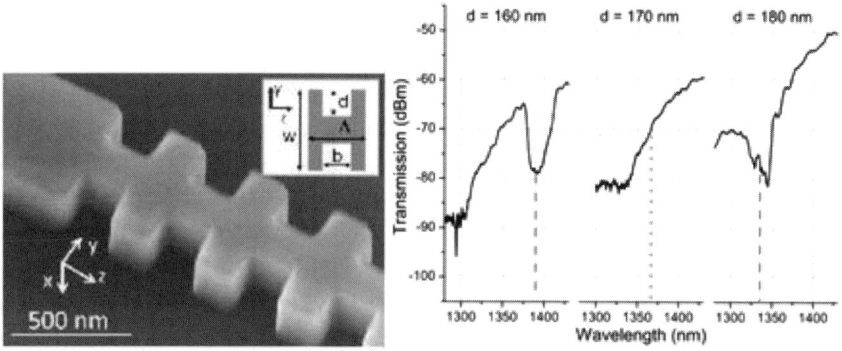

Fig. 2. (a) Micrograph of photonic wire Bragg grating, with inset showing characteristic dimensions that define a periodic unit cell. (b) Transmission spectra for Bragg grating at three different indentation sizes.

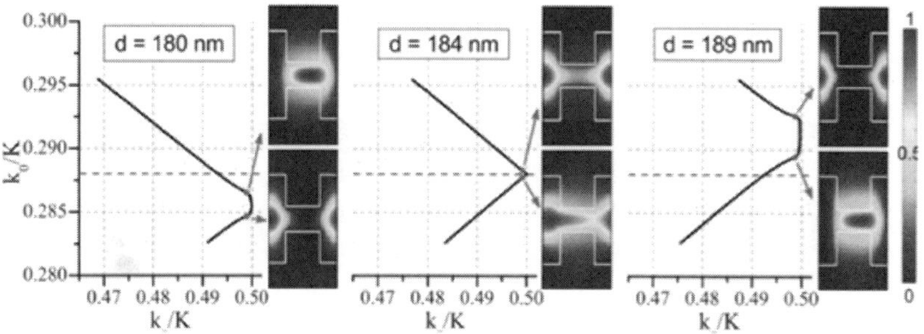

Fig. 3. Bragg-mode optical intensity plots at three different indentation values - and corresponding dispersion curves, from Ref. 4.

2.3. *Photonic crystal/photonic wire micro-/nano-cavities in SOI*

Photonic wires (PhWs) of silicon supported by a lower cladding of silica provide strong optical confinement. Combining photonic wires with one-dimensionally periodic, hole-based, photonic crystal (PhC) structures turns out to make it possible to achieve high quality-factor (Q-factor) resonators. Fig. 4 shows a scanning electron-micrograph of such a PhC/PhW micro-/nano-cavity. Resonance Q-factor values achieved experimentally in silica cladding supported structures[5] have begun to be comparable with those obtained in micro-resonators based on two-dimensionally periodic photonic-crystal based micro-cavities - which have so far required unsupported membrane guides. Q/V values that define the Purcell factor are even more favourable for the PhC/PhW microcavity, because its effective length is small. Optimisation using tapering within and outside the cavity makes it possible both to increase greatly the resonance Q-factor and simultaneously maintain useful transmission levels through the cavity, on-resonance. High-resolution 2D simulations for cavities in which each mirror has $N = 5$ periodic hole mirrors, together with partially optimised aperiodic tapered hole regions, have yielded Q-factor values as high as 177,000, together with transmission of nearly 50%. The corresponding experimental values obtained, a Q-factor of 147,000 and transmission of 34%, give reasonable confidence in the design approach and simulation accuracy. Variations in hole position and diameter at the few nanometre size scale are sufficient to produce noticeable differences in the resonator characteristics.

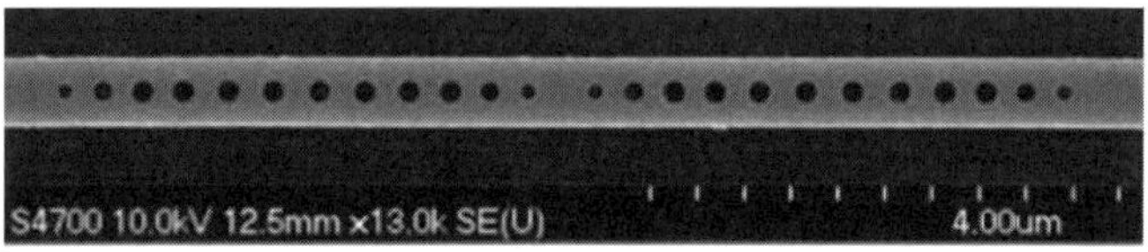

Fig. 4. Scanning electron micrograph of PhW/PhC micro-cavity in SOI.

Very high quality factor and a very small cavity volume are appealing properties for optical signal processing and filtering devices. 1D PhC micro-/nano-cavities embedded in photonic wires (PhWs) are potentially a fundamental building block for wavelength selective devices, especially for coarse

Wavelength Division Multiplexing (WDM) applications, due to the reduced footprint and low propagation losses that can now be obtained in photonic wire waveguides, with values below 1 dB/cm having been demonstrated. More recently, the high resonance Q-factor of the PhW/PhC micro-/nano-cavity resonators has been exploited to demonstrate low energy all-optical switching behaviour.[6] In these experiments, moderately short pulses of light in the green part of the spectrum (from a Q-switched Nd:YAG laser) were focussed onto the centre of a PhC/PhW micro-cavity for which the waveguide resonance was at an infra-red wavelength close to 1.451 μm. Tuning the infra-red probe-beam (PB) with respect to the cavity mode wavelength, made it possible to obtain either OFF/ON/OFF or ON/OFF/ON behaviour for the probe light propagating in the photonic wire waveguide - and to modulate fully this optical signal. Fig. 5 shows how very small changes in the wavelength of the probe beam propagating along the wire and through the resonant cavity lead to it behaving logically as an AND gate when tuned slightly away from the resonance peak, but providing a NOT operation when the infra-red waveguide light is tuned to coincide with the cavity resonance.

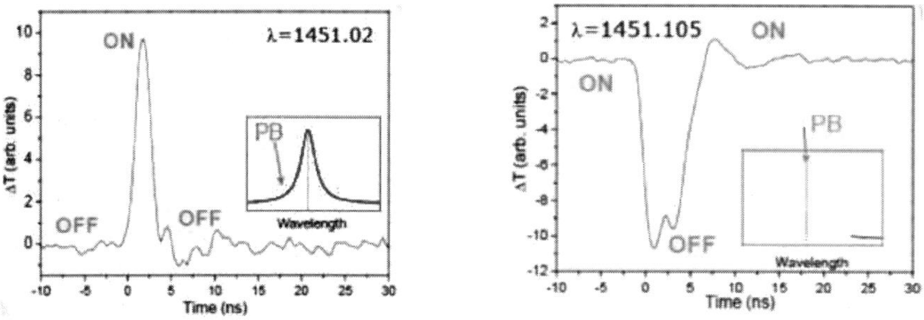

Fig. 5. (a) All-optical switching of infra-red light into the ON condition. (b) All-optical switching of infra-red light into the OFF condition.

2.4. *Photonic wire ring-resonators and ring-resonator based CROWs*

The scanning electron micrograph in Fig. 6 shows a silicon-based photonic-wire ring-resonator based coupled-resonator optical waveguide (CROW) formed from six strongly coupled race-track rings. The radius of the bends in the rings was 20 m, the free-spectral range (FSR) was 450 GHz, giving a system bandwidth of 110 GHz - and a slowing ratio S = 2.6. The estimated propagation loss of the constituent photonic-wire waveguides was estimated to be about 1.5 dB/cm.

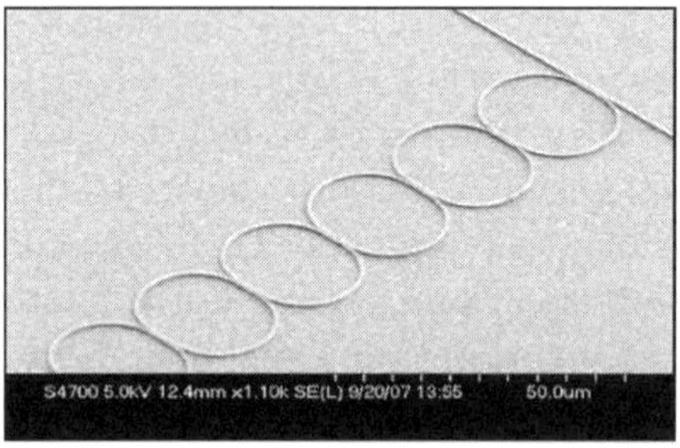

Fig. 6. Six coupled-ring CROW structure fabricated in SOI.

Four-wave mixing experiments have been carried out in CROWs similar to that shown in the micrograph, with the objective of assessing their potential application for wavelength conversion of differential phase-shift keying (DPSK) encoded data pulses.[7] Measurements showed that frequency translation by as much as 6 THz was possible while maintaining fidelity in a 10 Gb/s data stream. In a forthcoming paper,[8] the behaviour of individual bus-coupled ring resonators has been examined from the point-of-view of increased levels of back-scattered light - and it emerges that there are strong (and undesirable) coherence effects associated with multiple transits around the ring.

3. Novel Photonic Crystal Channel Waveguide Structures for Slow Light

Recently there has been much interest in using photonic-crystal based channel waveguides to obtain slow light behaviour. The reward for successful use of slow light can be very large enhancements in the magnitude of such things as non-linear optical effects and electro-optic effects. While it is correct to say that such enhancements are simply due to the greater device compactness that slow light makes possible, 'foot-print' is a centrally important concern for photonic and optoelectronic integration. So slow light

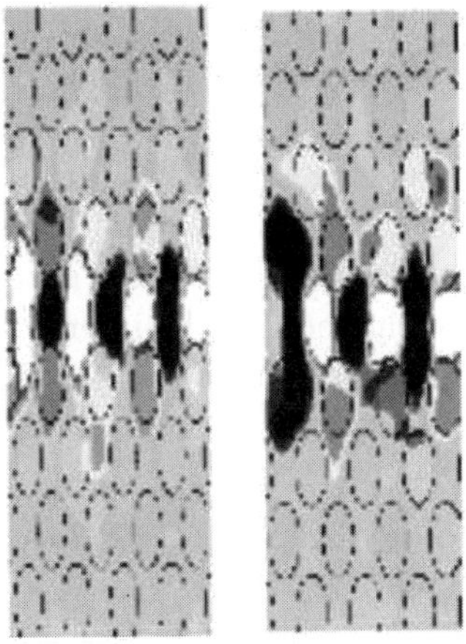

Fig. 7. (a) Modal Distribution at = 1.55 μm. (b) Modal Distribution at = 1.555 μm.

must be taken seriously. It is also intrinsic to quadratic and higher-order nonlinearity that the impact of slowing by a factor S gives appropriate power-law scaling, S^n. But propagation losses, whether due to in-plane backscattering, absorption or of out-of-plane scattering are also enhanced when the propagation is slowed - and the negative effects of sub-nanometre roughness and fabrication-related variations in hole shape and size are all 'enhanced'. In another recent paper,[9] it has been demonstrated that, with an appropriate thickness (450 nm) of silicon waveguide core, useful slow light behaviour can be obtained in photonic crystal channel waveguides in which the core is supported by a silica lower-cladding layer in an SOI configuration. As compared with the prevalent approach of using unsupported membrane 2D photonic crystal waveguides, an approximately twice larger thickness of silicon waveguide core layer enables a fundamental mode that is below the light line for the regions exterior to the core layer. Slow propagation with dispersion control in channel waveguides through 2D photonic crystal (PhC) regions of elliptical holes turns out to produce a desirable situation where slow propagation - as indicated by the group velocity (or, equivalently, group refractive index) - can be obtained while simultaneously suppressing higher order dispersion terms, to a large extent. The degree of ellipticity of the holes can be varied so as to optimise the suppression of the higher order dispersion terms. A channel waveguide formed in a 2D-periodic photonic crystal region is a particular kind of defect - and it has a dispersion relationship that can be plotted in the gap regions of the $-k$ diagram for the band-structure of the photonic crystal. The slope of the dispersion curve determines the group velocity, while the curvature determines the first higher order term - the GVD (group velocity dispersion). It is typical of slow light propagation in photonic crystal channel waveguides that the guided light spreads out more widely into the photonic crystal regions on either side of the channel. As the intensity plots of Figs. 7(a) and 7(b) illustrate, quite modest changes in the (free-space) wavelength of the light may then cause a substantial change in the modal distribution.

4. Fabrication Technology Aspects in Photonic Quasi-Crystals (PQCs)

A fabrication technique that seems likely to become of predominant importance in some areas of optoelectronic device production is 'Nano-Imprint Lithography' (NIL). Work on nano-imprint lithography has been carried out in the James Watt Nano-fabrication Centre (JWNC) at Glasgow University - and two distinct approaches have been investigated and characterised. Different approaches to NIL-based fabrication have been developed specifically for the realisation of photonic quasi-crystal (PQC) patterns and their use for enhanced extraction of blue light from light-emitting diodes (LEDs) fabricated on gallium nitride (GaN) based epitaxial structures.

Heat and Pressure NIL has been carried out in a sequence of steps, as follows: (i) preparation of a stamp with an array of correctly positioned and sized nano-pillars - formed in a section of silicon wafer. The stamp was fabricated using a combination of high resolution electron-beam lithography (EBL) in hydrogen silsesquioxane (HSQ) resist and inductively-coupled plasma (ICP) dry-etching. A micrograph of part of a silicon stamp is shown in Fig. 8(a). (ii) This stamp was then impressed into a deposited layer of mr-i-9020 NIL polymeric resist - at a temperature of 120oC and pressure of 40 Bar for 15 minutes. A moderate cost dedicated machine from the Obducat company was used for this part of the process. The polymeric resist has a low viscosity at high temperatures - and therefore is able to conform closely to the pattern on the stamp before pressure induced cross-linking. (iii) The NIL patterns were then transferred by dry-etching with appropriate gas combinations, flow-rates and machine pressure into both silicon and GaN wafer sections. Transfer of the photonic quasi-crystal pattern back into silicon was only carried out for demonstration purposes. Fig. 8(b) shows the end-result of the complete process of NIL transfer into a GaN LED wafer - and indicates that the fabrication process can produce excellent results, even when the wafer surface has a significant level of irregularity - as is a common situation. The central aim of the work carried out was to realise PQC structures that would both enhance the LED light extraction efficiency and direct the emitted light into a beam-shape (approximately cone-shaped) that is optimal for application in liquid-crystal display (LCD) back-lighting applications.

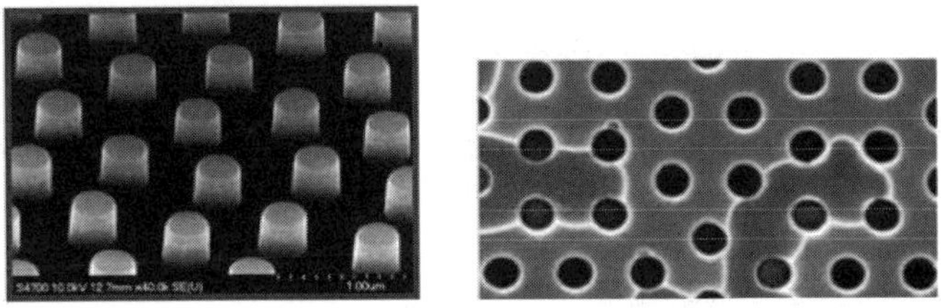

Fig. 8. (a) Stamp for heat-and-pressure NIL realised in single-crystal silicon. (b) PQC hole-pattern transferred into blue LED wafer surface by inductively coupled plasma (ICP) etching, after heat-and-pressure NIL.

Flash-imprint NIL has also been carried out in a sequence of steps, as follows: a) preparation of a fused-quartz stamp that is transparent to ultra-violet light. The process used for forming the stamp in fused-quartz was similar in general principles to that used for forming stamps in silicon, as described above. But the quality of the stamp produced was typically inferior to that obtained in single-crystal silicon. More work on the fabrication process to improve the fidelity of the stamp to the desired shape is clearly desirable. For this process, a special-purpose attachment for one of their mask-aligners was developed by the Karl Seuss company, allowing accurate location of the NIL generated photonic crystal region with respect to the rest of the device structure. A forthcoming paper[10] describes process details for several approaches to nano-imprint lithography for GaN LED applications, in some detail.

5. Planar Metamaterials Based on Split-Ring Resonator (SRR) Arrays

Planar metamaterial structures have been realised on both silicon and silica substrates.[11] The metamaterial 'surface' has been formed from arrays of metallic split-ring resonators (SRRs) realised in both gold and aluminium films. In these single-split SRR arrays (see Fig. 5(a)), it is clear that there is a fundamental resonance in the SRR structures, which may be identified as a 'magnetic' resonance - and its excitation requires appropriate polarisation in the incident light. That polarisation is often labelled, for linearly polarised light in normal incidence, as transverse electric (TE). The magnetic resonance is also commonly known as an LC resonance because the resonance frequency is determined by the combination of the (distributed) capacitance that is centred on the small gap between the ends of the ring pattern at the point where it is split, together with the (distributed) inductance produced by magnetic field lines enclosing the metallic conductor forming the ring. TE polarised light has its electric field component oriented across the gap in the SRR. The magnetic resonance can be considered as the fundamental resonance of the SRR, but, at higher frequencies, there may be several plasmonic resonances - the resonant frequency of which is determined by the velocity of plasmonic propagation along the SRR electrode and its length between the end points formed by the split. Clearly the end-point boundary conditions for SRRs with the geometry of Fig. 5(a) are significantly different from those for SRRs with the geometry of Fig. 5(b). The consequence is that, for otherwise identical size and geometry, there will be a difference in the resonance frequencies because of the substantially different size of the terminating gap. The plasmonic resonances will also have some measure of shape/form dependence - and the ability to excite them with linearly polarised light will depend on the detailed field distribution of each mode.

Fig. 9. (a) Scanning electron micrograph of part of closely packed array of rectangular geometry SRRs fabricated on silicon substrate. Note the small size of the gaps providing the splits in the rings. (b) Scanning electron micrograph of part of array of rectangular U-shaped SRRs having LC (magnetic) resonance at a wavelength in the mid infra-red. (Taken at an inclined angle with respect to the plane and to the array axes.

The schematic of Fig. 10 represents schematically the reflection spectrum of a regular array of single-split SRRs, with the broader LC peak occurring at a longer wavelength than the first plasmonic resonance observable with light having the same (TE) polarisation (black curve) - while a higher frequency plasmonic resonance is observed with the other (TM) polarisation (grey curve).

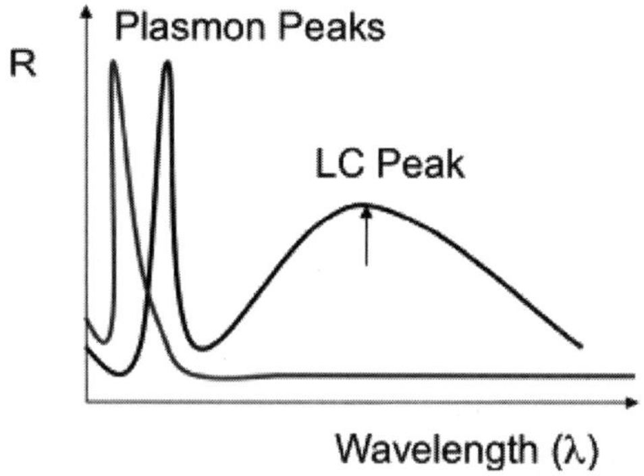

Fig. 10. Schematic reflection spectra of single-split SRR array.

The SRR may be considered as forming the resonant 'atomic' unit of the metamaterial. For some of our work, the basic form of SRR has been an asymmetric split-ring resonator (A-SRR) with a circular basic geometry[12] - in which, in general, there were two splits in each metallic ring pattern and the lengths of the two distinct metallic arcs forming the A-SRR pattern are different. Arrays of such A-SRRs (Fig. 5) characteristically have two resonant reflection peaks that may be identified as plasmon modes of the whole structure, with the frequency of the longer wavelength peak being mainly determined by the length of the longer arc - and the frequency of the shorter wavelength peak being mainly determined by the length of the shorter arc.

One methodology for applications in sensing is surface-enhanced infrared absorption (SEIRA) spectroscopy. Arrays of A-SRRs , with their combination of two reflectance peaks and an absorption resonance minimum between them, provide an interesting base for SEIRA type sensing of organic materials in small amounts, with the potential to show single molecule

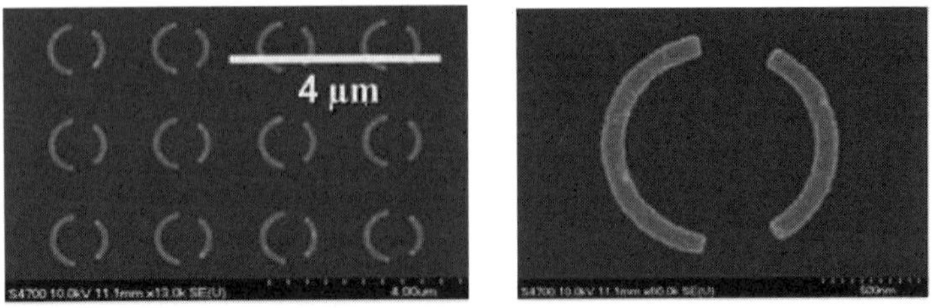

Fig. 11. (a) Scanning electron micrograph of part of array of A-SRRs. (b) Scanning electron micrograph of single A-SRR element-on larger magnification than for (a).

detection capability.[12] In the course of experiments in which the effect on the positions of the resonant features was changed by deposition of thin layers of a commonly used electron-beam resist, poly-methyl-methacrylate (PMMA), it was recognised that the signature of resonance in a characteristic bond of the PMMA (the carbonyl bond) was strongly present. By increasing the diameter of the A-SRR base units from 1.2 μm (Fig. 12(a)) to 1.35 μm, the impact of the resonance of the carbonyl bond could be considerably enhanced, as shown by the strongly oscillatory feature in the results of Fig. 12(b), at a wavelength of 5.8 μm.

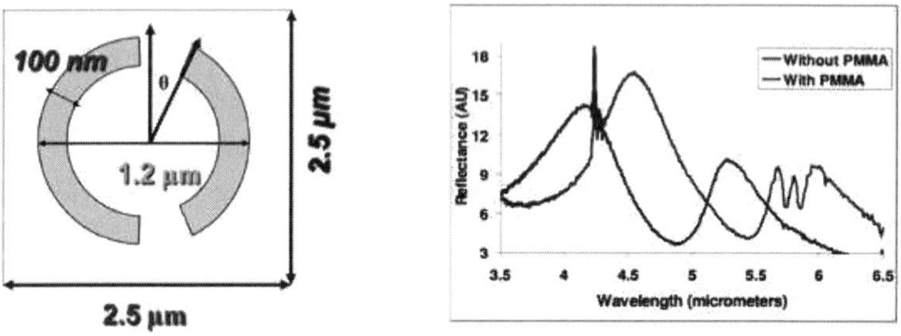

Fig. 12. (a) Schematic of A-SRR showing pattern dimensions and cell size. (b) Experimental reflection spectrum for array of A-SRRs with diameter increased to 1.35 μm.

Before concluding, mention will be made of the demonstration of visible wavelength magnetic resonance in arrays of single-split SRRs having a nominally rectangular U-geometry.[13] The choice of aluminium metallisation and silica substrates has made it possible to 'push' the resonance down to wavelengths as short as 530 nm. The choice of aluminium, in place of the gold that has been more prevalent, has made it possible to avoid the early onset of an overwhelming kinetic inductance effect.

6. Conclusions

Nanophotonic concepts, materials and technologies are poised to make a major impact across much of Photonics and Optoelectronics - enhancing the performance of the devices used in a wide variety of applications. Silicon waveguide photonics that exploits the 'nano' scale shows considerable promise for enhanced functionality in integrated switches, modulators, filters and sensors. Both fast light and slow light will have a role to play, particularly where non-linearity is exploited. Applications such as the next generation of 'active cable' are likely to use silicon-based integrated nanophotonics. Nano-imprint lithography (NIL) could be a key part of the mass-production of photonic crystal (PhC) and photonic quasi-crystal (PQC) structures, e.g. for LED applications. In general, mass-production processes, whether NIL or deep ultra-violet (DUV) lithography, will be an essential feature of the widespread application of 'top-down' nano-photonic structures. Primary pattern generation is however likely to remain with electron-beam lithography (EBL). Optical frequency metamaterials continue to excite interest - and will surely find large-scale application before long.

Acknowledgements

The author is pleased to acknowledge the support of the European Community, the U.K. Engineering and Physical Sciences Research Council (EPSRC) and the Royal Academy of Engineering (RAEng) for research activity in which he has been involved. The following list of people who have carried out the research is not exhaustive: F. Rahman, A.Z. Khokhar, D.S. Macintyre, B. Lahiri, R. Dylewicz, A.C. Bryce, A. Melloni, F. Morichetti, A. Canciamilla, C. Ferrari, M. Torregiani , M Gnan, Paolo Bassi, R.K. Sinha, S. Rawal, A. Samarelli, M. Belotti, M. Galli, D. Gerace, L.C. Andreani, S.G. Mc Meekin, M. Sorel, M. Strain, A.R. Md Zain, N.P. Johnson, H.M.H. Chong, G. Bellanca, R.M. de Ridder, W. Hopman, I. Watson, E. Gu, M. Dawson, W. Ding, C. Benton, A.V. Gorbach, W.J. Wadsworth, J.C. Knight and D.V. Skryabin.

References

1. W. Ding, C. Benton, A.V. Gorbach, W.J. Wadsworth, J.C. Knight, D.V. Skryabin, M. Gnan, M. Sorel, and R.M. De La Rue, Opt. Expr. **14 (20)**, 3310-3318, (2008).
2. G.A. Siviloglou, S. Suntsov, R. El-Ganainy, R. Ivanow, G.I. Stegeman, D.N. Christodoulides, R. Morandotti, D. Modotto, A. Locatelli, C. De Angelis, F. Pozzi, C.R. Stanley and M. Sorel, Opt. Expr.**16(5)**, 9377-9384 (2006).
3. M. Gnan, S. Thoms, D.S. Macintyre, R.M. De La Rue and M. Sorel, Electron. Letts.**44(2)**, 115 - 116, (2008).
4. M. Gnan, W.C.L. Hopman, G. Bellanca, R.M. de Ridder, R.M. De La Rue, and P. Bassi, Opt. Expr. **17(11)**, 8830-88412, (2009).
5. Ahmad Rifqi Md Zain, Nigel P. Johnson, Marc Sorel and Richard M. De La Rue, Opt. Expr. **16(16)**, 12084-12089, (2008).
6. Michele Belotti, Matteo Galli, Dario Gerace, Lucio Claudio Andreani, Giorgio Guizzetti, Ahmad R. Md Zain, Nigel P. Johnson, Marc Sorel and Richard M. De La Rue, Opt. Expr. **18(2)**, 1450 - 1461, (2010).
7. F.Morichetti et al, PD paper, OFC, San Diego, (2009).
8. F. Morichetti, A. Canciamilla, M. Martinelli, A. Melloni, A. Samarelli, R.M. De La Rue, and M. Sorel, Appl. Phys. Letts. **96**, accepted for publication, (2010).
9. Swati Rawal, R.K. Sinha and Richard M. De La Rue, Opt. Expr. **17(16)**, 13315-13325, (2009).
10. Ali Z. Khokhar et al, Microel. Engin. **87**, in press, (2010).
11. N.P. Johnson, A.Z. Khokhar, H.M.H. Chong, R.M. De La Rue and S. McMeekin, Elect. Letts. **42(19)**, 1117-1119, (2006).
12. Basudev Lahiri, Ali Z. Khokhar, Richard M. De La Rue, Scott G. McMeekin and Nigel P. Johnson, Opt. Expr. **17(2)**, 1107-1115, (2009).
13. Basudev Lahiri, Scott G. McMeekin, Ali Z. Khokhar, Richard M. De La Rue and Nigel P. Johnson, Opt. Expr. **18(3)**, 3210 - 3218, (2010).

ONE-, TWO-ELECTRONIC AND EXCITONIC STATES IN A QUANTUM DOTS WITH NONTRIVIAL GEOMETRIES: ADIABATIC DESCRIPTION

K. G. DVOYAN and E. M. KAZARYAN

Physical and Technical Department, Russian-Armenian (Slavonic) University, Yerevan, 0051, Armenia

H. A. SARKISYAN

Physical and Technical Department, Russian-Armenian (Slavonic) University, Yerevan, 0051, Armenia
Department of Physics, Yerevan State University, Yerevan, 0025, Armenia
E-mail: shayk@ysu.am
www.rau.am

The adiabatic description of the electronic and excitonic states in ellipsoidal and ring-like quantum dots is presented. It is shown that nontrivial geometry of considered quantum dots makes it possible to apply procedure of separation of the Hamiltonian on the "fast" and "slow" parts. The analytically solvable models of one- and two-electronic states in strongly prolate ellipsoidal quantum dot and quantum ring are presented.

Keywords: Adiabatic description, Ellipsoidal quantum dot, Quantum ring.

1. Introduction

Quantum dots (QDs) belong to the most intensively investigated objects of nanophysics. The uniqueness of these systems consists in the fact that the motion of particles present in them is quantized in all three directions. Therefore it is not occasional, that these systems are called as "artificial atoms".[1] The outstanding property of QD is the possibility to rule its energy spectrum during the growth of concrete sample.[2] State-of-the-art methods of nanostructure growth allow obtaining QDs of different geometrical shapes and sizes. Up to date, spherical, pyramidal and cylindrical QDs are obtained (see for example Refs. 3–5). Physical properties of QDs are investigated both theoretically and experimentally. As the result of these investigations, the strong interdependence between the character of QD energy spectrum and its geometrical parameters (size and shape) has been found. These physical-chemical properties of QD and surrounding medium form the character of potential well. On the other side, QD shape and size conditions the height and symmetry of this potential. From a theoretical point of view, the most convenient objects are spherical QDs, because the symmetry of such systems often allows constructing exactly solvable Hamiltonians for charge carriers present in them. In its turn, it gives the possibility to perform analytical investigation of some QD characteristics such as absorption coefficient, charge carrier mobility, etc.[6]

On the other hand, one can employ different approximate methods for the analytical description of the properties of QDs with more complicated geometry – perturbation theory, variational method, exact diagonalization method, etc.. In some cases complicated geometry allows application of adiabatic approximation for the description of the electronic properties of QDs, when one can divide Hamiltonian of the system under study into the "slow" and "fast" parts.[7] The coordinates of the "slow" subsystem in Hamiltonian play the role of constant parameters for the "fast" subsystem. A striking example of system, in which one can describe the electronic states within the framework of adiabatic approximation, is strongly prolate ellipsoidal QD (SPEQD). Supposing that major semiaxis of SPEQD directed along Z-axis one can average slow motion along this axis by fast motion in transversal plane XOY. In Refs. 8 and 9, the authors studied electronic and optical properties of SPEQD. Particularly, it was shown that the energy of the "fast" subsystem plays the role of effective potential energy in the "slow" subsystem. At the same time, the coordinates of the "slow" subsystem enter into this energy in the capacity of parameters. In the first approximation, one can consider the effective potential energy as the oscillator potential. A more detailed determination of this potential although is desirable. Second section is devoted to the discussion of this problem. It is shown in this section that effective potential of the "slow" subsystem can be approximated by the Poschl-Teller potential. In addition to that recently ring-like nanostructures have been fabricated,[10] where the radial motion of carriers is confined both on the inner and outer borders. The importance of detailed theoretical study of physical properties of such systems consists in the generalizing character of obtained results as one can realize quantum wells and wires as well as QDs by the limiting processes. The construction of the realistic Hamiltonian of the system as far as it possible is necessary for the description of the processes in quantum rings (QR), which allows us calculate certain characteristics of QR. One- and two-electronic states in QR first have been investigated theoretically in pioneering work of Chakraboty and Pietilainen,[11] where authors have modeled the confining potential of QR by the off-center two-dimensional oscillator. In frames of this model Halonen et al.[12] discussed the optical absorption in QR in a magnetic field in the same off-center oscillator approximation, in presence of repulsive scattering centers. The authors of[13] calculated the energy band structure in QR in the presence of magnetic flux using the Wigner molecule model. Szafran and Peeters[14] discussed the few-electron eigenstates confined in coupled concentric double QRs by the exact diagonalization technique. They show that the magnetic field suppresses the tunnel coupling between the rings, localizing the single electron states in the internal ring and the few-electron states in the external ring. The effect of spin-orbit coupling in two-dimensional circular QR was discussed in Ref. 15. In Ref. 16 it was shown that in thin ring-like QDs one can again apply adiabatic approximation and examine one-electron states within the framework of plane rotator. A question appears: can one take an advantage of the same approximation for description of two-electronic or excitonic states in QR? Third and fourth sections of this article are devoted to this problem.

2. Electronic States in the SPEQD

Consider the motion of a particle (electron, hole) in a SPEQD (Fig. 1). The particle potential energy in cylindrical coordinates can be written as

$$U(\rho, \varphi, Z) = \begin{cases} 0, & \frac{\rho^2}{a_1^2} + \frac{Z^2}{c_1^2} \le 1 \\ \infty, & \frac{\rho^2}{a_1^2} + \frac{Z^2}{c_1^2} > 1 \end{cases}, \quad a_1 \ll c_1, \tag{1}$$

where a_1 and c_1 are the minor and major semiaxes of the SPEQD, respectively. In the strong SQ regime, the electron-hole Coulomb interaction energy is much less than the confinement energy of the SPEQD walls. In this regime one can neglect the Coulomb interaction. Thus, the energy states of the electron and the hole should be determined independently. The particular shape of SPEQD suggests that the particle motion along the radial direction occurs more rapidly than along the Z-direction, what enables one to apply adiabatic approximation. The Hamiltonian of the system in the cylindrical coordinates has the form:

$$\hat{\mathcal{H}} = -\frac{\hbar^2}{2\mu_p}\left[\frac{\partial^2}{\partial\rho^2} + \frac{1}{\rho}\frac{\partial}{\partial\rho} + \frac{1}{\rho^2}\frac{\partial^2}{\partial\varphi^2}\right] - \frac{\hbar^2}{2\mu_p}\frac{\partial^2}{\partial Z^2} + U(\rho, \varphi, Z). \tag{2}$$

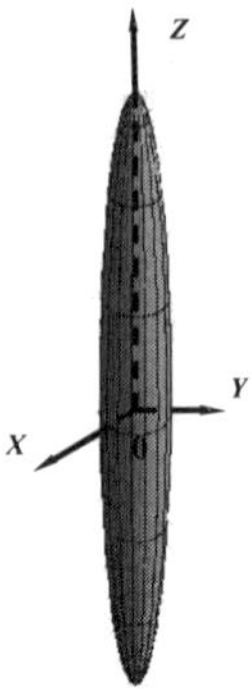

Fig. 1. Strongly prolated ellipsoidal QD (SPEQD).

Being expressed through dimensionless variables, the Hamiltonian (2) may be represented as the sum of the "fast" and "slow" subsystems' operators, $\hat{H}_1$ and $\hat{H}_2$, respectively:

$$\hat{H} = \hat{H}_1 + \hat{H}_2 + U\left(r, \varphi, z\right), \tag{3}$$

where

$$\hat{H}_1 = -\left[\frac{\partial^2}{\partial r^2} + \frac{1}{r}\frac{\partial}{\partial r} + \frac{1}{r^2}\frac{\partial^2}{\partial \varphi^2}\right], \tag{4}$$

$$\hat{H}_2 = -\frac{\partial^2}{\partial z^2}. \tag{5}$$

Here $\hat{H} = \hat{\mathcal{H}}/E_{\mathrm{R}}$, $r = \rho/a_{\mathrm{B}}$, $z = Z/a_{\mathrm{B}}$, μ_p is the effective mass of the particle, $E_{\mathrm{R}} = \hbar^2/2\mu_p a_{\mathrm{B}}^2$ is the effective Rydberg energy, $a_{\mathrm{B}} = \kappa\hbar^2/\mu_p e^2$ is the effective Bohr radius of the particle, e is the particle charge, κ is the dielectric constant. The wave function can be sought in the form

$$\psi\left(r, \varphi, z\right) = e^{im\varphi} R\left(r; z\right) \chi\left(z\right). \tag{6}$$

At a fixed value of the slow subsystem z-coordinate the particle motion is localized in a two-dimensional potential well with the effective variable width

$$L\left(z\right) = a\sqrt{1 - \frac{z^2}{c^2}}, \tag{7}$$

where $a = a_1/a_B$ and $c = c_1/a_B$. From the fast subsystem Schrödinger equation we obtain particle energy spectrum

$$\varepsilon_1(z) = \frac{\alpha_{n+1,m}^2}{L^2(z)}, \quad n = 0, 1, 2..., \quad m = 0, \pm 1, \pm 2..., \tag{8}$$

where $\alpha_{n+1,m}$ are the zeros of the first-kind $J_m(r)$ Bessel function. For the lower levels of the spectrum the particle is mainly localized in the region $|z| \ll a$.

Expression (8) takes the place of the potential in the Schrödinger equation for the "slow" subsystem, but the Schrödinger equation with such effective potential is not analytically solvable. The adiabatic approximation is applicable for calculation of lower levels of the energy spectrum. Parabolic potential, obtained by use of Taylor series of the energy expression for the "fast" subsystem,[7] gives rise to a set of equidistant energy levels in spectrum. It is notable that each energy level of the "fast" subsystem has its own set of equidistant levels with gaps depending on the quantum number of the particular "fast" subsystem. However, only two or three lower energy levels are split into equidistant level subsystems; for higher levels of the "fast" subsystem the sublevels are not equidistant any more.

We suggest a more realistic model of one-dimensional effective potential which we represent in the form of modified Pöschl-Teller potential (see Fig. 2).[17] In dimensionless quantities this potential has the following form:

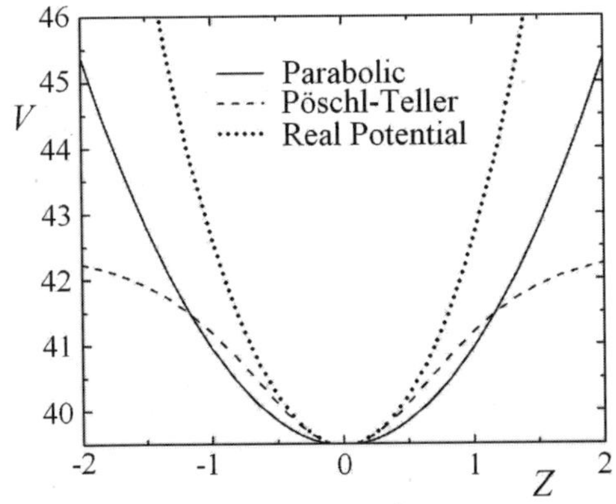

Fig. 2. Dependence of one-dimensional effective potentials on coordinate Z.

$$\varepsilon_1(z) = V_{\mathrm{PT}}(z) = \frac{\alpha^2_{n+1,m}}{a^2} - \frac{V_0}{ch^2(z/\gamma)} + \frac{V_0}{\gamma}. \tag{9}$$

Here V_0 and γ are parameters describing the depth and width of corresponding quantum well, respectively. Note that they depend on the quantum numbers n and m of the "fast" subsystem. Choice of this particular potential is explained by the fact that the Taylor expansion of potential (8) for small values of the z-coordinate is parabolic as it is the case for (9) also. On the other hand, at higher values of the z-coordinate the discrepancy of the Pöschl-Teller potential from parabolic one is increasing. Thus, violation of equidistance of energy levels of "slow" subsystem can be taken into account.

The one-dimensional Schrödinger equation with the Pöschl-Teller potential reads:

$$\chi''(z) + \left(\varepsilon - \frac{\alpha^2_{n+1,m}}{a^2} + \frac{V_0}{ch^2(z/\gamma)} - \frac{V_0}{\gamma}\right)\chi(z) = 0 \tag{10}$$

A series of transformations results in the following expressions for the wave function and energy spectrum of charge carriers:

$$\chi(z) = C\left(1 - th^2\left(\frac{z}{\gamma}\right)\right)^{\frac{k}{2}} \times {}_2F_1\left(k - s, k + s + 1, k + 1; \frac{1}{2} - \frac{1}{2}th\left(\frac{z}{\gamma}\right)\right) \tag{11}$$

$$\varepsilon = \frac{\alpha^2_{n+1,m}}{a^2} - \frac{\left(-(1+2n) + \sqrt{1 + 4V_0\gamma^2}\right)}{4\gamma^2} + V_0, \tag{12}$$

where $k^2 = -\varepsilon + \pi^2 n^2/a^2 + \lambda(\lambda - 1)/\gamma^2$ and $s = (-1 + \sqrt{1 + 4\lambda(\lambda - 1)})/2$, C is normalization constant, ${}_2F_1(a, b, c; z)$ is the hypergeometric function. For small values of the coordinate z, the potential (9) takes the form

$$V_{\mathrm{PT}}(z) \approx \frac{\alpha^2_{n+1,m}}{a^2} + \frac{V_0}{\gamma}z^2. \tag{13}$$

For the Schrödinger equation for the "slow" subsystem with the potential (13) we arrive at the following expression for the equidistant energy spectrum of a charge carriers:

$$\varepsilon = \frac{\alpha^2_{n+1,m}}{a^2} + \frac{2\sqrt{V_0}}{\gamma}(N + 1/2), \quad N = 0, 1, 2, ... \tag{14}$$

which perfectly agrees with the result.[18]

As is seen from formula (14), the energy spectrum of charge carriers in SPEQD is equidistant. More precisely, each level of the "fast" subsystem has its own family of equidistant energy levels created by the "slow" subsystem. As a consequence of the adiabatic approximation, this result is valid only for the low spectrum levels (i.e. small quantum numbers). It is also important that approximation of "fast" subsystem energy expression by a modified Pöschl-Teller potential enables one to take into account the energy levels which are non-equidistance at higher energy values. Numerical calculations performed for a $GaAs$ QD with the following parameters: $\mu_e = 0.067m_e$, $a_e = 104\text{Å}$ is the effective Bohr radii of the electron. The frequency of transition between the equidistant levels (for the value $n = 0$), at fixed values $a_1 = 0.5a_e$ and $c_1 = 2.5a_e$, is equal to $\omega_{00} = 3.32 \times 10^{13}s^{-1}$, which corresponds to the infrared region of the spectrum. For the same values of quantum numbers, but with the values $a_1 = 0.4a_e$ and $c_1 = 2a_e$, we obtain $\omega_{10} = 5.19 \times 10^{13}s^{-1}$, which is half as much against the preceding case. As it is seen from formula (12), with increasing semiaxes the particle energy is lowered. Note that this energy is more "sensitive" to changes of the minor semiaxis, which is a consequence of the higher contribution of SQ into the particle energy in the direction of the axis of ellipsoid revolution. With increasing semiaxes the energy levels come closer together, but remain equidistant.

One can see from Fig. 2 that the effective one-dimensional potential is well approximated by the modified Pöschl-Teller potential. As the z-coordinate grows, the discrepancy between the exact and approximate potential becomes evident for the modified Pöschl-Teller potential. For small values of z- coordinate the modified Pöschl-Teller potential is well approximated with parabolic potential.

Fig. 3 illustrates the dependence of the first equidistant family of charge carriers energy in SPEQD on the minor semiaxis of the SPEQD at a fixed value of the major semiaxis for both modified Pöschl-Teller and parabolic approximation cases of the one-dimensional potential approximation. In other words, we compare results obtained from relations (12) with those from (14). From Fig. 3 it is easily seen that the charge carriers energy levels are equidistant in both cases since for small values of the z-coordinate it is sufficient to keep only quadratic terms in the Taylor development of the modified Pöschl-Teller potential what leads to practical coincidence with parabolic potential.

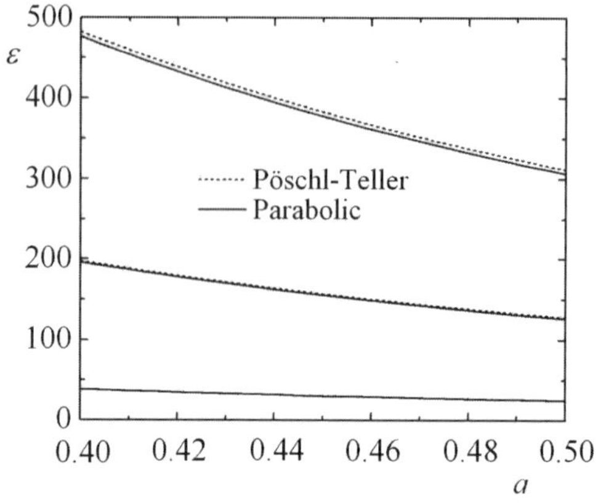

Fig. 3. Dependence of the first equidistant family of charge carriers energy in SPEQD on the minor semiaxis of the SPEQD at a fixed value of the major semiaxis $c = 2.5\,a_{\mathrm{B}}$ for both parabolic and Pöschl-Teller cases.

3. Two-Electronic States in QR

Let us consider a QR with a small width ($R_1 - R_2 \ll \{R_1, R_2\}$), which contains two electrons (Fig. 4). The influence of the QR borders in radial direction is assumed to be very strong. Then for both electrons in future we will consider one particle states and neglect the Coulomb interaction.

The assumption of strong radial quantization allows to consider the electrons on the same fixed distance from the QR center and the energy of Coulomb interaction between the electrons is a function only of polar angles φ_1 and φ_2. In this case the two particle Hamiltonian is the following:

$$\hat{\mathcal{H}}(1,2) = \sum_{i=1}^{2} \left\{ -\frac{\hbar^2}{2\mu} \left(\frac{1}{\rho_i} \frac{\partial}{\partial \rho_i} \left(\rho_i \frac{\partial}{\partial \rho_i} \right) + \frac{1}{\rho_i^2} \frac{\partial^2}{\partial \varphi_i^2} \right) + V_{\mathrm{conf}}(i) \right\} + V_{\mathrm{Coul}}(\varphi_1, \varphi_2) \quad (15)$$

where V_{conf} describes the confining potential in radial direction:

$$V_{\mathrm{conf}} = \begin{cases} 0, & R_1 \le \rho \le R_2, \\ \infty, & [\rho < R_1,\ \rho > R_2] \end{cases}.$$

The expression of Coulomb interaction $V_{\mathrm{Coul}}(\varphi_1, \varphi_2)$ will be discussed later.

Because the radial wave functions of electrons are expressed by the Bessel functions of the first and second kinds, then in the adiabatic approach the total wave function can be presented in the form

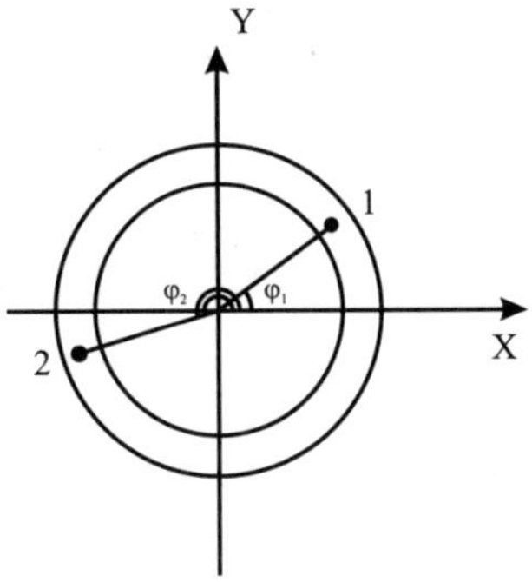

Fig. 4. Quantum ring (QR) with the two electrons.

$$\Psi\left(r_1, r_2, \varphi_1, \varphi_2\right) = f_0\left(r_1\right) f_0\left(r_2\right) \Phi\left(\varphi_1, \varphi_2\right), \tag{16}$$

where

$$f_0\left(r_i\right) = C_1 I_0\left(\kappa_0 r_i\right) + C_2 Y_0\left(\kappa_0 r_i\right), \tag{17}$$

$I_0\left(x\right), Y_0\left(x\right)$ are the Bessel functions of the first and the second kinds respectively,[19] $\kappa_0 = \sqrt{2\mu E_0/\hbar^2}$, E_0 is the energy of the one particle ground state, which is determined from the following condition

$$\begin{vmatrix} I_0\left(\kappa_0 R_1\right) & Y_0\left(\kappa_0 R_1\right) \\ I_0\left(\kappa_0 R_2\right) & Y_0\left(\kappa_0 R_2\right) \end{vmatrix} = 0. \tag{18}$$

Note, that the function $f_0\left(r\right)$ satisfies to radial equation

$$-\frac{\hbar^2}{2\mu}\left(\frac{1}{\rho}\frac{d}{d\rho}\left(\rho\frac{d}{d\rho}\right)\right) f_0\left(\rho\right) + V_{\text{conf}}\left(\rho\right) = E_0 f_0\left(\rho\right). \tag{19}$$

Substituting (16) into (15) and using (19), we obtain for the angular wave function the following equation:

$$\begin{aligned} &-\frac{\hbar^2}{2\mu R_{\text{eff}}^2}\left(\frac{\partial^2}{\partial\varphi_1^2} + \frac{\partial^2}{\partial\varphi_2^2}\right)\Phi\left(\varphi_1, \varphi_2\right) + V_{\text{Coul}}\left(\varphi_1, \varphi_2\right)\Phi\left(\varphi_1, \varphi_2\right) \\ &\quad = \left(E - 2E_0\right)\Phi\left(\varphi_1, \varphi_2\right) = \varepsilon\Phi\left(\varphi_1, \varphi_2\right), \end{aligned} \tag{20}$$

where

$$R_{\text{eff}} = \frac{R_1 + R_2}{2}. \tag{21}$$

Let us turn to the definition of the expression of the electrons' interaction potential $V_{\text{Coul}}(\varphi_1, \varphi_2)$. Note that the system represents two particles on a circle interacting with the repulsive potential.[17] On the other hand it is clear that this potential should depend on the relative angle of the electrons, i.e. on $\varphi_1 - \varphi_2$, and it should be an even function, because the replacement of electrons does not change the physical situation. Meanwhile if $\varphi_1 - \varphi_2 = \pi$ then $V_{\text{Coul}}(\varphi_1, \varphi_2)|_\pi = e^2/(2\varepsilon_{\text{d}} R_{\text{eff}})$, and if $\varphi_1 - \varphi_2 = \pi/2$ then $V_{\text{Coul}}(\varphi_1, \varphi_2)|_{\pi/2} = e^2/(\sqrt{2}\varepsilon_{\text{d}} R_{\text{eff}})$ (ε_{d} is dielectric constant). By a direct check one can see that the function

$$V_{\text{Coul}}(\varphi_1, \varphi_2) = \frac{e^2}{R_{\text{eff}}\varepsilon_{\text{d}}} \left\{ \frac{1}{\sqrt{2}} + \frac{\sqrt{2}-1}{2} \cos(\varphi_1 - \varphi_2) \right\} \tag{22}$$

has the required characteristics.

Thus, for the radial wave function we obtain the following equation:

$$-\frac{\hbar^2}{2\mu R_{\text{eff}}^2} \left(\frac{\partial^2}{\partial \varphi_1^2} + \frac{\partial^2}{\partial \varphi_2^2} \right) \Phi(\varphi_1, \varphi_2) \tag{23}$$

$$+ \frac{e^2}{\varepsilon_{\text{d}} R_{\text{eff}}} \left\{ \frac{1}{\sqrt{2}} + \frac{\sqrt{2}-1}{2} \cos(\varphi_1 - \varphi_2) \right\} \Phi(\varphi_1, \varphi_2) = \varepsilon \Phi(\varphi_1, \varphi_2).$$

Let us enter new variables α and β:

$$\alpha = \varphi_1 - \varphi_2, \ \beta = (\varphi_1 + \varphi_2)/2. \tag{24}$$

In this case the variables in (23) separate and due to this the wave function $\Phi(\alpha, \beta)$ can be expressed in the form

$$\Phi(\alpha, \beta) = u(\alpha) v(\beta). \tag{25}$$

Applying the separation procedure for defining of the functions $u(\alpha)$ and $v(\beta)$ one can obtain the following equations:

$$-\frac{\hbar^2}{\mu R_{\text{eff}}^2} \frac{d^2 u(\alpha)}{d\alpha^2} + \frac{e^2}{\varepsilon_{\text{d}} R_{\text{eff}}} \left\{ \frac{\sqrt{2}-1}{2} \cos\alpha \right\} u(\alpha) = E_\alpha u(\alpha), \tag{26}$$

$$-\frac{\hbar^2}{4\mu R_{\text{eff}}^2} \frac{d^2 v(\beta)}{d\beta^2} = E_\beta v(\beta), \tag{27}$$

where

$$(E_\alpha + E_\beta) = \varepsilon - \frac{e^2}{\sqrt{2}R_{\text{eff}}}. \tag{28}$$

Inserting in the equation (27) the electrons total moment of inertia $J_{\text{eff}} = 2\mu R_{\text{eff}}^2$, for the eigenfunctions and the eigenvalues one can write

$$v(\beta) = \frac{\exp\{im\beta\}}{\sqrt{2\pi}}, \quad E_\beta = \frac{\hbar^2 m^2}{2J_{\text{eff}}} \quad (m = 0, \pm1, \pm2, ...), \tag{29}$$

Let us now consider equation (26). Introducing the following notifications

$$\alpha = 2\phi,\ \lambda = \frac{4\mu R_{\text{eff}}^2}{\hbar^2} E_\alpha,\ 2q = \left(\frac{\sqrt{2}-1}{2}\right)\frac{4\mu e^2 R_{\text{eff}}}{\varepsilon_{\text{d}}\hbar^2},$$

we can transform equation (26) to the Mathieu equation:

$$\frac{d^2u(\phi)}{d\phi^2} + (\lambda - 2q\cos 2\phi)\,u(\phi) = 0. \tag{30}$$

As the one dimensional equation (30) contains symmetrical potential energy with respect to the point $\phi = 0$, its' solutions have to be either even or odd periodic functions. One can write these solutions in the form of Fourier expansion:[20]

$$u_+(\phi) = \sum_{j=0}^{\infty} A_{2j}\cos(2j\phi), \quad u_-(\phi) = \sum_{j=1}^{\infty} B_{2j}\sin(2j\phi), \tag{31}$$

where A_{2j}, B_{2j} are the expansion coefficients. It is known from the Mathieu equation theory that even and odd solutions and hence the energy levels alternate. If $\{a_{2j}\}_{j=0}^{\infty}$ is the ensemble of the even, and $\{b_{2j}\}_{j=1}^{\infty}$ is the ensemble of the odd solutions with respect to λ, hence the following inequalities take place:

$$a_0 < b_2 < a_2 < b_4 < a_4 < .... \tag{32}$$

Finally for the even and odd energy levels we obtain the expressions:

$$E_{2j}^{+} = \frac{\hbar^2 a_{2j}}{4\mu R_{\text{eff}}^2}, \quad E_{2j}^{-} = \frac{\hbar^2 b_{2j}}{4\mu R_{\text{eff}}^2}. \tag{33}$$

Taking into account electron spin, we obtain the singlet states corresponding to the even states, as the spin part of the wave function is odd. In z−presentation by the Pauli matrix

$$\hat{\sigma}_z = \begin{pmatrix} 1 & 0 \\ 0 & -1 \end{pmatrix}, \quad \xi = \begin{pmatrix} 1 \\ 0 \end{pmatrix} \text{ and } \varsigma = \begin{pmatrix} 0 \\ 1 \end{pmatrix},$$

the asymmetric spin wave function can be written as

$$\chi_{\mathrm{A}} = \frac{1}{\sqrt{2}} \left\{ \xi(1)\varsigma(2) - \xi(2)\varsigma(1) \right\}. \tag{34}$$

For the odd levels spin part of the wave function is symmetric and, hence, the total wave function is triplet with the spin wave functions

$$\chi_{\mathrm{S}} = \left\{ \xi(1)\xi(2), \ \varsigma(1)\varsigma(2), \ \frac{1}{\sqrt{2}} \left\{ \xi(1)\varsigma(2) + \xi(2)\varsigma(1) \right\} \right\}. \tag{35}$$

Thus in the considered system the singlet and triplet levels alternate and we obtain finally for the energy spectrum the following expression:

$$E_{2j,m}^{+(-)} = \left(2E_0 + \frac{\hbar^2 a_{2j}(b_{2j})}{4\mu R_{\mathrm{eff}}^2} + \frac{\hbar^2 m^2}{2J_{\mathrm{eff}}} + \frac{e^2}{\sqrt{2}\varepsilon_{\mathrm{d}} R_{\mathrm{eff}}} \right). \tag{36}$$

Now let us consider the behavior of the system depending on the effective radius of QR. According to (30) the effective "depth" of the one dimensional well increases with the increase of R_{eff} as $q \sim R_{\mathrm{eff}}$. It can be shown from the Mathieu equation theory, that for $\alpha = \pi$ Mathieu equation reduces to the harmonic oscillator equation, with the eigenvalues:

$$E_n = \frac{\hbar^2 \sqrt{q}}{\mu R_{\mathrm{eff}}^2} \left(n + \frac{1}{2} \right) - \frac{q\hbar^2}{2\mu R_{\mathrm{eff}}^2}. \tag{37}$$

It is due to the small oscillations of diametrically located electrons we get the decrease of repulsive force with the increase of QR effective radius. That is why these oscillations will be harmonic. As follows from (37) the oscillations effective frequency decreases with the increase of R_{eff} due to weakening of the oscillating strength.

4. Excitonic States in QR with Smorodinsky–Winternitz Confining Potential

It should be mentioned that in QR may also arise double-particle problems of another type – excitonic states. It is clear that in the general case such problem is extremely difficult to treat mathematically. However in the case of ring-like structures with small width, size quantization in the direction of ring axis OZ is so strong that one may ignore Coulomb interaction in this direction. In other words, here one may again use adiabatic approximation and separate Z-part of Hamiltonian considering Coulomb problem purely two-dimensional. Let us investigate magnetoexcitonic states in QR with Smorodinsky–Winternitz confinement potential[21] assuming that Coulomb interaction is weak enough. The analytical expression of Smorodinsky–Winternitz potential has the form

$$V_{\text{conf}} = \alpha/\rho^2 + \beta\rho^2 - 2\sqrt{\alpha\beta},$$

where α and β are constants characterizing the nanostructure.[22] Choosing (in the polar coordinate system) a vector potential gauge in the form $\mathbf{A} = \{A_\varphi = H\rho/2,\ A_\rho = A_z = 0\}$, for Schrödinger equation in one-particle approximation we obtain the following equation:

$$\left\{-\frac{\hbar^2}{2\mu}\left(\frac{1}{\rho}\frac{\partial}{\partial\rho}\left(\rho\frac{\partial}{\partial\rho}\right)+\frac{1}{\rho^2}\frac{\partial^2}{\partial\varphi^2}\right)-\frac{i\hbar\omega_H}{2}\frac{\partial}{\partial\varphi}+\frac{\mu\omega_H^2\rho^2}{8}\right.$$
$$\left. + V(\rho)\right\}\psi(\rho,\varphi) = E_{n_\rho,m}\psi(\rho,\varphi), \tag{38}$$

where $\omega_H = |e|\,H/\mu c$ is the cyclotron frequency, μ is the effective mass of charge carrier. Equation (38) is exactly solvable. The analytical expressions for wave function and energy spectrum are[23]:

$$\psi(\rho,\varphi) = \frac{1}{\sqrt{2\pi}}e^{im\varphi}\frac{1}{a_\Omega^{M+1}}\left[\frac{(n_\rho+M)!}{2^M n_\rho! M!^2}\right]^{1/2}$$
$$\times\,\rho^M e^{-\frac{\rho^2}{4a_\Omega^2}}\,{}_1F_1\left[-n_\rho, M+1, \frac{\rho^2}{2a_\Omega^2}\right] \tag{39}$$

$$E_{n_\rho,m} = \hbar\omega_H\left(\frac{\Omega}{\omega_H}\left(n_\rho+\frac{M+1}{2}\right)+\frac{m}{2}\right) - 2\sqrt{\alpha\beta}. \tag{40}$$

Here $n_\rho = 0, 1, 2, \ldots$ is the radial quantum number, m is the magnetic quantum number, $\Omega = \sqrt{\omega_H^2 + 8\beta/\mu}$, $M = \sqrt{m^2 + 2\mu\alpha/\hbar^2}$, $a_\Omega = \sqrt{\hbar/\mu\Omega}$, ${}_1F_1$ is the confluent hypergeometric function.

Let us now consider excitonic effects on interband transitions. The Hamiltonian of a two-dimensional magnetoexciton takes the following form:

$$\hat{\mathcal{H}}_{\mathrm{ex}} = \hat{\mathcal{H}}_{\mathrm{e}} + \hat{\mathcal{H}}_{\mathrm{h}} - \frac{e^2}{\varepsilon_{\mathrm{d}} \left| \boldsymbol{\rho}_{\mathrm{e}} - \boldsymbol{\rho}_{\mathrm{h}} \right|} \tag{41}$$

where $\hat{\mathcal{H}}_i = (\hat{\mathbf{p}} - \mathbf{A} e_i / c)^2 / 2\mu_i + V(\rho_i)$ and $\hat{\mathcal{H}}_i \psi_i = E_i \psi_i$, $i = e, lh, hh$; ε is dielectric constant. We consider the third term of Eq. (41) as a small perturbation for the Hamiltonian $\hat{\mathcal{H}}_{\mathrm{e}} + \hat{\mathcal{H}}_{\mathrm{h}}$. Therefore, the wave function of the magnetoexciton, in a first order approximation, may be presented in the following form:

$$\Psi^0_{\mathrm{ex}}(\rho_{\mathrm{e}}, \varphi_{\mathrm{e}}, \rho_{\mathrm{h}}, \varphi_{\mathrm{h}}) = \psi_{\mathrm{e}}(\rho_{\mathrm{e}}, \varphi_{\mathrm{e}}) \psi_{\mathrm{h}}(\rho_{\mathrm{h}}, \varphi_{\mathrm{h}}). \tag{42}$$

The correction for the energy is given by the following expression

$$\Delta E_0 = \int \Psi^{0*}_{\mathrm{ex}} \left(- \frac{e^2}{\varepsilon_{\mathrm{d}} \sqrt{\rho_{\mathrm{e}}^2 + \rho_{\mathrm{h}}^2 - 2\rho_{\mathrm{e}}\rho_{\mathrm{h}} \cos(\varphi_{\mathrm{e}} - \varphi_{\mathrm{h}})}} \right) \Psi^0_{\mathrm{ex}} dS_{\mathrm{e}} dS_{\mathrm{h}}. \tag{43}$$

Thus, for the energy we have:

$$E = E_{\mathrm{e}} + E_{\mathrm{h}} + \Delta E_0. \tag{44}$$

Let us analyze the obtained results. Note that the parameters of restriction potentials α and β are connected with inner and outer radii according to the following relations (see Ref. 22):

$$\beta R_i^2 + \frac{\alpha}{R_i^2} - 2\sqrt{\alpha\beta} = U_i, \; i = 1; 2, \tag{45}$$

where U_i is defined by a relation of $U_i = 1.247 Q x_i$: for an electron $Q_{\mathrm{e}} = 0.4$, for a hole $Q_{\mathrm{h}} = 0.6$. Note that for calculation we used the following material parameters for $Ga_{1-x_1}Al_{x_1}As/GaAs/Ga_{1-x_2}Al_{x_2}As$ QR: $\mu_{\mathrm{e}} = 0.067 m_0$, $\mu_{\mathrm{hh}} = 0.45 m_0$, $\mu_{\mathrm{lh}} = 0.082 m_0$, $\varepsilon = 12.9$, $x_1 = x_2 = 0.4$).The energies of $lh - e$ and $hh - e$ pairs dependent on inner radius R_1 of QR at the fixed value of outer radius R_2 are presented on Fig. 5 ($R_2 = 35$nm, $a_H = 32nm$ ($H \approx 6400$Gs)). The solid lines correspond to the case when the Coulomb interaction between the electron and hole is absent, while dashed lines describe the same dependencies in the presence of this interaction. The increase of radius R_1 leads to enhancing of size quantization and therefore the energy levels go up. On the other hand, the more effective is the size quantization, the Coulomb interaction becomes weaker. It is the main reason why the dashed lines become closer to the solid lines. Let us note that

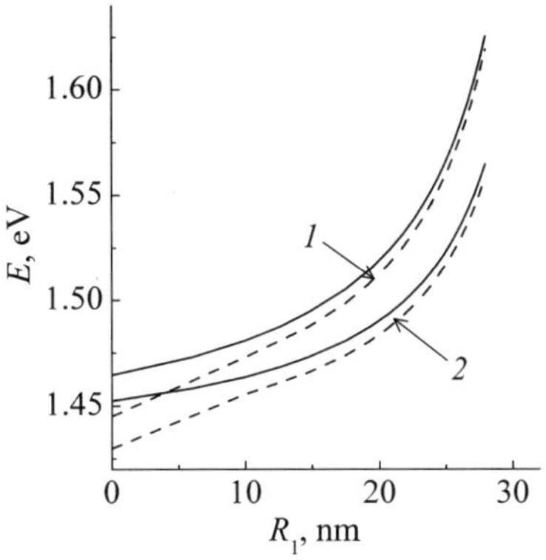

Fig. 5. Dependencies of energy for (1) $lh - e$ and (2) $hh - e$ pairs on inner radius of QR: without accounting of Coulomb interaction (solid curves) and with its accounting (dashed curves). $a_H = 32$ nm.

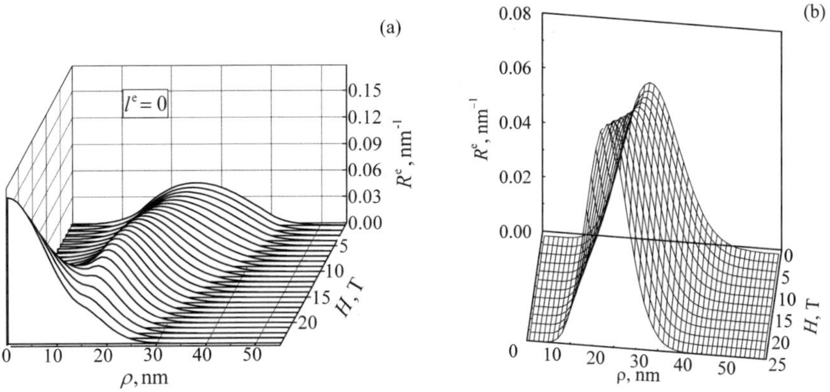

Fig. 6. The evolution of radial eigenfunction for the singlet electron, with increasing magnetic field: (a) figure taken from Ref. 24, (b) our calculations.

the lower curves correspond to $hh - e$ pair, which is direct a consequence of hh large effective mass. The opposite picture appears, when the outside radius is increased at fixed value of inner radius. Finally, for the comparison, we present the evolution of radial eigenfunction for the singlet electron from the Ref. 2 (Fig. 6(a)) and for our case (Fig. 6(b)). With the increase of magnetic field, the maximum of eigenfunction drifts to the inner radius side. However, in contrast to the finite-height confinement potential,[24] when charge carrier can exude and appear at the centre of QR, for the system under consideration the maximum of radial eigenfunction (which is real) can only shift to the side of inner radius, because we consider impenetrable confinement potential.

5. Conclusion

In this article we applied the adiabatic procedure for description of electronic and excitonic states in zero-dimensional nanostructures with nontrivial geometry. For ellipsoidal and ring-like QDs the analytical expression for wave function and energy spectrum of one- and two-electronic states are determined. For excitonic states in the QR with a small height we show that the model of two-dimensional Coulomb system is applicable, because it is possible to separate the Z-part of Hamiltonian.

Acknowledgments

The authors are thankful to Dr. A.V. Meliksetyan and Dr. D.B. Hayrapetyan for collaboration and useful discussions.

References

1. P.A. Maksym, T. Chakraborty, Phys. Rev. Lett., **65**, 108 (1990).
2. Zh.I. Alferov, Semiconductors, **32**, 1 (1998).
3. D.S. Chuu, C.M. Hsiao, W.N. Mei, Physical Review B (Condensed Matter), **46**, 3898 (1992).
4. M. Grundmann, O. Stier, D. Bimberg, Physical Review B (Condensed Matter), **52**, 11969 (1995).
5. M. Tadić, F.M. Peeters, K.L. Janssens, M. Korkusiński, P. Hawrylak, Journal of Applied Physics, **92**, 5819 (2002).
6. D. Bimberg, M. Grundmann, N.N. Ledentsov, Quantum dot heterostructures, (John Wiley & Sons, New York, 1999).
7. V.M. Galitsky, B.M. Karnakov, V.I. Kogan, Practical Quantum Mechanics, (Nauka, Moscow, 1981).
8. K.G. Dvoyan, D.B. Hayrapetyan, E.M. Kazaryan, A.A. Tshantshapanyan, Nanoscale Research Letters, **2**, 601 (2007).
9. K.G. Dvoyan, D.B. Hayrapetyan, E.M. Kazaryan, Nanoscale Research Letters, **4**, 106 (2009).
10. A. Lorke, R.J. Luyken, A.O. Govorov, J.P. Kotthaus, J.M. Garcia, P.M. Petroff, Phys. Rev. Lett., **84**, 2223 (2000).
11. T. Chakraborty, P. Pietiläinen, Phys. Rev. B (Condensed Matter), **50**, 8460 (1994).
12. V. Halonen, P. Pietiläinen, T. Chakraborty, Europhysics Letters, **33**, 377 (1996).

13. L. Wendler, V.M. Fomin, Phys. Rev. B (Condensed Matter), **51**, 17814 (1995).
14. B. Szafran, F.M. Peeters, Phys. Rev. B, **72**, id. 155316 (2005).
15. M.P. Nowak, B. Szafran, Phys. Rev. B, **80**, id. 195319 (2009).
16. V.A. Harutyunyan, E.M. Kazaryan, A.A. Kostanyan, H.A. Sarkisyan, Physica E, **36**, 114 (2007).
17. S. Flügge, Practical Quantum Mecannics, (Springer-Werlag, 1971).
18. D.B. Hayrapetyan, J. Contemp. Phys., **42**, 293 (2007).
19. N.S. Koshlyakov, E.B. Gliner, M.M. Smirnov, Equations in partial derivatives of mathematical physics, (Vysshaya Shkola, Moscow, 1970).
20. M. Abramovitz, I.A. Stegun, Handbook of mathematical functions, (Dover , New York, 1972).
21. P. Winternitz, Ya.A. Smorodinsky, M. Uhlir, I. Fris, Sov. Nucl. Phys., **4**, 444 (1966).
22. A.K. Atayan, E.M. Kazaryan, A.V. Meliksetyan, H.A. Sarkisyan, Journal of Computational and Theoretical Nanoscience, **7**, 1 (2010).
23. W.-C. Tan, J.C. Inkson, Semicond.Sci.Technol., **11**, 1635 (1996).
24. T.V. Bandos, A. Cantarero, A. García-Cristóbal, The European Physical Journal B, **53**, 99 (2006).

PLANAR PLASMONIC STRUCTURES AND NONLINEAR METAL-DIELECTRIC SUBWAVELENGTH WAVEGUIDES

A. R. DAVOYAN*, I. V. SHADRIVOV, and YU. S. KIVSHAR

Nonlinear Physics Center, Research School of Physics and Engineering,
Australian National University,
Canberra, ACT 0200, Australia
**E-mail: ard124@physics.anu.edu.au*
http://physics.anu.edu.au/nonlinear

We overview our recent studies of the properties of planar metal-dielectric waveguides for surface plasmon polaritons. For linear slot waveguides, we revisit the problem of the guided mode structure and dispersion properties of surface waves. We study the effect of losses and show a significant change of the mode properties in the presence of losses. For nonlinear waveguides, we study both second- and third-order nonlinear effects. In particular, we show that a plasmonic beam launched at a metal-dielectric interface can experience self-focusing accompanied by the soliton formation due to the Kerr-type nonlinearity of the dielectric. We also show that quadratic nonlinear response may lead to plasmon-to-plasmon parametric frequency conversion, and analyze the frequency conversion in the presence of losses.

Keywords: Surface plasmon-polaritons, metal-dielectric structures, plasmon soliton, slot waveguide, second-harmonic generation.

1. Introduction

One of the most active fields of research in modern optics is associated with manipulating light at the nanoscale. It was shown that optical systems with metal inclusions can allow miniaturizing optical components through excitation of surface plasmon-polaritons, or plasmon modes. Metal-dielectric structures exhibit a wide range of novel physical phenomena which can be employed for various applications such as sensing, imaging, and waveguiding (for general overview, see Refs. 1–3). Most importantly, plasmonic optical elements provide possibilities to overcome diffraction limit, and they allow squeezing light to subwavelength dimensions. One may think of designing plasmonic integrated circuits which would allow miniaturizing optical devices to increase their functionalities in signal processing.[4] One of the key components of any integrated circuit is a waveguide, which transfers a signal between the circuit elements. Thus, the study of plasmonic waveguides are of a fundamental importance for the design of the nanoscale structures and circuits for future applications.

Optical structures supporting surface plasmon polaritons have been a subject of extensive studies, and different types of metal-dielectric waveguides have been suggested theoretically and demonstrated experimentally (see, e.g., Refs. 5–8). The simplest plasmonic waveguide is an interface between metal and insulator which supports plasmon polaritons; however, due to losses in metal, an excited plasmon can propagate for only a very short distance.[5,9] Introducing more complex structures, such as a three-layer system, helps to increase the propagation distance due to the coupling of plasmons at the neighboring interfaces and the field localization in dielectric rather than in metal.[2] There are two basic geometries for plasmon guiding, namely, dielectric-metal-dielectric and metal-dielectric-metal structures. In past decades, rigorous analysis of these structures has been presented.[9–11] In Ref. 12 it was suggested that not all regimes of the plasmon propagation in waveguides were analyzed previously.

In order to increase functionality and efficiency of plasmonic effects in integrated circuit elements, one should employ *nonlinear effects* which may enhance substantially capabilities of plasmonic nanoguides and circuits, also leading to a range of novel effects such as plasmon self-focusing, frequency conversion, parametric amplification, etc. To study the plasmon propagation in nonlinear media, we will analyze nonlinear Maxwell's equations for the transverse magnetic (TM) waves in the presence of a metal-dielectric interface. Relatively small number of studies has been devoted to the analysis of corresponding nonlinear problems. In 1980-s, several groups suggested to employ simple approximations for the analysis of Kerr nonlinearities in a single metal-dielectric interface geometry.[13,14] Most of the earlier studies considered nonlinear guided waves in metal-dielectric structures localized in the transverse dimension. The temporal effects and the formation of temporal surface-polariton solitons were first discussed by Boardman *et al.*,[15] whereas recently Feigenbaum and Orenstein[16] considered spatial localization of plasmon waves in the planar metal-dielectric waveguide geometry. However the detailed analysis of nonlinear plasmon localization should be carried out taking into account boundary conditions, and, more importantly, losses which affect soliton propagation.

Second-order nonlinear processes can be employed for the plasmon harmonic generation and nonlinear parametric amplification. However, until now, many studies have demonstrated the excitation of plasmons through the second-order optical processes.[17,18] And the study plasmon-to-plasmon frequency conversion in subwavelength structures has not been addressed yet.

In this Chapter we address the propagation of plasmons in linear and nonlinear subwavelength metal-dielectric structures. In Sec. 2 we study the mode structure and dispersion properties of plasmons in the presence of absorbtion, and reveal the significance of the effect of losses. In Sec. 3 we analyze the propagation of plasmons in nonlinear systems predicting the possibility of plasmon-to-plasmon frequency conversion (Sec. 3.1) and generation of plasmon solitons in lossy plasmonic systems (Sec. 3.2).

2. Linear Metal-Dielectric-Metal Waveguides

The properties of metals in optics can be described within classical electrodynamics, where dielectric response of metal comes from the free electron gas, or plasma.[2] This model was suggested by Drude and extended later by Lorentz to take into account the transitions of electrons between conduction bands at upper visible and ultraviolet range of frequencies. We will use the Drude-Lorentz model for describing linear dispersive response of the metals. At the same time, in the region of electron transitions the decay of electromagnetic wave propagating in metal is very strong, and the nonlinearities associated with metallic properties are large. We will be interested in the infrared and visible frequency ranges, thus we can neglect nonlinear properties of metals, and assume that nonlinear effects arise due to the inclusion of nonlinear dielectrics.

Planar metal-dielectric interface supports only TM-polarized surface waves - *surface plasmons*, which have two components of electric field and one component of magnetic field. The dispersion relations are derived in a standard way by applying the continuity conditions for the tangential field components, and they were obtained for both insulator-metal-insulator and metal-dielectric-metal structures.[9–11] Detailed analysis of plasmonic waveguide modes, including radiative waves, in thin metallic films sandwiched between two semi-infinite dielectrics was presented in Ref. 10. Later, Prade et al.[11] developed a comprehensive analysis of dispersion relations for three-layer structures and studied plasmonic modes in the slot waveguide geometry (metal-dielectric-metal) as well as in a thin metal film (dielectric-metal-dielectric). Nevertheless, *the effect of losses* in metallic layers was not studied rigorously, and the dispersive nature of dissipation in metals was not addressed either. Only recent studies provided first analysis of waveguides with real metal losses, see Refs. 5,19,20. However, not all properties of slot waveguides have been revealed and effect of loss on the dispersion properties of surface plasmons has not been addressed detailed enough yet.

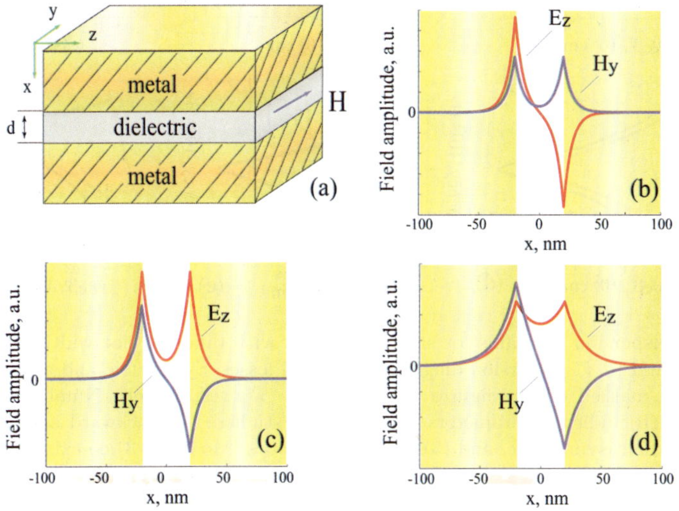

Fig. 1. (a) Schematic of a symmetric plasmonic slot waveguide; (b-d) Magnetic and electric field profiles calculated in a lossless 40 nm wide slot waveguide for the frequency corresponding to the free space wavelength of 465 nm for forward propagating (b) symmetric mode, (c) antisymmetric modes, and (d) backward antisymmetric mode.

2.1. *Lossless slot waveguides*

We start our analysis with symmetric metal-dielectric-metal plasmonic waveguide of the width d, which is schematically shown on Fig. 1(a).

For the linear slot waveguide the dispersion relation is:[10,11]

$$\tanh\left(\frac{1}{2}\kappa_d d\right) = -\begin{cases} \dfrac{\varepsilon_d}{\varepsilon_m}\dfrac{\kappa_m}{\kappa_d}, & (symmetric), \\ \dfrac{\varepsilon_m}{\varepsilon_d}\dfrac{\kappa_d}{\kappa_m}, & (antisymmetric), \end{cases} \tag{1}$$

where $\kappa_m = \sqrt{\beta^2 - \varepsilon_m}$ and $\kappa_d = \sqrt{\beta^2 - \varepsilon_d}$, β normalized mode wavenumber. Note, that the coordinates normalized to c/ω.

The dispersion relation (1) splits into two separate equations describing symmetric and antisymmetric modes with respect to the magnetic field, due to the symmetry of the structure.

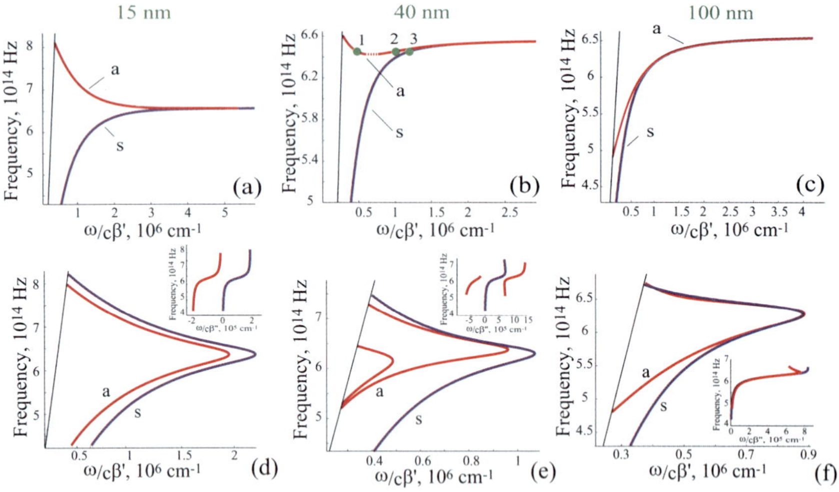

Fig. 2. Dispersion diagrams for slot waveguides with different slot widths. (a–c) lossless case, and (d–f) dispersion diagrams with real losses. Symmetric and antisymmetric modes are marked on the figures with "a" and "s", respectively. Numbered dots on the figure (b) indicate parameters of the modes studied: (1) backward antisymmetric, (2) forward antisymmetric, and (3) symmetric modes. Modes (1-3) co-exist at the same frequency.

Since the dielectric permittivity of metals is complex, $\varepsilon_m = \varepsilon'_m + i\varepsilon''_m$, with imaginary part describing losses in metal, the solutions of dispersion relation (1) are complex as well, $\beta = \beta' + i\beta''$, with β'' corresponding to the plasmon decay rate.

In a standard analysis of the waveguide modes, the losses are usually neglected,[11,12] $\varepsilon''_m \to 0$, and only real parts of the propagation constants, $\beta = \beta'$, are taken into account. Then the losses considered to be small (which is often true in dielectric waveguides), and as a result the absorption leads to the decay of the plasmons, leaving the real part of the wavenumber mostly unchanged. Here we will show that significant losses not only introduce the decay rate for the guided waves, but also dramatically change the dispersion properties of the modes. To demonstrate this, we analyze the simplest case of nonabsorbing metal, i.e. $\varepsilon''_m \equiv 0$, and discuss only propagating waves $\beta = \beta'$. Prade et al.[11] have presented a detailed study of the dispersion relation (1) in this case. Curiously, our analysis of the well-known dispersion relation (1) for different slot widths shows novel features not discussed in the previous works.

To be more specific, we consider dielectric with refractive index $n = 2.5$ sandwiched between thick silver plates having permittivity, ε_m, with zero losses, $\varepsilon''_m \to 0$. We solve dispersion relation taking into account only propagating waves, $\beta'' = 0$, and plot dispersion curves for this structure, see Fig. 2(a-c). For very narrow slot widths, $d = 15$ nm, we observe the regime predicted by Prade et al., where for $\delta \equiv \varepsilon_d/|\varepsilon_m| > 1$ only one antisymmetric mode exists and for $\delta < 1$ only symmetric mode is observed, see Fig. 2(a). For relatively wide slot waveguides, e.g. for 100 nm, we found symmetric and antisymmetric modes [see Fig. 2(c)] which exist only for $|\varepsilon_m| > \varepsilon_d$, as it was predicted earlier.[11] Note that for all slot widths, in case of lossless metal, when the $|\varepsilon_m| = \varepsilon_d$ the plasmon propagation constant for both modes becomes infinitely large, $\beta \to \infty$, corresponding to surface plasmon-polariton resonance, $\omega = \omega_{spp}$,[2]

In the case of a waveguides with the slot width varying in range $35 \lesssim d \lesssim 50$nm we reveal a new regime where three modes, symmetric, antisymmetric and *backward antisymmetric*, coexist as shown in Fig. 2(b), for frequencies below surface plasmon resonance, $\delta > 1$. The backward mode is characterized by a negative slope of the corresponding dispersion curve. The profiles of all three modes, for the frequency corresponding to the free-space wavelength $\lambda = 465$ nm, are presented on the Figs. 1(b-d). The backward antisymmetric mode is less confined to the interfaces, since it has smaller propagation constants, β, and its field is concentrated mainly in metal, rather than in dielectric. The backward nature of this mode is due to the fact that the energy flow in metallic slabs is dominant in relation to dielectric layer energy flow, the analogous condition was discussed in.[20]

Even for zero losses, $\varepsilon''_m = 0$, the dispersion relation generally has not only propagating mode solutions, but also solutions corresponding to *evanescent modes* with $\beta'' \neq 0$ (thus exponentially decaying from the source). Due to a rapid decay of these solutions in lossless structures, corresponding modes are usually not taken into account in the mode analysis of plasmonic waveguides. In plasmonic systems, on the other hand, such modes may play a significant role since their decay rate may become comparable to the decay of the 'main' modes.

As an example, on the Fig. 3(a,b) we present the real and imaginary wavenumbers of the plasmons as a function of frequency for several lower order evanescent modes for the case of 40 nm wide slot waveguide. It is clearly seen that both backward and forward antisymmetric modes have solutions continuing into the complex space of propagation constants. Moreover, the solutions are complex conjugates, and the imaginary propagation constant

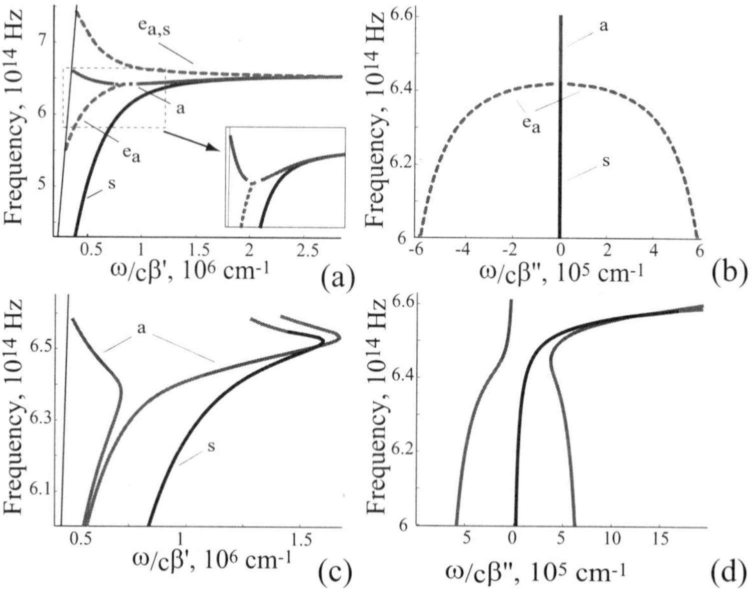

Fig. 3. Dispersion diagrams for 40nm slot waveguide: (a) Dispersion of real part of propagation constant in lossless case, $\varepsilon_m'' = 0$. Dashed line corresponds to evanescent mode marked e_a. Inset shows the magnified dashed region with merging propagating and evanescent modes; (b) Decay factor (imaginary part of wavenumber) dispersion, $\varepsilon_m'' = 0$; (c,d) Same as (a) and (b) but with small losses described by reduced collision frequencies $\gamma_i/10$.

is negative for backward antisymmetric branch and positive for forward branch, which defines backward and forward nature of evanescent waves, see Fig. 3(b). We also observe the increase of decay factors, $|\beta''|$, of evanescent plasmons with the decrease of frequency. Introduction of small losses in the structure clearly shows evolution of corresponding modes, as can be seen in Fig. 3(c,d). All modes now have complex wavenumbers, and antisymmetric mode dispersion curves merge with corresponding evanescent modes.

2.2. *Plasmonic slot waveguides with realistic losses*

Continuing our analysis, we would like to consider realistic losses in the structure. In this case all solutions of the dispersion relation become complex, $\beta = \beta' + i\beta''$, and thus we cannot distinguish the propagating and evanescent modes.

As discussed above, introduction of small losses leads to the merging of the propagating mode solutions with the evanescent branches. One of the antisymmetric branches of evanescent modes has negative β'', see Fig. 3(d). In our convention, negative β'' corresponds to the wave decaying (and thus the energy flowing too) in negative-z direction.

Figure 2(d-f) shows the dispersion curves for the structure with realistic losses. Only lower order modes with smaller imaginary parts of the wavenumber are shown. In contrast to lossless case, when the propagating waves existed either below or above the surface plasmon resonance frequency ω_{spp}, the modes are now spread across the whole frequency domain. In narrow waveguides, i.e. $d \lesssim 35$ nm, for lossy systems two coexisting modes are observed - forward symmetric and backward antisymmetric. For relatively wide slot waveguides, with $d \sim 100$ nm, forward symmetric and antisymmetric modes coexist. For moderate slot widths [Fig. 2(e)] three coexisting modes can be excited: forward symmetric and antisymmetric and backward antisymmetric waves. Note that at lower frequencies counter-propagating antisymmetric waves have similar real parts of wavenumbers, β', which may be used in plasmonic couplers.

For the forward waves above the surface plasmon resonance frequency the losses increase dramatically, whereas the inverse behavior is observed for backward antisymmetric wave - losses are higher at low frequencies, see insets on the Fig. 2(d-f).

3. Nonlinear Effects in Plasmonic Waveguides

Above we have studied only linear properties of plasmons, however nonlinear plasmonic waveguides offer additional possibilities for the control of light at subwavelength scales. Nonlinear plasmons can be excited at the metal-dielectric interfaces where the dielectric possesses the nonlinear response.[12,13,15,17,18] Since only TM electromagnetic waves are supported by a metal-dielectric interface, the corresponding nonlinear Maxwell equations involve two components of the electric field; they cannot be solved analytically in a general case and several approximations and simplifications have been suggested. Among nonlinear problems of interest are quadratic processes, which can be employed for parametric amplification, and cubic nonlinear effects, leading to possible applications in all-optical light control. In this section we study plasmon-to-plasmon frequency conversion in a plasmonic waveguide filled with quadratically nonlinear dielectric, and the processes of light localization in plasmonic structure with Kerr nonlinearity.

3.1. *Plasmonic waveguides with a quadratic nonlinear response*

Parametric processes of frequency conversion are of great interest for optical applications. The main requirement for parametric processes is phase matching which should be satisfied either in bulk or waveguiding structures. Several approaches to achieve efficient phase matching in crystals and waveguides have been suggested and demonstrated, including the use of birefringence and quasi-phase matching. Many of such methods have physical limitations.

One of the recently discussed novel possibilities of phase matching is based on tailoring the mode dispersion in nanoscale waveguides.[21,22] In particular, it was shown that collinear phase matching between TM and TE modes becomes possible for the subwavelength dimensions of a slot waveguide[21] or a high-index guiding slab.[22] However, since a change of polarization is implemented, the effective mode size can not be less than a half of the wavelength. We apply a novel phase-matching scheme[21,22] to the case of plasmonic waveguiding and to study the possibility of plasmon-to-plasmon conversion[23] in planar metal-dielectric-metal nonlinear waveguide, schematically shown on Fig. 1.

To be more specific, we consider silver for the metal (as the material extensively used for plasmonic applications due to its relatively low losses), and lithium niobate $LiNbO_3$ as a nonlinear dielectric (also widely used for second-order optical parametric processes). The dielectric constants of $LiNbO_3$ are $\varepsilon_{dx} = 5.52$ and $\varepsilon_{dz} = 5.06$. Note that in our geometry $LiNbO_3$ has three components of second order nonlinearity tensor, which give rise to nonlinear polarizations leading to frequency conversion.

To reveal the possibility of plasmonic frequency conversion in such structure, first we should analyze the phase matching conditions. We calculate the plasmonic guide indices for the fundamental frequency (FF) and second harmonic (SH) depending on the slot width, Fig. 4(a-b). In Fig. 4(a) we plot the guide index vs. the slot width for free space FF wavelength $\lambda_0 = 840$ nm. For this case, we observe that the exact phase matching between the symmetric and antisymmetric modes becomes possible for a slot of 40 nm. Moreover, in this case the phase matching occurs between the *forward symmetric* and *backward antisymmetric* mode (typical mode profiles are shown on Fig. 1). For the fundamental frequency at $\lambda_0 = 1550$ nm, the linear phase matching occurs for the slot width $a \simeq 320$ nm, and in this case both FF and SH are forward waves.

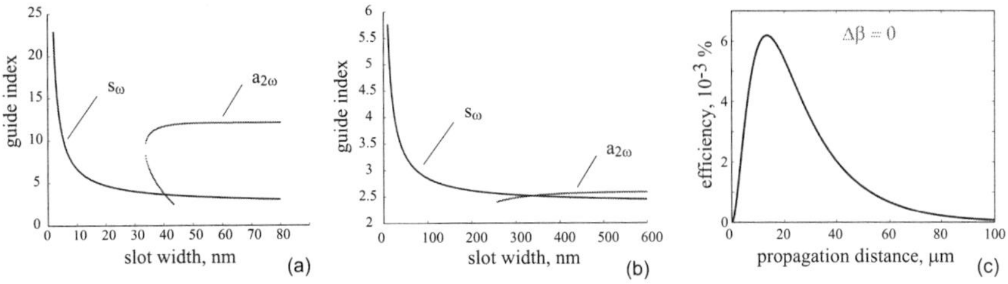

Fig. 4. Plasmonic guide index versus the slot width for different FF wavelengths: (a) $\lambda_0 = 840$ nm and (b) $\lambda_0 = 1550$ nm. Notations "s" and "a" stand for the symmetric and antisymmetric modes, respectively, and the indices ω and 2ω correspond to the fundamental and second harmonics, respectively. And the conversion efficiency with propagation (c),[23]

To describe the interaction of the mode at the fundamental frequency with the mode of the second harmonics, we follow the approach suggested in Ref. 24, and derive the equations for the slowly varying amplitudes of corresponding modes:

$$\frac{dA_1}{dz} = -\alpha_1 A_1 + i\Gamma_1 A_1^* A_2 e^{i\Delta\beta z},$$

$$\frac{dA_2}{dz} = -\alpha_2 A_2 + i\Gamma_2 A_1^2 e^{-i\Delta\beta z} \tag{2}$$

where A_m are the amplitudes of FF and SH modes defined through the expression $\mathbf{E_m} = A_m(z)[\mathbf{x}E_{x0}^{(m)}(x) + \mathbf{z}E_{z0}^{(m)}(x)]e^{i\beta_m z - im\omega t}$, where $E_{x0,z0}^{(m)}$ are the profiles of the linear modes in the guiding structure at the corresponding frequencies, $m = 1, 2$; $\Delta\beta = \beta_2 - 2\beta_1$ is the phase mismatch, and

$$\alpha_m = m\frac{\int \varepsilon''(m\omega)|\mathbf{E}_0^{(m)}|^2 dx}{2\int E_{x0}^{(m)} H_{y0}^{(m)} dx},$$

$$\Gamma_m = 2\pi m\frac{\int [P_{x0}^{(m)} E_{x0}^{(m)} + P_{z0}^{(m)} E_{z0}^{(m)}] dx}{\int E_{x0}^{(m)} H_{y0}^{(m)} dx},$$

where ε'' is the imaginary part of the dielectric permittivity, and $\mathbf{P}_0^{(m)}$ are the corresponding nonlinear polarizations due to mode overlap.

In a sharp contrast to dielectric waveguides, the conversion between the FF symmetric mode and SH antisymmetric mode becomes possible for plasmonic waveguides,[23] Within the undepleted pump approximation, when the amplitude at fundamental frequency is not affected by nonlinearity, and also for the exact phase matching, $\Delta\beta = 0$, we obtain

$$A_1 = A_1(0)\exp(-\alpha_1 z),$$
$$A_2 = \frac{i\Gamma_2 A_1(0)^2}{(\alpha_2 - 2\alpha_1)}\left[\exp(-2\alpha_1 z) - \exp(-\alpha_2 z)\right]$$

Now we introduce the normalized conversion efficiency as follows,

$$\eta = \frac{\text{Power}(2\omega)}{\text{Power}(\omega)} = \frac{|A_2|^2 \int E_{x0}^{(2)} H_{y0}^{(2)}\, dx}{|A_1|^2 \int E_{x0}^{(1)} H_{y0}^{(1)}\, dx}.$$

Dependence of the conversion efficiency on the propagation distance (or the device length) in the case of the exact phase matching is shown in Fig. 4(c). The FF pump power is 10 MW/m, and the free-space wavelength is $\lambda = 1550$ nm. The efficiency curve displays a characteristic interplay between the SH generation and losses, which manifests itself a maximum of the conversion efficiency at $\sim$15 μm.

We note that equations similar to Eqs. (2) can also be derived for the second-order nonlinear processes with plasmons guided by a thin metallic film embedded into nonlinear dielectric media. We have analyzed the phase-matching conditions for this case as well, and have found that, in contrast to the slot waveguide, the lower branch of the dispersion curves describes an antisymmetric mode, whereas the upper branch describes a symmetric mode.[11]

3.2. *Self-focusing in plasmonic structures. Plasmon-soliton generation*

Now we turn our attention to the influence of cubic nonlinearity on surface plasmons. Recently Feigenbaum and Orenstein[16] have presented spatial localization of plasmon waves in the planar metal-dielectric waveguide geometry. However, the surface plasmon polaritons are usually introduced for a single metal-dielectric interface,[2,3] where the analysis of nonlinear localization is more complicated, and it should be carried out by taking into account the boundary conditions, and more importantly, losses which cannot be neglected in plasmonic structures.[25]

We study propagation of a surface plasmon-polariton beam along a metal-nonlinear dielectric interface, as shown in Fig. 5(a). We assume that dielectric is a Kerr-type nonlinear material with the intensity-dependent dielectric permittivity. The dielectric permittivity in the structure can be written in the form: $\varepsilon(x) = \varepsilon_{\text{metal}} + i\varepsilon''$, for $x < 0$, and $\varepsilon_d(x) = \varepsilon_{\text{nln}} \equiv$

$\varepsilon_{\text{lin}} + \alpha|\mathbf{E}|^2$, for $x > 0$, where $\varepsilon_{\text{metal}}$ is the real part of the dielectric permittivity of a metal, ε'' is the imaginary part of dielectric permittivity that describes losses in metal, ε_{lin} is the linear part of the dielectric permittivity of the dielectric, and α is a nonlinear coefficient.

To study the processes of plasmonic beam propagation we consider that the linear plasmon mode profile, not affected by the nonlinearity, and it propagation is described by the slowly varying amplitude, thus the full field can be presented as $\mathbf{E} = A(\mathbf{x}E_{x0} + \mathbf{z}E_{z0})e^{i\beta_m z - im\omega t}$, where A - slowly varying amplitude, β - plasmon propagation constant, and $\mathbf{E}_0 = \mathbf{x}E_{x0} + \mathbf{z}E_{z0}$ - linear mode profile.

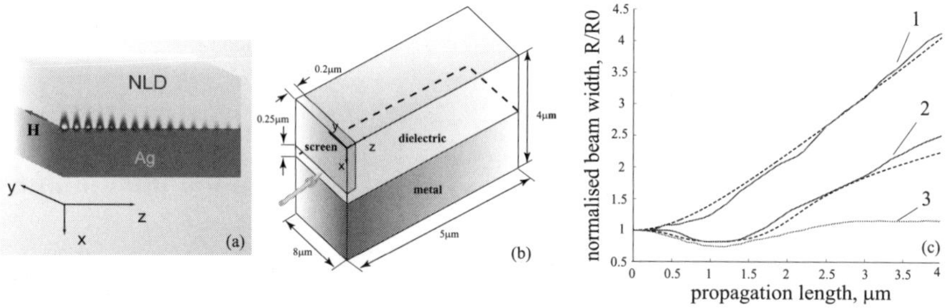

Fig. 5. (a) Schematic of a metal-dielectric interface shown with a transverse structure of the surface plasmon. (b) Geometry used in the numerical FDTD calculations. (c) Evolution of the beam width for: 1 – linear regime, 2 – nonlinear regime with realistic losses, 3 – soliton formation in nonlinear regime with low losses. Dashed lines correspond to the results obtained from Eq. (3).[25]

Substituting this solution into Maxwell equations and providing corresponding averaging over the transverse dimension $\mathbf{x}$, after some simple derivations it is possible to obtain effective nonlinear Schrödinger quation with losses,

$$-2i\beta D\frac{\partial A}{\partial z} + \frac{\partial^2 A}{\partial y^2} + 2I|A|^2 A + i\Gamma A = 0, \tag{3}$$

where $D = \int E_{x0}^2 dx / \int |\mathbf{E}_0|^2 dx$, $\Gamma = \int \varepsilon''(x)|\mathbf{E}_0|^2 dx / \int |\mathbf{E}_0|^2 dx$ and $I = 1/2 \int \alpha(x)|\mathbf{E}_0|^4 dx / \int |\mathbf{E}_0|^2 dx$, $\varepsilon(x)$ is the imaginary part of dielectric constant of metal, $\alpha(x)$ - Kerr nonlinear coefficient. Equation (3) describes the propagation of the plasmon polariton beam localized at the interface and expanding or focusing in the transverse direction $\mathbf{y}$. The similar equation but without losses was derived in Ref. 16 for a nonlinear slot waveguide.

In the linear regime ($I = 0$), Eq. (3) describes plasmon diffraction with the diffraction coefficient $D_{\text{diff}} = 1/2\beta D R_0^2$, where R_0 is the beam width.

In the lossless ($\Gamma = 0$) nonlinear ($I \neq 0$) regime, Eq. (3) has a stationary soliton solution. This solution can be presented as $A = \phi(y)e^{-i\triangle\beta z}$, where $\phi(y)$ satisfies the equation, $\phi_{yy} - 2\beta D\triangle\beta\phi + 2I\phi^3 = 0$, $\triangle\beta$ being a nonlinear correction to the plasmon wavenumber. Finally, the spatial plasmon-polariton soliton profile is $\phi(y) = qI^{-1/2}\text{sech}(q\,y)$, where $q = \sqrt{2\beta D\triangle\beta}$, and the soliton width is $\triangle y \approx 1/q$, which is much larger than the plasmon transverse width. In the lossless case, the soliton peak intensity is found as $|\mathbf{E}|^2 = q^2/I|\mathbf{E}_0|^2$.

In the presence of losses ($\Gamma \neq 0$), Eq. (3) has no stationary solutions, and the soliton can be treated in a generalized sense as a nonlinear localized wave with varying parameters.[26]

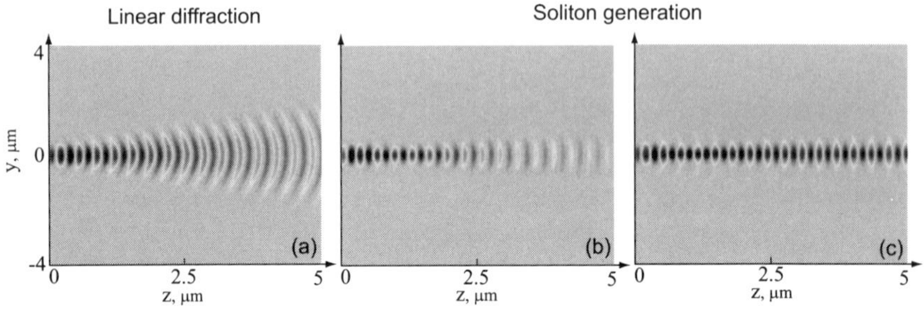

Fig. 6. Magnetic field distribution at the metal-dielectric interface for: (a) Linear diffraction (low power), $R_0 = 0.35$ μm. (b) Self-focusing and soliton generation for $|\mathbf{E}|^2 \simeq 3 \times 10^6 V^2/\mu\text{m}^2$. (c) Same as in (b) but for low losses,[25]

To evaluate the applicability of analytical results, we employ numerical FDTD simulations and study the propagation of plasmonic beams along a metal-dielectric interface for different powers. For the FDTD simulations, we use the commercial software RSoft with Fullwave package implemented at the Australian National Supercomputer Facility.

To excite plasmon-polaritons at the metal-dielectric interface we use a setup shown in Fig. 5(b). The dielectric is chosen to be nonlinear chalcogenite glass with $\varepsilon_{\text{lin}} = 4.84$ and $\chi^{(3)} = 1.4 \times 10^{-19}$ m^2/V^2. We illuminate the structure with a continuous TM polarized light ($\mathbf{H} \uparrow\uparrow \mathbf{y}$) with the free space wavelength $\lambda = 800$ nm. The source is placed in a close proximity to the structure in order to create a relatively narrow plasmonic beam with $R_0 \simeq 0.35$ μm. Plasmons are excited by the light scattered on the metal corner. To avoid exciting plane waves in the dielectric, we place a thin metal screen parallel to the front dielectric facet, 250nm above the metal surface, see Fig. 5(b).

To observe the plasmon beam propagation, we plot the field in the interface plane. Distribution of the magnetic field in this plane is shown in Figs. 6(a-c). Figure 5(a) shows the transverse plasmon structure in the linear regime, where the plasmon diffracts in the propagation direction. Due to losses in metal, the plasmon beam decays while propagating along the surface, the estimated propagation distance is about 5 μm.

At low powers, the plasmon beam diffracts, the diffraction length is about 2.5 μm. For larger powers, e.g. $|\mathbf{E}|^2 \simeq 2 \times 10^6 V^2/\mu\text{m}^2$, we observe the beam narrowing and self-focusing. Further increase of the beam intensity leads to the soliton generation, see Fig. 6(b); this corresponds to the maximum change of the nonlinear index $\alpha|\mathbf{E}|^2\varepsilon^{-1/2} = \varepsilon^{-1/2}3/4\chi^{(3)}|\mathbf{E}|^2 \approx 0.05$, being consistent with the previous nonlinear plasmonic results.[16] In our case, self-focusing is clearly observed at $z = 1$ μm. The generated spatial soliton propagates about 700–800 nm and then starts loosing its power due to dissipation, propagating for about 3 μm. Since the process of the soliton formation is overshadowed by the losses, solely for the presentation purposes we demonstrate the same dynamics at lower value of losses. The results of the corresponding FDTD simulations are presented in Fig. 6(c), where the generated soliton propagates preserving its shape without loosing energy, similar to spatial solitons in optics.[26]

To analyze the effect of losses on the plasmon self-focusing, we trace the variation of the effective beam width, R, and present the results in Fig. 5(c), similar to the physics of magnetic solitons.[27] In the linear regime (curve 1), the plasmon beam diffracts, this is indicated by a monotonic growth of the beam width with the propagation distance. In the nonlinear regime, the beam width decreases reaching a minimum (curve 2), where self-trapping takes place and the spatial plasmon soliton is formed. For longer propagation distances, the soliton broadens due to dissipation.

4. Conclusion

We have studied both linear and nonlinear effects in planar metal-dielectric waveguides. In linear case, we have shown that losses change dramatically the dispersion properties of the guided modes of the system. We have studied both quadratic and cubic nonlinear effects in plasmonic structures. We have analyzed the plasmon-to-plasmon frequency conversion in quadratic nonlinear waveguides, and revealed a possibility of the parametric conversion between the modes with different symmetry. For plasmons excited at the interfaces with the Kerr media, we have studied the process of plasmon self-focusing and plasmon-soliton generation in the presence of losses.

Acknowledgement

The authors acknowledge a support of the Australian Research Council and useful collaboration with S.I. Bozhevolnyi.

References

1. *Surface Plasmon Nanophotonics* Ed. M.L. Brongersma and P.G. Kik, (Springer-Verlag, New York, 2007).
2. S.A. Maier, *Plasmonics: Fundamentals and Applications* (Springer-Verlag, Berlin, 2007).
3. *Plasmonic Nanoguides and Circuits*, Ed. S.I. Bozhevolnyi (Pan Stanford, Singapore, 2009).
4. S.I. Bozhevolnyi, V.S. Volkov, E. Devaux, J.-Y. Laluet, and T.W. Ebbesen, *Nature* **440,** 508 (2006).
5. J.A. Dionne, L.A. Sweatlock, H.A. Atwater, and A. Polman, *Phys. Rev. B* **73,** 035407 (2006).
6. Y. Satuby and M. Orenstein, *Appl. Phys. Lett.* **90,** 251104 (2007).
7. N.-N. Feng and L. Dal Negro, *Opt. Lett.* **32,** 3086 (2007).
8. P. Neutens, P. Van Dorpe, I. De Vlaminck, L. Lagae, and G. Borghs, *Nature Photon.* **3,** 283 (2009).
9. E.N. Economou, *Phys. Rev.* **182,** 539 (1969).
10. J.J. Burke, G.I Stegeman, and T. Tamir, *Phys. Rev. B* **33,** 5186 (1986).
11. B. Prade, J.Y. Vinet, A. Mysyrowicz, *Phys. Rev. B* **44,** 13556 (1991).
12. A.R. Davoyan, I.V. Shadrivov, and Yu.S. Kivshar, *Opt. Express* **16,** 21209 (2008); and *Opt. Express* **17,** 4833 (2009).
13. D. Mihalache, G.I. Stegeman, C.T. Seaton, E.M. Wright, R. Zanoni, A.D. Boardman, and T. Twardowski, *Opt. Lett.* **12,** 187 (1987).
14. A.D. Boardman, A.A. Maradudin, G.I. Stegeman, T. Twardowski, and E.M. Wright, *Phys. Rev. A* **35,** 1159 (1987).
15. A.D. Boardman, G.S. Cooper, A.A. Maradudin, and T.P. Shen, *Phys. Rev. B* **34,** 8273 (1986).
16. E. Feigenbaum and M. Orenstein, *Opt. Lett.* **32,** 674 (2007).
17. J.C. Quail and H.J. Simon, *J. Appl. Phys.* **56,** 2589 (1984)
18. G.I. Stegeman, J.J. Burke, D.G. Hall, *Appl. Phys. Lett.* **41,** 906 (1982).
19. J.A. Dionne, L.A. Sweatlock, H.A. Atwater, and A. Polman, *Phys. Rev. B* **72,** 075405 (2005).
20. E. Feigenbaum, N. Kaminski, M. Orenstein, *Opt. Express* **17,** 18934 (2009).
21. A. Di Falco, C. Conti, and G. Assanto, *Opt. Lett.* **31,** 3146 (2006).
22. H. Ishikawa and T. Kondo, *Appl. Phys. Express* **2,** 042202 (2009).
23. A.R. Davoyan, I.V. Shadrivov, Yu.S. Kivshar, *Opt. Express* **17,** 20063 (2009).
24. Z. Ruan, G. Veronis, K.L. Vodopyanov, M.M. Fejer, and S. Fan, *Opt. Express* **17,** 13502 (2009).
25. A.R. Davoyan, I.V. Shadrivov, Yu.S. Kivshar, *Opt. Express* **17,** 21732 (2009).
26. Yu.S. Kivshar, G.P. Agrawal, *Optical Solitons. From Fibers to Photonic Crystals* (Elsevier Science, USA, 2003).
27. S.O. Demokritov, B. Hillebrands, A.N. Slavin, *Phys. Rep.* **348,** 441 (2001).

COMPUTER ALGEBRA STUDY OF STRUCTURAL AND SYMMETRY PROPERTIES OF DISCRETE DYNAMICAL SYSTEMS

V. V. KORNYAK

Laboratory of Information Technologies, Joint Institute for Nuclear Research, Dubna, Moscow Region 141980, Russia
E-mail: kornyak@jinr.ru

To study structural and symmetry properties of discrete dynamical systems of different types — *deterministic*, *mesoscopic statistical* and *quantum* — we develope various approaches based essentially on the computer algebra and computational group theory methods. We introduce the concept of a system of *discrete relations on an abstract simplicial complex* and develope algorithms for analysis of *compatibility* and construction of *canonical decompositions* of such systems. We describe application of these techniques to some cellular automata. Much attention is paid to study symmetries of the systems. In the case of deterministic systems we reveale some important relations between symmetries and dynamics. In particular, we demonstrate that moving *soliton-like structures* — like "*spaceships*" in cellular automata, "*traveling waves*" in mathematical physics and "*generalized coherent states*" in quantum physics — arise inevitably in *deterministic* dynamical system whose symmetry group splits the set of states into finite number of group orbits. We propose an approach to *quantization of discrete systems* based on introduction of gauge connection with values in unitary representations of some finite *quantizing* groups — the elements of the connection are interpreted as amplitudes of quantum transitions. To study properties of the suggested quantization we introduce a class of simple models — *local quantum models on regular graphs*.

Keywords: Discrete relations, Cellular automata, Symmetries of discrete systems, Discrete gauge principle, Quantization, Computer algebra.

1. Introduction

There are many reasons — physical, mathematical, and conceptual — to study discrete structures. Discrete systems are important in applications — *nanostructures*, for example, by their nature are discrete, not continuous, formations. From a fundamental point of view, there are many philosophical and physical arguments that discreteness better describes physics at small distances than continuity which arises only as approximation or as a logical limit in considering large collections of discrete structures.

We begin with introduction of the concept of a system of *discrete relations on an abstract simplicial complex*.[1,2] This general concept covers many discrete mathematical structures. In particular, it can be considered as generalization of cellular automata or as a set-theoretical analog of systems of polynomial equations. We describe algorithms for analysing *compatibility* and constructing *canonical decompositions* of discrete relations. As an illustration, we describe results of application of the algorithms to some cellular automata, namely, Conway's automaton *Game of Life* and Wolfram's *elementary cellular automata*. For many automata the canonical decomposition allows to obtain either general solutions in closed form or important information on their global behavior.

Symmetry is a property of fundamental importance for any mathematical or physical structure. Symmetries play essential role in the dynamics of the systems. In this paper we consider connection between symmetries of discrete dynamical systems on graphs — one-dimensional simplicial complexes — and their dynamics.[3,4]

In the case of *deterministic dynamical systems*, such as cellular automata, non-trivial connections between the lattice symmetries and dynamics are revealed. In particular, we show that formation of moving soliton-like structures — typical examples are "spaceships" in cellular automata — is a direct result of the existence of non-trivial symmetry.

For *mesoscopic lattice models* (omitted in the present paper) we developed algorithms exploiting symmetries to compute partition functions and search phase transitions.

We consider a class of discrete dynamical models allowing quantum description.[5] Our approach to quantization consists in introduction of gauge connection with values in unitary representation (not necessarily one-dimensional) of some group (of *internal symmetries*) Γ — the elements of the connection are interpreted as amplitudes of quantum transitions. The standard quantization is a special case of this construction — Feynman's path amplitude $\mathrm{e}^{i\int Ldt}$ can be interpreted as parallel transport with values in (one-dimensional) fundamental representation of the group $\Gamma = \mathrm{U}(1)$. For discrete systems it is natural to take a *finite* group as the *quantizing* group, in this case all manipulations — in contrast to the standard quantization — remain within the framework of constructive discrete mathematics requiring no more than the ring of *algebraic integers*. On the other hand, the standard quantization can be approximated by taking one-dimensional representations of large enough finite groups.

2. Discrete Relations on Abstract Simplicial Complexes

Let us consider the Cartesian product $\Sigma^n = \Sigma_1 \times \Sigma_2 \times \cdots \times \Sigma_n$, i.e., the set of ordered n-tuples $(\sigma_1, \sigma_2, \ldots, \sigma_n)$, with $\sigma_i \in \Sigma_i$ for each i, where Σ_i are finite sets of $q_i = |\Sigma_i|$ elements that we shall call *states*. By definition, *n-ary relation* is any subset of the n-dimensional hyperparallelepiped Σ^n.

We can treat n dimensions of the hyperparallelepiped Σ^n as elements of a set of points $X = \{x_1, x_2, \ldots, x_n\}$. To make this initially amorphous set into a "space" (or "space-time") we should provide X with a structure determining how "close" to each other are different points. The relevant mathematical abstraction of such a structure is an *abstract simplicial complex* (see, e.g., Ref. 6). The natural concept of space assumes the homogeneity of its points, i.e., existence of a symmetry group acting transitively on X. The homogeneity is possible only if all Σ_i are equivalent. We denote the equivalence class by Σ and represent it canonically in the form $\Sigma = \{0, \ldots, q-1\}$, $q = |\Sigma|$.

If the number of states is a prime power, $q = p^m$, we can equip the set Σ with the structure of the Galois field $\mathbb{F}_q$. Using the functional completeness of polynomials over finite fields,[7] we can represent any k-ary relation on Σ as a set of zeros of some polynomial belonging to the ring $\mathbb{F}_q[x_1, \ldots, x_k]$. Although this description is not necessary (and does not work, if Σ_i are different sets or q is not prime power), it is useful due to our habit to employ polynomials wherever possible and capability of applying advanced tools of polynomial algebra, such as the Gröbner bases.

2.0.1. *Relations*

We have:

Definition 1 (relation). A *relation* R^δ on the set of points $\delta = \{x_1, \ldots, x_k\}$ is any subset of the hypercube Σ^δ; i.e., $R^\delta \subseteq \Sigma^\delta$. The relation R^δ can be regarded as the Boolean-valued function $R^\delta : \Sigma^\delta \to \{0, 1\}$. We can think of x_i's as variables taking values in Σ and write the relation as

$$a = R^\delta(x_1, \ldots, x_k), \quad a \in \{0, 1\}.$$

An important special case of relations:

Definition 2 (functional relation). A *relation* R^δ on the set of points $\delta = \{x_1, \ldots, x_k\}$ is called *functional* if there is a position $i \in (1, \ldots, k)$ such that for any $\sigma_1, \ldots, \sigma_{i-1}, \sigma_{i+1}, \ldots, \sigma_k, \varsigma, \tau \in \Sigma$ from $(\sigma_1, \ldots, \sigma_{i-1}, \varsigma, \sigma_{i+1}, \ldots, \sigma_k) \in R^\delta$ and $(\sigma_1, \ldots, \sigma_{i-1}, \tau, \sigma_{i+1}, \ldots, \sigma_k) \in R^\delta$ it follows that $\varsigma = \tau$.

In terms of variables the functional relation R^δ can be written in the form

$$x_i = F\left(x_1, \ldots, x_{i-1}, x_{i+1}, \ldots, x_k\right), \quad \text{where} \quad F : \Sigma^{\delta \setminus \{x_i\}} \to \Sigma.$$

We need to be able to extend relations from subsets of points to larger sets:

Definition 3 (extension of relation). For given set of points δ, its subset $\tau \subseteq \delta$ and relation R^τ on the subset τ, we define the *extension* of R^τ as the relation

$$R^\delta = R^\tau \times \Sigma^{\delta \setminus \tau}.$$

This definition, in particular, allows the relations $R^{\delta_1}, \ldots, R^{\delta_m}$ defined on different domains to be extended to the common domain, i.e., to the union $\delta_1 \cup \cdots \cup \delta_m$.

Logical implications of the relations are defined in a natural way:

Definition 4 (consequence of relation). A relation Q^δ is called a *consequence* of the relation R^δ if $R^\delta \subseteq Q^\delta \subseteq \Sigma^\delta$; i.e., Q^δ is arbitrary *superset* of the set R^δ.

The relation R^δ may have many different consequences: their total number (including R^δ itself and the trivial relation Σ^δ) is evidently equal to $2^{|\Sigma^\delta| - |R^\delta|}$.

It is natural to single out the consequences that can be reduced to relations on smaller sets of points:

Definition 5 (proper consequence). A *nontrivial* relation Q^τ is called the *proper consequence* of the relation R^δ if τ is a *proper* subset of δ (i.e., $\tau \subset \delta$) and the relation $Q^\tau \times \Sigma^{\delta \setminus \tau}$ is a consequence of R^δ.

We call relations that have no proper consequences the *prime relations.*

2.0.2. *Compatibility of Systems of Relations*

The compatibility of a system of relations can naturally be defined by the intersection of their extensions to the common domain:

Definition 6 (base relation). The *base relation* of the system of relations $R^{\delta_1}, \ldots, R^{\delta_m}$ is the relation

$$R^\delta = \bigcap_{i=1}^{m} R^{\delta_i} \times \Sigma^{\delta_i \setminus \delta}, \quad \text{where} \quad \delta = \bigcup_{i=1}^{m} \delta_i.$$

Let us make two comments for the polynomial case $q = p^n$, where the standard tool for the compatibility analysis is the Gröbner basis method: (a) The compatibility condition determined by the *base relation* can be represented by a *single* polynomial, unlike the Gröbner basis, which is normally a system of polynomials. (b) Any possible Gröbner basis of polynomials representing the relations $R^{\delta_1}, \ldots, R^{\delta_m}$ corresponds to some combination of consequences of the *base relation*.

2.0.3. *Decomposition of Relations*

If a relation has proper consequences, we can try to express it as far as possible in terms of these consequences, i.e., relations on smaller sets of points. To this end we introduce

Definition 7 (canonical decomposition). The *canonical decomposition* of a relation R^δ with proper consequences $Q^{\delta_1}, \ldots, Q^{\delta_m}$ is the relation

$$R^\delta = P^\delta \bigcap \left(\bigcap_{i=1}^{m} Q^{\delta_i} \times \Sigma^{\delta \setminus \delta_i} \right), \qquad (1)$$

where the factor P^δ is defined by the following

Definition 8 (principal factor). The *principal factor* of the relation R^δ with proper consequences $Q^{\delta_1}, \ldots, Q^{\delta_m}$ is the relation

$$P^\delta = R^\delta \bigcup \left(\Sigma^\delta \setminus \bigcap_{i=1}^{m} Q^{\delta_i} \times \Sigma^{\delta \setminus \delta_i} \right).$$

The principal factor is the maximally "free" — i.e., the closest to the trivial — relation that, together with the proper consequences, makes it possible to recover the initial relation.

If the principal factor in the canonical decomposition is trivial, the relation is completely reduced to relations on smaller sets of points.

Definition 9 (reducible relation). A relation R^δ is said to be *reducible* if it can be represented as

$$R^\delta = \bigcap_{i=1}^{m} Q^{\delta_i} \times \Sigma^{\delta \setminus \delta_i}, \qquad (2)$$

where δ_i are *proper subsets* of δ.

This definition makes it possible to impose a "topology" — i.e., the structure of an abstract simplicial complex with the corresponding theories of homologies, cohomologies, etc. — on an *arbitrary* n-ary relation $R \subseteq \Sigma^n$. This is achieved by (1) naming the dimensions of the hypercube Σ^n as the "points" $x_1, \ldots, x_n \in X$; (2) decomposing R (which can now be denoted by R^X) into *irreducible* components; and (3) defining the *maximal simplices* of the set Δ as the *domains* of irreducible components of R^X.

2.0.4. *On Representation of Relations in Computer*

To specify a k-ary relation R^k we should mark its points within the k-dimensional hypercube Σ^k, i.e., define a *characteristic function* $\chi : \Sigma^k \to \{0,1\}$, with $\chi(\vec{\sigma}) = 1$ or 0 according as $\vec{\sigma} \in R^k$ or $\vec{\sigma} \notin R^k$. Here $\vec{\sigma} = (\sigma_0, \sigma_1, \ldots, \sigma_{k-1})$ is a point of the hypercube. The simplest way to implement the characteristic function is to enumerate all the q^k hypercube points in some, e.g., lexicographic, order. Then the relation can be represented by a string of q^k bits $\alpha_0\alpha_1 \cdots \alpha_{q^k-1}$ in accordance with the table below. We call this string *bit table* of relation. Symbolically $\mathrm{BitTable}\,[\,i_{\vec{\sigma}}\,] := \left(\vec{\sigma} \in R^k\right)$.

σ_0	σ_1	$\ldots$	σ_{k-2}	σ_{k-1}	$i_{\vec{\sigma}}$	$\chi(\vec{\sigma})$
0	0	$\ldots$	0	0	0	α_0
1	0	$\ldots$	0	0	1	α_1
$\vdots$	$\vdots$	$\ldots$	$\vdots$	$\vdots$	$\vdots$	$\vdots$
$q-1$	$q-1$	$\ldots$	$q-1$	$q-1$	q^k-1	α_{q^k-1}

Note that $\vec{\sigma}$ is the ("little-endian") representation of the number $i_{\vec{\sigma}}$ in the radix q:

$$i_{\vec{\sigma}} = \sigma_0 + \sigma_1 q + \cdots + \sigma_i q^i + \cdots + \sigma_{k-1} q^{k-1}.$$

The characteristic function (bit table) can be represented as the *binary* integer

$$\chi = \alpha_0 + \alpha_1 2 + \cdots + \alpha_i 2^i + \cdots + \alpha_{q^k-1} 2^{q^k-1}. \tag{3}$$

Manipulations with relations are reduced to very efficient bitwise computer commands.

2.1. *Illustration: application to some cellular automata*

2.1.1. *J. Conway's game of life*

The "Life family" is a set of 2-dimensional, binary (i.e., $\Sigma = \{0,1\}$; $q = 2$) cellular automata similar to ***Conway's Life***, which rule is defined on 9-cell (3×3) Moore neighborhood and is described as follows. A cell is "born" if it has exactly 3 "alive" neighbors, "survives" if it has 2 or 3 such neighbors, and "dies" otherwise. This rule is symbolized in terms of the "birth"/"survival" lists as B3/S23. Another examples of automata from this family are ***HighLife*** (the rule B36/S23), and ***Day&Night*** (the rule B3678/S34678).

Generalizing this type of local rules, we define a k*-valent Life rule* as a *binary* rule on a k-valent neighborhood (we adopt that x_{k+1} is the central cell and $x_1, \ldots, x_k$ are neighbors of x_{k+1}) described by two *arbitrary* subsets of the set $\{0, 1, \ldots, k\}$. These subsets $B, S \subseteq \{0, 1, \ldots, k\}$ contain conditions for the one-time-step transitions $x_{k+1} \to x'_{k+1}$ of the forms $0 \to 1$ and $1 \to 1$, respectively. Since the number of subsets of any finite set A is $2^{|A|}$, the number of different rules defined by two sets B and S is equal to

$$N_{B/S,\,k} = 2^{2k+2}. \tag{4}$$

There is another way to characterize this type of local rules. Let us consider k-valent rules symmetric with respect to the group $\mathrm{Sym}\,(k)$ of all permutations of k outer points of the neighborhood. We shall call such rules k-symmetric. It is not difficult to count the total number of different q-ary k-symmetric rules:

$$N_{q,\,\mathrm{Sym}(k)} = q^{\binom{k+q-1}{q-1}q}. \tag{5}$$

We see that (5) evaluated at $q = 2$ coincides with (4), i.e., $N_{2,\,\mathrm{Sym}(k)} = N_{B/S,\,k}$. Since k-valent Life rules are obviously k-symmetric we have the following:

Proposition. *The sets of* k*-symmetric binary rules and* k*-valent Life rules coincide.*
I.e., any k-symmetric binary rule can be represented as B/S lists.

The local relation of ***Conway's Life*** automaton R^{δ}_{CL} is defined on the 10-set $\delta = \{x_1, \ldots, x_{10}\}$, where the point $x_{10} \equiv x'_9$ is the next time step of the point x_9. By construction, elements of the 10-dimensional hypercube Σ^{δ} belong to the relation of ***Conway's Life*** automaton, i.e., $(x_1, \ldots, x_{10}) \in R^{\delta}_{CL}$, in the following cases:
(1) $\left(\sum_{i=1}^{8} x_i = 3\right) \wedge (x_{10} = 1)$; (2) $\left(\sum_{i=1}^{8} x_i = 2\right) \wedge (x_9 = x_{10})$; (3) $x_{10} = 0$, otherwise. The number of elements of the relation R^{δ}_{CL} is $\left|R^{\delta}_{CL}\right| = 512$. The relation R^{δ}_{CL}, as is the case for any cellular automaton, is *functional*: the state of x_{10} is uniquely determined by the states of other points. We accompany the below analysis of the structure of R^{δ}_{CL} by description in terms of polynomials from $\mathbb{F}_2\left[x_1, \ldots, x_{10}\right]$. This is done only for illustrative purposes and for comparison with the Gröbner basis method. In fact, we transform the relations into polynomials only for output.

The polynomial representing R^{δ}_{CL} takes the form

$$P_{CL} = x_{10} + x_9\left(\Pi_7 + \Pi_6 + \Pi_3 + \Pi_2\right) + \Pi_7 + \Pi_3, \tag{6}$$

where $\Pi_k \equiv \Pi_k(x_1, \ldots, x_8)$ is the kth *elementary symmetric polynomial* defined as

$$\Pi_k(x_1, \ldots, x_n) = \sum_{1 \le i_1 < i_2 < \cdots < i_k \le n} x_{i_1} x_{i_2} \cdots x_{i_k}.$$

Hereafter, we will use the following notation: $\Pi_k \equiv \Pi_k(x_1, \ldots, x_8)$, $\Pi_k^i \equiv \Pi_k(x_1, \ldots, \widehat{x_i}, \ldots, x_8)$, $\Pi_k^{ij} \equiv \Pi_k(x_1, \ldots, \widehat{x_i}, \ldots, \widehat{x_j}, \ldots, x_8)$. Applying the computer program to R_{CL}^{δ}, we find that the relation R_{CL}^{δ} is *reducible* and has the decomposition

$$R_{CL}^{\delta} = R_2^{\delta \setminus \{x_9\}} \times \Sigma^{\{x_9\}} \bigcap \left(\bigcap_{k=1}^{7} R_1^{\delta \setminus \{x_{i_k}\}} \times \Sigma^{\{x_{i_k}\}} \right), \tag{7}$$

where $(i_1, \ldots, i_7)$ is arbitrary 7-element subset of the set $(1, \ldots, 8)$.
The eight relations $R_1^{\delta \setminus \{x_i\}}$ have the following polynomial form:

$$x_9 x_{10}(\Pi_6^i + \Pi_5^i + \Pi_2^i + \Pi_1^i) + x_{10}(\Pi_6^i + \Pi_2^i + 1) + x_9(\Pi_7^i + \Pi_6^i + \Pi_3^i + \Pi_2^i) = 0.$$

The relation $R_2^{\delta \setminus \{x_9\}}$ has the form: $x_{10}(\Pi_7 + \Pi_6 + \Pi_3 + \Pi_2 + 1) + \Pi_7 + \Pi_3 = 0$.
The relations $R_1^{\delta \setminus \{x_i\}}$ and $R_2^{\delta \setminus \{x_9\}}$ are *not prime*, and can be decomposed. Continuing the decomposition iterations, we finally obtain the following system of relations:

$$x_9 x_{10}\left(\Pi_2^i + \Pi_1^i\right) + x_{10}\left(\Pi_2^i + 1\right) + x_9\left(\Pi_7^i + \Pi_6^i + \Pi_3^i + \Pi_2^i\right) = 0, \tag{8}$$

$$x_{10}(\Pi_3 + \Pi_2 + 1) + \Pi_7 + \Pi_3 = 0, \tag{9}$$

$$(x_9 x_{10} + x_{10})\left(\Pi_3^{ij} + \Pi_2^{ij} + \Pi_1^{ij} + 1\right) = 0, \tag{10}$$

$$x_{10}\left(\Pi_3^i + \Pi_2^i + \Pi_1^i + 1\right) = 0, \tag{11}$$

$$x_{10} x_{i_1} x_{i_2} x_{i_3} x_{i_4} = 0. \tag{12}$$

One can easily interpret the simplest relations (12): if the point x_{10} is in the state 1, then at least one point in any set of four points surrounding x_9 must be in the state 0.

The above analysis of the relation R_{CL}^{δ} takes < 1 sec on a 1.8GHz AMD Athlon notebook with 960Mb.

To compute the Gröbner basis we must add to polynomial (6) ten polynomials

$$x_i^2 + x_i, \; i = 1, \ldots, 10$$

corresponding to the relation $x^q = x$ that holds for all elements of any finite field $\mathbb{F}_q$.

Computation of the Gröbner basis over $\mathbb{F}_2$ with the help of **Maple 9** gives the following: (a) pure lexicographic order with variable ordering $x_{10} \succ x_9 \succ \cdots \succ x_1$ does not provide any new information leaving initial polynomial (6) unchanged; (b) pure lexicographic order with variable ordering $x_1 \succ x_2 \succ \cdots \succ x_{10}$ reproduces relations (8)—(12) (modulo several polynomial reductions violating the symmetry of polynomials); the computation takes 1 h 22 min; (c) degree-reverse-lexicographic order also reproduces system (8)—(12) (same comment as above); the times are: 51 min for the variable ordering $x_1 \succ x_2 \succ \cdots \succ x_{10}$, and 33 min for the ordering $x_{10} \succ x_9 \succ \ldots \succ x_1$.

2.1.2. *Elementary cellular automata*

Simplest binary, nearest-neighbor, one-dimensional cellular automata were named *elementary cellular automata* by S. Wolfram, who has extensively studied their properties.[8] A large collection of results concerning these automata is presented in Wolfram's online atlas.[9] In the exposition below we use Wolfram's notations and terminology. The elementary cellular automata are simpler than ***Conway's Life***, and we may pay more attention to the topological aspects of our approach.

Local rules of the elementary cellular automata are defined on the 4-set $\delta = \{p, q, r, s\}$ which can be pictured by the icon 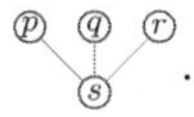. A local rule is a binary function of the form $s = f(p, q, r)$. There are totally $2^{2^3} = 256$ such functions.

Our computation with relations representing the local rules shows that the total number 256 of them is divided into 118 reducible and 138 irreducible relations. Only two of the irreducible relations appeared to be prime, namely, the rules 105 and 150 in Wolfram's numeration. This numeration is based on the "big-endian" — i.e., opposite to our convention (3) — representation of binary numbers.

We consider the elementary automata on a space-time lattice with integer coordinates (x, t), i.e., $x \in \mathbb{Z}$ or $x \in \mathbb{Z}_m$ (spatial m-periodicity), $t \in \mathbb{Z}$. We denote a state of the point on the lattice by $u(x, t) \in \Sigma = \{0, 1\}$. Generally the points are connected as is shown:

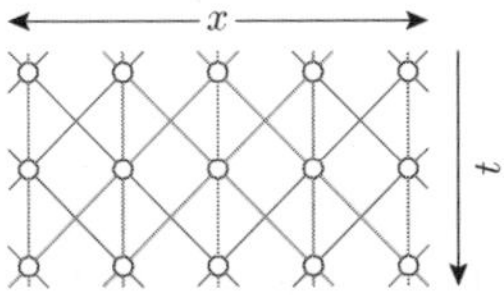

.

Reducible Automata. Some automata with reducible local relations can be represented as unions of automata defined on disconnected subcomplexes:

- Two automata 0 and 255 are determined by unary relations $s = 0$ and $s = 1$ on the disconnected set of points. Note that unary relations are usually called *properties*.
- Six automata 15, 51, 85, 170, 204 and 240 are, in fact, disjoint collections of spacially zero-dimensional automata, i.e., single cells evolving in time. As an example, let us consider the automaton 15. The local relation is defined on the set 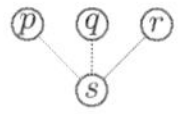 and its bit table is 0101010110101010. This relation is reduced to the relation on the face 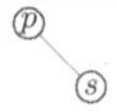 with bit table 0110. The spacetime lattice is split in the following way:

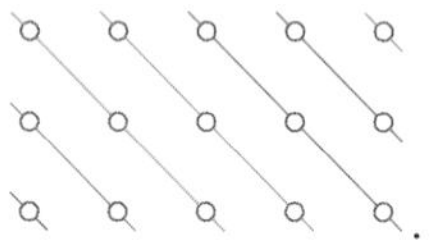

The bit table 0110 means that the points p and s can be only in opposite states, and we can write immediately the general solution for the automaton 15:

$$u(x,t) = a(x-t) + t \mod 2,$$

where $u(x,0) \equiv a(x)$ is an arbitrary initial condition.
- Each of the ten automata 5, 10, 80, 90, 95, 160, 165, 175, 245, 250 is decomposed into two identical automata. As an example let us consider the rule 90. This automaton is distinguished as producing the fractal of topological dimension 1 and Hausdorff dimension $\ln 3/\ln 2 \approx 1.58$ known as the *Sierpinski sieve* (or *gasket* or *triangle*). Its local relation on the set 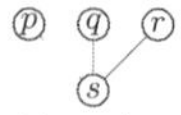 is represented by the bit table 1010010101011010. The relation is reduced to the relation with the bit table

$$10010110 \quad \text{on the face} \quad \tag{13}$$
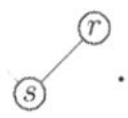

It follows from the structure of face (13) that the spacetime lattice is split into two identical independent complexes as is shown

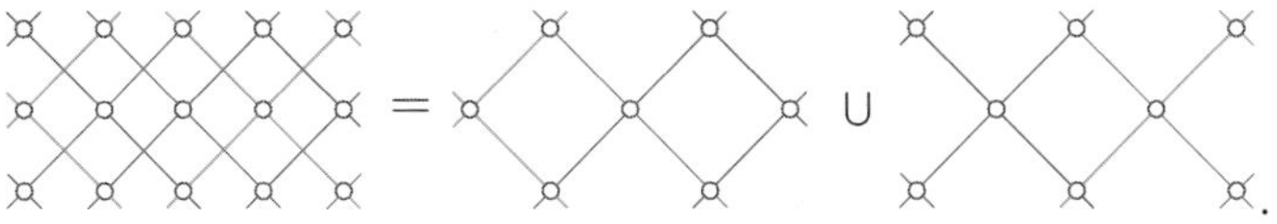

To find a general solution of the automaton 90 it is convenient to use the polynomial form of relation (13) $s+p+r=0$. With this linear expression, the general solution is easily constructed:

$$u(x,t)=\sum_{k=0}^{t}\binom{t}{k}a(x-t+2k) \mod 2, \qquad u(x,0)\equiv a(x).$$

Using Proper Consequences. Proper consequences — even if they are not functional — can provide useful information on the behavior of a cellular automaton.

For example, 64 automata (with both reducible and irreducible local relations) have proper consequences with the bit table

$$1101 \tag{14}$$

on, at least, one of the faces

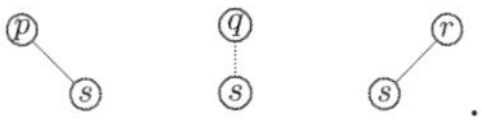

(15)

The algebraic forms of relation (14) on faces (15) are $ps+s=0$, $qs+s=0$, $rs+s=0$, respectively. Relation (14) is *not functional*, and hence can not describe any deterministic evolution. Nevertheless, it imposes severe restrictions on the behavior of the automata. The features of the behavior resulting from relation (14) are clearly seen from many of computational results presented in the atlas.[9] A typical pattern from this atlas is reproduced in Fig. 1, where several evolutions of the automaton 168 are presented. In the figure, 0's and 1's are denoted by the empty and filled square cells, respectively. Note that the authors of the figure use a spatially periodic condition: $x\in\mathbb{Z}_{30}$.

The local relation of the automaton 168 — its polynomial form is $pqr+qr+pr+s=0$ — has the proper consequence $rs+s=0$. Relation (14) means that if, say r, as for the rule 168, is in the state 1 then s may be in both states 0 or 1, but if the state of r is 0, then the state of s must be 0:

$$r=1\Rightarrow s=0\vee s=1,$$

$$r=0\Rightarrow s=0.$$

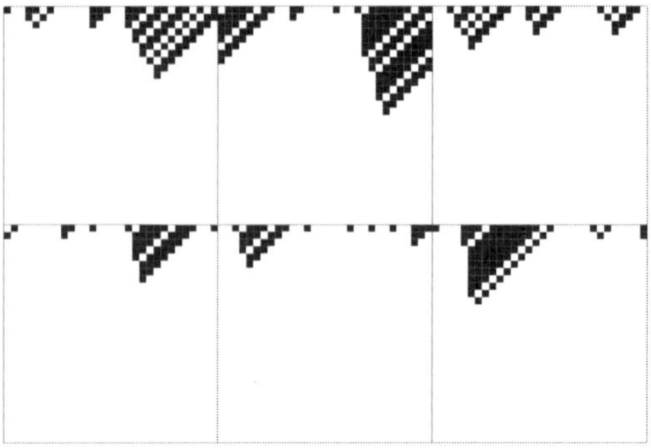

Fig. 1. Rule 168. Several random initial conditions.

One can see that all evolutions in Fig. 1 consist of diagonals $x = x_0 - t$ directed leftward and downward. Each diagonal begins with a several units, but after the first appearance of zero all subsequent points along the diagonal are zeros.

3. Discrete Dynamical Systems with Symmetries

Generally, a dynamical system is a set S of distinguishable states evolving in time. Very often the set S has a special structure $\mathrm{S} = \Sigma^X$, i.e., it decomposes into a 'space' X and a set Σ of 'values' of functions on the space. We will denote by G and Γ the symmetry groups of the sets X and Σ, respectively. These groups are obviously subgroups of the full — i.e., without any restrictions imposed by possible structures of the sets X and Σ — symmetric groups of the sets: $\mathrm{G} \leq \mathrm{Sym}(X)$ and $\Gamma \leq \mathrm{Sym}(\Sigma)$. Further we will refer to the groups G and Γ as the *space* and *internal* symmetries, respectively.

3.1. *Symmetries of lattices*

Let the space X be a finite set of points: $X = \{x_1, \ldots, x_{N_X}\}$. We provide this set with the structure of one-dimensional complex, i.e., *graph*. Graphs — we shall call them also *lattices* — are sufficient for all our purposes. In particular, they are adequate to introduce the gauge and quantum structures. The symmetry group of the space X is the *graph automorphism group* $\mathrm{G} = \mathrm{Aut}(X)$.

Very often dynamics of a model is expressed in terms of rules defined on the neighborhoods of lattice vertices. For models of this sort with locally defined evolution rules — typical examples are cellular automata and Ising model — the group of *local symmetries* $\mathrm{G_{loc}}$ is essential. This group is a subgroup of the space symmetry group: $\mathrm{G_{loc}} \leq \mathrm{G}$, and is defined as the *stabilizer* of a vertex x in the space group G: $g \in \mathrm{G_{loc}} = \mathrm{Stab_G}(x)$ means $xg = x$. Local rules are defined on *orbits* of $\mathrm{G_{loc}}$ on *edges* from the *neighborhoods* of points x.

Let us consider the role of the local group $\mathrm{G_{loc}}$ in more detail using the *buckyball* 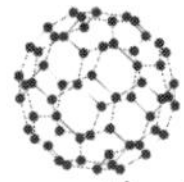 as an example. The incarnations of this 3-valent graph include in particular:

- the *Caley graph of the icosahedral group* Alt(5) (in mathematics);
- the *molecule of fullerene* C_{60} (in carbon chemistry).

The symmetry group of the buckyball is $\mathrm{G} = \mathrm{Aut}(X) = \mathbb{Z}_2 \times \mathrm{Alt}(5)$. The neighborhood of a vertex x_i takes the form 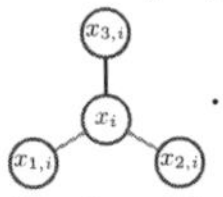

. The stabilizer of x_i is $\mathrm{G_{loc}} = \mathrm{Stab_G}(x_i) = \mathbb{Z}_2$. The set of neighborhood edges contains three elements:

$$E_i = \{e_{1,i} = (x_i, x_{1,i}), e_{2,i} = (x_i, x_{2,i}), e_{3,i} = (x_i, x_{3,i})\}.$$

The set of orbits of $\mathrm{G_{loc}}$ on E_i consists of two orbits:

$$\Omega_i = \{\omega_{1,i} = \{e_{1,i}, e_{2,i}\}, \omega_{2,i} = \{e_{3,i}\}\},$$

i.e., the stabilizer does not move the edge $(x_i, x_{3,i})$ and swaps $(x_i, x_{1,i})$ and $(x_i, x_{2,i})$. This asymmetry results from different roles the edges play in the structure of the buckyball: $(x_i, x_{1,i})$ and $(x_i, x_{2,i})$ are edges of a pentagon adjacent to x_i , whereas $(x_i, x_{3,i})$ separates two hexagons; in the carbon molecule C_{60} the edge $(x_i, x_{3,i})$ corresponds to the double bond, whereas others are the single bonds.

Naturally formulated local rules determining behavior of a system must respect decompositions of neighborhoods into group orbits. For example, the Hamiltonian of the Ising model on the buckyball must depend on two, generally different, coupling constants J_{12} and J_3. Moreover, the coupling constants may be of different types — ferromagnetic or antiferromagnetic — and this may lead to interesting behavior of the model. Such natural Hamiltonian should take the form

$$H_{bucky} = -\frac{1}{2}\sum_i s_i \left[J_{12}(s_{1,i} + s_{2,i}) + J_3 s_{3,i}\right] - B\sum_i s_i\,, \tag{16}$$

where $s_i, s_{1,i}, s_{2,i}, s_{3,i} \in \Sigma = \{-1, 1\}$. In a similar way the local rule for a cellular automaton on the buckyball must have the form

$$x_i' = f\left(x_i, x_{1,i}, x_{2,i}, x_{3,i}\right) ,$$

where function f must be symmetric with respect to variables $x_{1,i}$ and $x_{2,i}$, i.e.,

$$f\left(x_i, x_{1,i}, x_{2,i}, x_{3,i}\right) \equiv f\left(x_i, x_{2,i}, x_{1,i}, x_{3,i}\right) .$$

3.2. *Unification of space and internal symmetries*

Having the groups G and Γ acting on X and Σ, respectively, we can combine them into a single group $\mathrm{W} \leq \mathrm{Sym}\left(\Sigma^X\right)$ which acts on the states Σ^X of the whole system. The group W can be identified, as a set, with the *Cartesian product* $\Gamma^X \otimes \mathrm{G}$, where Γ^X is the set of Γ-valued functions on X. That is, every element $u \in \mathrm{W}$ can be represented in the form $u = (\alpha(x),\ a)$, where $\alpha(x) \in \Gamma^X$ and $a \in \mathrm{G}$. *A priori* there are many possible ways to combine G and Γ into a single group. So selection of possible combinations should be guided by some natural (physical) considerations. Such considerations convince that the required combination W is described by an equivalence class of *split extensions*

$$1 \to \Gamma^X \to \mathrm{W} \to \mathrm{G} \to 1 \tag{17}$$

determined by an *antihomomorphism* $\mu : \mathrm{G} \to \mathrm{G}$. The equivalence is described by *arbitrary* function $\kappa : \mathrm{G} \to \mathrm{G}$. The explicit formulas for main group operations — *action* on Σ^X, *multiplication* and *inversion* — are

$$\sigma(x)\left(\alpha\left(x\right),\ a\right) = \sigma\left(x\mu(a)\right)\alpha\left(x\kappa(a)\right), \tag{18}$$

$$\left(\alpha\left(x\right),\ a\right)\left(\beta\left(x\right),\ b\right) = \left(\alpha\left(x\kappa(ab)^{-1}\mu(b)\kappa(a)\right)\beta\left(x\kappa(ab)^{-1}\kappa(b)\right),\ ab\right), \tag{19}$$

$$\left(\alpha(x),\ a\right)^{-1} = \left(\alpha\left(x\kappa\left(a^{-1}\right)^{-1}\mu(a)^{-1}\kappa(a)\right)^{-1},\ a^{-1}\right). \tag{20}$$

3.3. *Basic constituents of discrete dynamical models*

The following constructions form the basis for all types of dynamical systems we consider in this paper:

(1) *Space* X is a k-valent graph with symmetry group $\mathrm{G} = \mathrm{Aut}(X)$ — *space symmetries*.
(2) Vertices x of X take values in a set Σ with symmetry group $\Gamma \leq \mathrm{Sym}\left(\Sigma\right)$ — *internal symmetries*.

(3) *States of the whole system* are functions $\sigma(x) \in \Sigma^X$.
(4) We define *whole system symmetry* groups W — unifying space G and internal Γ symmetries — as equivalence classes of split group extensions of the form (17).
(5) Action of the group W splits the set of states Σ^X into *orbits* of different sizes:

$$\Sigma^X = \bigsqcup_i O_i.$$

(6) Evolution proceeds in *discrete time* $t \in \mathbb{Z} = \{\ldots, -1, 0, 1, \ldots\}$.
(7) *Dynamics* is determined by some *evolution rule* connecting the current state $\sigma_t(x)$ of the system with its prehistory $\sigma_{t-1}(x)$, $\sigma_{t-2}(x)$, $\sigma_{t-3}(x), \ldots$

3.4. *Deterministic and non-deterministic dynamics*

Evolution rule of *deterministic* (or *causal*) dynamical system is *functional relation*. This means that the current state is a function of the prehistory:

$$\sigma_t\left(x\right) = F\left(\sigma_{t-1}\left(x\right), \sigma_{t-2}\left(x\right), \sigma_{t-3}\left(x\right), \ldots\right). \tag{21}$$

Cellular automaton is a typical example of deterministic dynamical system. The causality imposes several restrictions on the system dynamics.[4] In particular, for the first order functional relations:

- *dynamical trajectories* pass group orbits in *non-decreasing* order of orbit sizes,
- *periodic* trajectories lie within orbits of the *same size*.

Mesoscopic lattice model is a sort of non-deterministic dynamical system. This is a special case of *Markov chain*. In mesoscopic lattice model transition from one state to any other is possible with probability controlled by a Hamiltonian.

Quantum system is another important type of non-deterministic dynamical system. The probabilities of transitions between states are expressed in terms of complex-valued transition amplitudes.

4. Soliton-like Structures in Deterministic Dynamics

Let us begin with a simple example. Consider a cube X whose vertices take values in two-element set, say $\Sigma = \{0, 1\}$. The 48-element symmetry group of a cube has the structure $G = \mathbb{Z}_2 \times \mathrm{Sym}\left(4\right)$. The group is generated by 3 elements:

- p — 120° rotation around diagonal of the cube;
- q — 90° rotation around axis passing through the centers of opposite cube faces;
- r — reflection interchanging opposite faces of the cube.

Total number of states of the model is $\left|\Sigma^X\right| = 2^8 = 256$. If we assume that the group Γ is trivial, then $\mathrm{W} = \Gamma^X \rtimes \mathrm{G} = 1 \rtimes \mathrm{G} \cong \mathrm{G}$. The group W decomposes the set Σ^X into 22 orbits in accordance with the table:

Size of orbits	1	2	4	6	8	12	24
Number of orbits	2	1	2	2	5	4	6

.

Let us consider a *deterministic* dynamical system on the cube, namely, symmetric binary 3-valent cellular automaton with rule 86. The number 86 is the "little endian" representation of the bit string 01101010 taken from the last column of the rule table with Sym (3)-symmetric combinations of values for $x_{1,i}, x_{2,i}, x_{3,i}$

$x_{1,i}$	$x_{2,i}$	$x_{3,i}$	x_i	x'_i
0	0	0	0	0
0	0	0	1	1
1	0	0	0	1
1	0	0	1	0
1	1	0	0	1
1	1	0	1	0
1	1	1	0	1
1	1	1	1	0

.

Here x_i is value of ith vertex of the cube; $x_{1,i}, x_{2,i}, x_{3,i}$ are values of the cube vertices adjacent to the ith one and x'_i is the next time value of ith vertex. The rule can also be represented in ***Conway's Life*** style "Birth"/"Survival" notation as B123/S0, or as polynomial over the field $\mathbb{F}_2$

$$x'_i = x_i + \Pi_3 + \Pi_2 + \Pi_1,$$

where $\Pi_1 = x_{1,i} + x_{2,i} + x_{3,i}$, $\Pi_2 = x_{1,i}x_{2,i} + x_{1,i}x_{3,i} + x_{2,i}x_{3,i}$, $\Pi_3 = x_{1,i}x_{2,i}x_{3,i}$ are elementary symmetric functions.

The phase portrait of the automaton is shown in Fig. 2. In the figure the group orbits are represented by circles. The ordinal numbers[a] of orbits are placed within these circles. The numbers over orbits and within cycles are sizes of the orbits (recall that all orbits belonging to one cycle have the same size — see Section 3.4). The rational number p indicates the *weight* of

[a]These ordinal numbers are specified by the computer program.

the corresponding element of the phase portrait. In fact, p is a probability for randomly chosen state to appear in an isolated cycle or to be caught by an attractor: $p = (\textit{size of basin})/(\textit{total number of states})$. Here *size of basin* is sum of sizes of orbits involved in the struture. The structures in Fig. 2 are placed in the decreasing order of their weights.

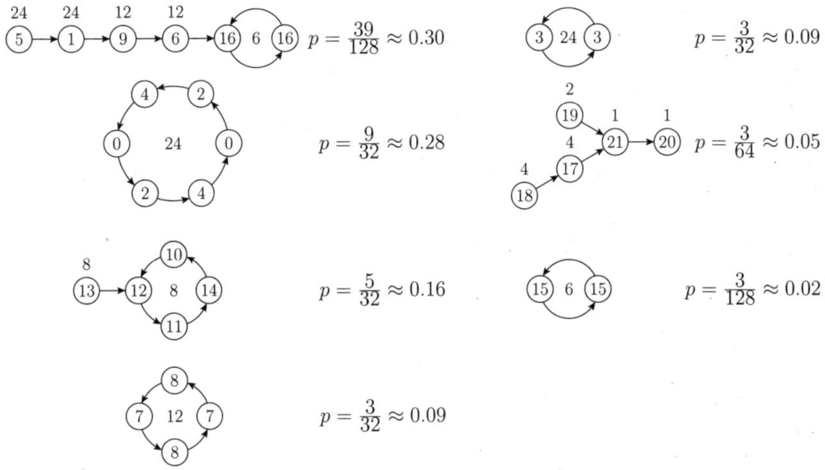

Fig. 2. Rule 86. Equivalence classes of trajectories on hexahedron.

Generalizing this example of *deterministic* dynamical system, we see that if the symmetry group W splits the state set Σ^X into *finite* number of orbits — this is the case for all systems we consider here — then after some lapse of time *any* trajectory comes *inevitably* to a cycle over some finite sequence of orbits. This just means formation of *soliton-like structures*. Namely, let us consider evolution

$$\sigma_{t_0}(x) \rightarrow \sigma_{t_1}(x) = A_{t_1 t_0}\left(\sigma_{t_0}(x)\right). \tag{22}$$

If the states at the moments t_0 and t_1 belong to the *same orbit*: $\sigma_{t_0}(x) \in O_i$ and $\sigma_{t_0}(x) \in O_i$, $O_i \subseteq \Sigma^X$; then evolution (22) can be replaced by the *group action*

$$\sigma_{t_1}(x) = \sigma_{t_0}(x)w, \quad w \in W,$$

i.e., the initial state ("shape") $\sigma_{t_0}(x)$ is reproduced after some "*movement*" in the space Σ^X.

Let us give several examples (including continuous cases) of cycles over group orbits:

- *traveling waves* $\sigma(x - vt)$ in mathematical physics — Galilei group;

- "*generalized coherent states*" in quantum physics — unitary representations of compact Lie groups;
- "*spaceships*" in cellular automata — lattice symmetries.

Fig. 3 illustrates formation of "glider" — one of the "spaceships" in ***Conway's Life*** automaton. In the general case of $n \times n$ square lattice closed into torus — the space of ***Conway's Life*** — the symmetry group is the semidirect product (excepting the case 4×4) of 2D translations and dihedral group D_8: $\mathrm{W} = \mathrm{G} = \mathrm{T}^2 \rtimes \mathrm{D}_8$.

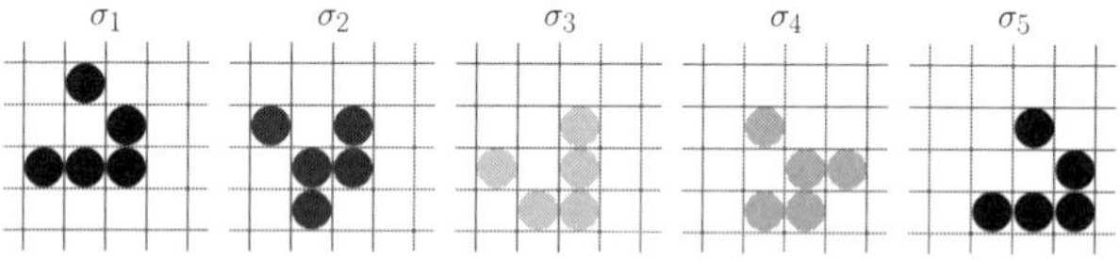

Fig. 3. Example of soliton-like structure. "Glider" in ***Conway's Life*** is cycle in *two* orbits of square lattice symmetry group $\mathrm{G} = \mathrm{T}^2 \rtimes \mathrm{D}_8$: configurations σ_3 and σ_4 are obtained from σ_1 and σ_2, respectively, by *the same* combination of downward *shift*, 90° clockwise *rotation* and *reflection* in respect to vertical.

Note that typical deterministic dynamical system is *irreversible* — it's phase portrait *modulo* group orbits looks like in Fig. 2. We see several isolated and limit cycles (fixed points are regarded as cycles of unit length) accompanied by influxes flowing into the limit cycles. After some time any system "forgets" influxes and appears in either isolated or limit cycles. After loss of information about influxes both types of cycles became physically indistinguishable and the system behaves just like reversible. There might be a hint for explanation of observable reversibility of the fundamental laws of nature.

5. Gauge Principle and Quantization

5.1. *Discrete gauge principle*

In fact, the gauge principle expresses the very general idea that any observable data can be presented in different "frames" at different points of space and time, and there should be some way to compare these data. At the set-theoretic level, i.e., in the form suitable for both discrete and continuous cases, the main concepts of the gauge principle can be reduced to the following elements

- a set X, space or space-time;

- a set Σ, local states;
- the set Σ^X of Σ-valued functions on X, the set of states of *dynamical system*;
- a group $\mathrm{W} \leq \mathrm{Sym}\left(\Sigma^X\right)$ acting on Σ^X, *symmetries of the system*;
- identification of data describing dynamical system with states from Σ^X makes sense only modulo symmetries from W;
- having no *a priori* connection between data from Σ^X at different points x and y in time and space we impose this *connection* (or *parallel transport*) explicitly as W-valued functions on edges of *abstract* graph:

$$P(x, y) \in \mathrm{W}, \ \varsigma(y) = \sigma(x)P(x, y);$$

  connection $P(x, y)$ has obvious property $P(y, x) = P(x, y)^{-1}$;
- connection $P(x, y)$ is called *trivial* if it can be expressed in terms of a function on *vertices* of the graph: $P(x, y) = p(x)^{-1}p(y), \ p(x), p(y) \in \mathrm{W}$;
- invariance with respect to gauge symmetries depending on time or space $u(x), u(y) \in \mathrm{W}$ leads to transformation rule for connection

$$P(x, y) \rightarrow u(x)^{-1}P(x, y)u(y); \tag{23}$$

- the *curvature* of connection $P(x, y)$ is defined as the conjugacy class of the *holonomy* along a cycle of a graph:

$$P(x_1, x_2, \ldots, x_k) = P(x_1, x_2)P(x_2, x_3) \cdots P(x_k, x_1)$$

  (the conjugacy means $P'(x_1, \ldots, x_k) \sim u^{-1}P(x_1, \ldots, x_k)u$ for any $u \in \mathrm{W}$);
  the curvature of trivial connection is obviously trivial: $\widetilde{P}(x_1, \ldots, x_k) \equiv 1$;
- the gauge principle does not tell us anything about the evolution of the connection itself, so gauge invariant relation describing dynamics of connection (*gauge field*) should be added.

5.2. *Gauge connection and quantization*

The Aharonov–Bohm effect (Fig. 4) is one of the most striking illustrations of interplay between quantum behavior and gauge connection. Charged particles moving through the region containing perfectly shielded thin solenoid produce different interference patterns on a screen depending on whether the solenoid is turned on or off. There is no electromagnetic force acting on the particles, but working solenoid produces U(1)-connection adding

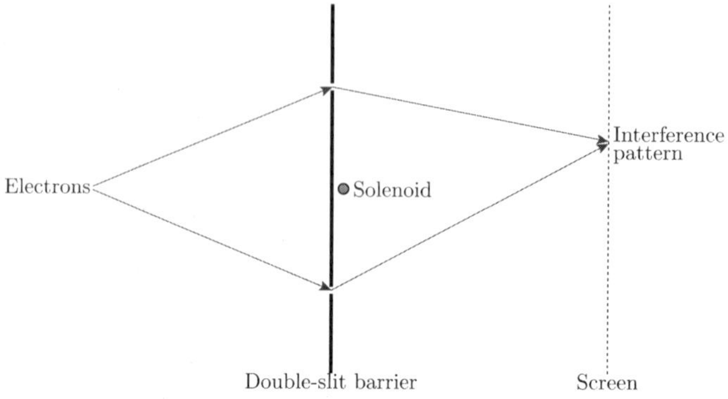

Fig. 4. Aharonov–Bohm effect. Magnetic flux is confined within the perfectly shielded solenoid; interference pattern is shifted in spite of absence of electromagnetic forces acting on the particles.

or subtracting phases of the particles and thus changing the interference pattern.

In the discrete time Feynman's path amplitude[10] decomposes into the product of elements of the fundamental representation of the group $\Gamma = \mathrm{U}(1)$:

$$A_{\mathrm{U}(1)} = \exp\left(iS\right) = \exp\left(i\int Ldt\right) \longrightarrow \mathrm{e}^{iL_{0,1}}\dots\mathrm{e}^{iL_{t-1,t}}\dots\mathrm{e}^{iL_{T-1,T}}. \tag{24}$$

By the notation $L_{t-1,t}$ we emphasize that the Lagrangian is in fact a function defined on pairs of points (graph edges) — this is compatible with physics where the typical Lagrangians are depend on the *first order* derivatives. Thus the expression $P(t-1,t) = e^{iL_{t-1,t}} \in \Gamma = \mathrm{U}(1)$ can be interpreted as U(1)-parallel transport.

A natural generalization of this is to suppose that:

- group Γ may differ from U(1),
- dimension of unitary representation $\rho\left(\Gamma\right)$ may differ from 1.

We can introduce quantum mechanical description of a discrete system interpreting states $\sigma \in \Sigma$ as basis elements of a Hilbert space Ψ. This allows to describe statistics of observations of σ's in terms of the *inner product* in Ψ.

Now let us replace expression (24) for Feynman's path amplitude by the following parallel transport along the path

$$A_{\rho(\Gamma)} = \rho\left(\alpha_{0,1}\right)\dots\rho\left(\alpha_{t-1,t}\right)\dots\rho\left(\alpha_{T-1,T}\right).$$

Here $\alpha_{t-1,t}$ are elements of a *finite group* Γ — we shall call Γ *quantizing group* — and ρ is an unitary representation of Γ on the space Ψ.

Recall that the characters and eigenvalues of unitary (for finite groups all linear representations are unitary) representations of finite groups are elements of the ring $\mathbb{A}$ of *algebraic integers*.[11] Thus algebraic integers are sufficient for all our computations except for normalization of probabilities requiring the quotient field of the ring $\mathbb{A}$.

5.3. *Simple model inspired by free particle*

In quantum mechanics — as is clear from the *never vanishing* expression $\exp\left(\frac{i}{\hbar}S\right)$ for the path amplitude — transitions from one to any other state are possible in principle. However we shall consider computationally more tractable models with restricted sets of possible transitions.

Let us consider quantization of a free particle moving in one dimension. Such a particle is described by the Lagrangian $L = \frac{m\dot{x}^2}{2}$. Keeping only transitions to the closest points in the discretized space we come to the following rule for the one-time-step transition amplitudes

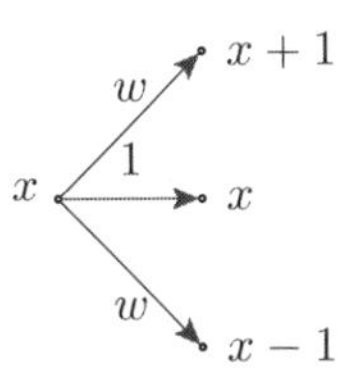

$$e^{\frac{i}{\hbar}\frac{m\{(x+1)-x\}^2}{2}} = e^{i\frac{m}{2\hbar}}$$

$$e^{\frac{i}{\hbar}\frac{m(x-x)^2}{2}} = 1$$

$$e^{\frac{i}{\hbar}\frac{m\{(x-1)-x\}^2}{2}} = e^{i\frac{m}{2\hbar}}.$$

That is, we have evolution rule as an U(1)-valued function R defined on pairs of points (graph edges). Symbolically:

$$R\,(x \to x) = 1 \quad \in \quad \mathrm{U}(1),$$
$$R\,(x \to x-1) = R\,(x \to x+1) = w = e^{i\frac{m}{2\hbar}} \quad \in \quad \mathrm{U}(1). \qquad (25)$$

Now let us assume that w in (25) is an element of some representation of a finite group: $w = \rho\,(\alpha)\,,\ \alpha \in \Gamma = \{\gamma_0 = 1, \ldots, \gamma_{M-1}\}$. Rearranging *multinomial coefficients* — *trinomial* in this concrete case — it is not difficult to write the sum amplitude over all paths of the form $(0,0) \longrightarrow (x,t)$

$$A_x^t\,(w) = \sum_{\tau=0}^{t} \frac{\tau!}{\left(\frac{\tau-x}{2}\right)!\left(\frac{\tau+x}{2}\right)!} \times \frac{t!}{\tau!\,(t-\tau)!}\, w^{\tau}. \qquad (26)$$

Note that x must lie in the limits determined by t: $x \in [-t, t]$.

One of the most expressive peculiarities of quantum-mechanical behavior is the *destructive interference* — cancellation of non-zero amplitudes attached to different paths converging to the same point. By construction, the sum of amplitudes in our model is a function $A(w)$ depending on distribution of sources of the particles, their initial phases, gauge fields acting along the paths, restrictions — like, e.g., "slits" — imposed on possible paths, etc. In the case of one-dimensional representation the function $A(w)$ is a polynomial with algebraic integer coefficients and w is a root of unity. Thus the condition for destructive interference can be expressed by the system of polynomial equations: $A(w) = 0$ and $w^M = 1$. For concreteness let us consider the cyclic group $\Gamma = \mathbb{Z}_M = \{\gamma_0, \cdots, \gamma_k, \cdots, \gamma_{M-1}\}$. Any of its M irreducible representations takes the form $\rho(\gamma_k) = w^k$, where w is one of the Mth roots of unity. For simplicity let w be the *primitive root*: $w = \mathrm{e}^{2\pi i/M}$.

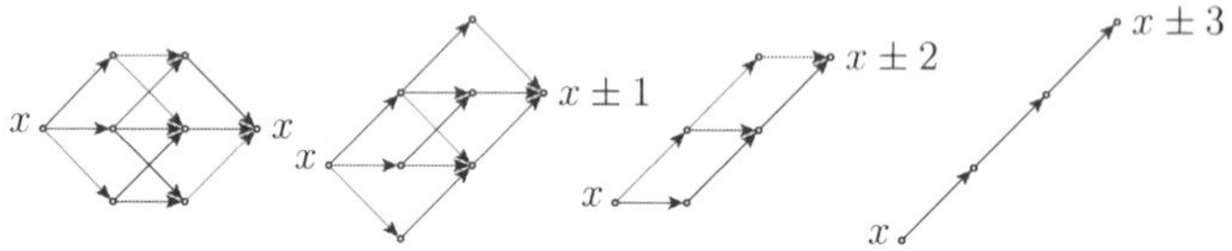

Number of paths	7	6	3	1
Amplitude	$A_0^3 = 1 + 6w^2$	$A_{\pm1}^3 = 3w + 3w^3$	$A_{\pm2}^3 = 3w^2$	$A_{\pm3}^3 = w^3$

Fig. 5. Amplitudes for all possible paths in three time steps.

Fig. 5 shows all possible transitions (with their amplitudes) from the point x in three time steps. We see that the polynomial $A_{\pm1}^3 = 3w + 3w^3 = 3w\left(w^2+1\right)$ contains the *cyclotomic polynomial* $\Phi_4(w) = w^2 + 1$ as a factor. The smallest group associated to $\Phi_4(w)$ — and hence providing the destructive interference — is $\mathbb{Z}_4$. Thus, $\mathbb{Z}_4$ is the natural quantizing group for the model under consideration.

Fig. 6 shows interference patterns — normalized squared amplitudes ("probabilities") — from two sources placed in the positions $x = -4$ and $x = 4$ for 20 time steps. The upper and lower graph show interference pattern when sources are in the same ($\Delta\phi = 0$) and in the opposite ($\Delta\phi = \pi$) phases, respectively.

5.4. *Local quantum models on regular graphs*

The above model — with quantum transitions allowed only within the neighborhood of a vertex of a one-dimensional lattice — can easily be

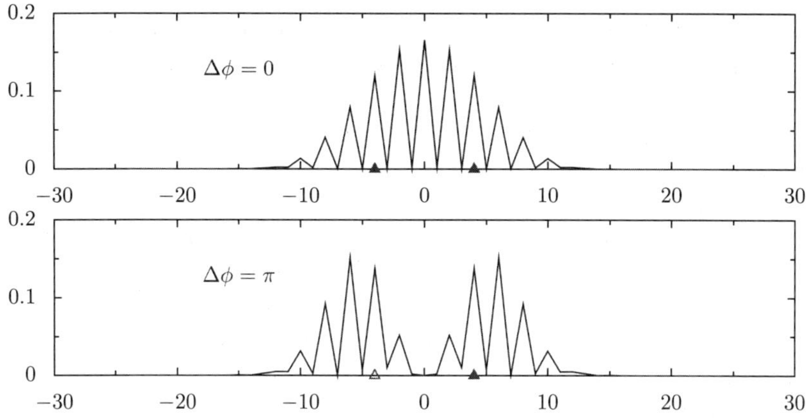

Fig. 6. Group $\mathbb{Z}_4$. Interference from two sources at points -4 and 4. Number of time steps $T = 20$. Phase differences $\Delta\phi = \phi_4 - \phi_{-4}$ between sources are 0 and π.

generalized to arbitrary regular graph. Our definition of *local quantum model on k-valent graph* uncludes the following:

(1) *Space* $X = \{x_1, \cdots, x_{N_X}\}$ is a $k-$valent graph.
(2) *Set of local transitions* $E_i = \{e_{0,i}, e_{1,i}, \cdots, e_{k,i}\}$ is the set of k adjacent to the vertex x_i edges $e_{m,i} = (x_i \to x_{m,i})$ completed by the edge $e_{0,i} = (x_i \to x_i)$.
(3) We assume that the *space symmetry* group $\mathrm{G} = \mathrm{Aut}\,(X)$ acts transitively on the set $\{E_1, \cdots, E_{N_X}\}$.
(4) $\mathrm{G_{loc}} = \mathrm{Stab_G}\,(x_i) \leq \mathrm{G}$ is the *stabilizer* of x_i.
(5) $\Omega_i = \{\omega_{0,i}, \omega_{1,i}, \cdots, \omega_{h,i}\}$ is the *set of orbits* of $\mathrm{G_{loc}}$ on E_i.
(6) *Quantizing group* Γ is a finite group: $\Gamma = \{\gamma_0, \cdots, \gamma_{M-1}\}$.
(7) *Evolution rule* R is a function on E_i with values in some representation $\rho\,(\Gamma)$. The rule R prescribes $\rho\,(\Gamma)$-weights to the one-time-step transitions from x_i to elements of the neighborhood of x_i. From the symmetry considerations R must be a function on orbits from Ω_i, i.e., $R\,(e_{m,i}g) = R\,(e_{m,i})$ for $g \in \mathrm{G_{loc}}$.

6. Conclusion

We developed and implemented algorithms for *analyzing compatibility* of systems of discrete relations and for constructing *canonical decompositions* of discrete relations.

Applying these algorithms to some cellular automata — a particular case of discrete relations — we obtained a number of new results. The most interesting among them, in our opinion, is the demonstration of how the presence of non-trivial *proper consequences* may determine the global behavior of an automaton.

We suggested an algorithmic approach — based on discrete symmetry analysis and implemented in C — for construction and investigation of discrete dynamical models — *deterministic*, *mesoscopic* and *quantum*.

We constructed a family of groups unifying space and internal symmetries in a natural way. This construction generalizes the standard *direct* and *wreath* products.

We demonstrated that soliton-like moving structures — like "*spaceships*" in cellular automata, "traveling waves" in mathematical physics and "generalized coherent states" in quantum physics — arise inevitably in *deterministic* dynamical systems whose symmetry group splits the set of states into finite number of group orbits.

We formulated the gauge principle in the form most suitable for discrete and finite systems. We also proposed a method — based on introduction of unitary gauge connection of a special kind — for quantizing discrete systems and constructed simple models for studying properties of suggested quantization.

Acknowledgments

This work was supported by the grant 10-01-00200 from the Russian Foundation for Basic Research and by the grant 1027.2008.2 from the Ministry of Education and Science of the Russian Federation.

References

1. Kornyak V.V., On Compatibility of Discrete Relations, *Lect. Notes Comp. Sci.* **3718**, Springer-Verlag Berlin Heidelberg, (2005), 272–284.
2. Kornyak V.V., Discrete Relations On Abstract Simplicial Complexes, *Programming and Computer Software.* **32**, No 2, (2006) 84–89.
3. Kornyak V.V., Cellular Automata with Symmetric Local Rules, *Lect. Notes Comp. Sci.* **4194**, Springer-Verlag Berlin Heidelberg, (2006), 240–250.
4. Kornyak V.V., Discrete Dynamical Systems with Symmetries: Computer Analysis, *Programming and Computer Software.* **34**, No 2, (2008) 84–94.
5. Kornyak V.V., Discrete Dynamics: Gauge Invariance and Quantization, *Lect. Notes Comp. Sci.* **5743**, Springer-Verlag Berlin Heidelberg, (2009), 180–194.
6. Hilton P. and Wayle S., *Homology Theory: An Introduction to Algebraic Topology,* Cambrige Univ. Press, 1960.
7. Lidl R. and Niederreiter H., *Finite Fields,* Reading, Mass.: Addison-Wesley, 1983.

8. Wolfram, S.: *A New Kind of Science.* Wolfram Media, Inc (2002).
9. http://atlas.wolfram.com/01/01/
10. Feynman R.P., Hibbs A.R., *Quantum Mechanics and Path Integrals.* McGraw-Hill (1965).
11. Serre J.-P., *Linear Representations of Finite Groups.* Springer-Verlag (1977).

EXOTIC FEW-BODY BOUND STATES IN A LATTICE

D. PETROSYAN* and M. VALIENTE

Institute of Electronic Structure and Laser,
Foundation for Research and Technology–Hellas,
71110 Heraklion, Crete, Greece
**E-mail: dap@iels.forth.gr*

Strongly-interacting ultra-cold atoms in tight-binding optical lattice potentials provide an ideal platform to realize the fundamental Hubbard model. Here, after outlining the elementary single particle solution, we review and expand our recent work on complete characterization of the bound and scattering states of two and three bosonic atoms in a one-dimensional optical lattice. In the case of two atoms, there is a family of interaction-bound "dimer" states of co-localized particles that exists invariantly for either attractive or repulsive on-site interaction, with the energy below or above the two-particle scattering continuum, respectively. Adding then the third particle — "monomer" — we find that, apart from the simple strongly-bound "trimer" corresponding to all three particles occupying the same lattice site, there are two peculiar families of weakly-bound trimers with energies below and above the monomer–dimer scattering continuum, the corresponding binding mechanism being an effective particle exchange interaction.

Keywords: Cold atoms; optical lattices; Hubbard model; boson systems.

1. Introduction

Among the tight-binding lattice models of condensed matter physics,[1,2] the Hubbard model[3] plays a fundamental role. It describes particle tunneling between adjacent lattice sites as well as short range (contact) interaction between the particles on the same lattice site. Despite apparent simplicity, this model is very rich in significance and implications for the many body physics on a lattice.[4] This is perhaps most profoundly manifested with numerous important experimental and theoretical achievements with cold neutral atoms trapped in deep optical lattice potentials,[5–7] wherein the Hubbard model is being realized with unprecedented accuracy.

A remarkable Hubbard model phenomenon is the existence of stable repulsively-bound pairs of atoms in an optical lattice, as was experimentally demonstrated in Ref. 8. This seminal achievement has led to several theoretical studies of the properties of interaction-bound atom pairs — "dimers" — in periodic potentials.[9–19]

In a one-dimensional (1D) tight-binding lattice, two bosons can form a bound dimer[8–13] for any finite strength of the on-site interaction, be it an attraction or a repulsion. Next level of complexity corresponds to three bosons, which obviously can form a strongly-bound "trimer" with all three particles occupying the same lattice site. For large enough on-site interaction strength, however, there are two more kinds of weakly bound trimers.[20] The corresponding binding mechanism turns out to be an effective particle exchange interaction between the dimer and the third particle — "monomer" — leading to symmetric and antisymmetric trimer states with energies slightly above and below the continuum of scattering states of (asymptotically) free dimer and monomer.

Below, after introducing the Bose-Hubbard model and outlining its elementary single particle solution, we review and expand our recent work[12,20] on the bound and scattering states of two and three bosonic atoms in a 1D optical lattice.

2. The Model

Cold bosonic particles in a 1D tight-binding lattice can be accurately described by the (second-quantized) Hubbard Hamiltonian[3–7]

$$H = -J\sum_j (\hat{b}_j^\dagger \hat{b}_{j+1} + \hat{b}_{j+1}^\dagger \hat{b}_j) + \frac{U}{2}\sum_j \hat{n}_j(\hat{n}_j - 1), \tag{1}$$

where $\hat{b}_j^\dagger$ ($\hat{b}_j$) is the particles creation (annihilation) operator and $\hat{n}_j = \hat{b}_j^\dagger \hat{b}_j$ the number operator at jth lattice site; $J (> 0)$ is the inter-site tunneling, or hopping, rate; and U is the on-site interaction, which can be attractive or repulsive.

Single particle solution: Denoting by $|j\rangle$ the state with a single particle at the jth lattice site, the Hamiltonian reduces to

$$H^{(1)} = -J\sum_j (\,|j\rangle\langle j+1| + |j+1\rangle\langle j|\,). \tag{2}$$

Expanding the single-particle state vector as $|\psi\rangle = \sum_j \psi(j)\,|j\rangle$, the stationary Schrödinger equation $H^{(1)}\,|\psi\rangle = E^{(1)}\,|\psi\rangle$ leads to the difference equation

$$-J\,[\psi(j+1) + \psi(j-1)] = E^{(1)}\psi(j), \tag{3}$$

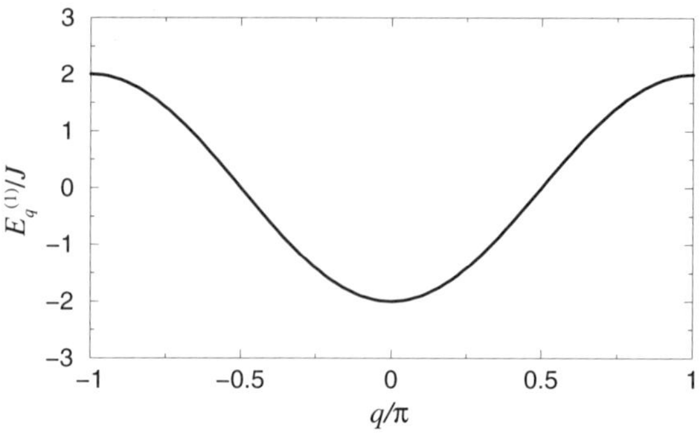

Fig. 1. Single particle Bloch band in a 1D tight-binding lattice.

which is satisfied by the (discrete) plane wave ansatz $\psi(j) = \psi_q(j) = \exp(iqj)$ for the wave function with quasimomentum $q \in \Omega$ restricted to the first Brillouin zone $\Omega \equiv [-\pi, \pi]$, while for the corresponding eigenenergy we obtain $E_q^{(1)} = \epsilon(q) \equiv -2J\cos(q)$. The single-particle energy (Bloch) band, shown in Fig. 1, has, therefore, a width of $4J$.

Having reviewed the trivial single-particle solution,[1] in the following Sections we present complete solutions of the much richer two- and three-body problems in a 1D lattice.

3. Two Particles in a Lattice

Although the Bose-Hubbard Hamiltonian (1) corresponds to identical bosons, which is our main concern here, it is nevertheless instructive to tackle the more general problem of two distinguishable particles in a lattice,[11] a simple limit of which produces the solution for two indistinguishable bosons.

We thus consider two particles A and B having, in general, different hopping rates J_A and J_B, respectively. The Hamiltonian describing their dynamics is the non-symmetrized (first-quantized) version of the Hubbard Hamiltonian (1) acting on the two particle subspace:

$$\begin{aligned} H^{(2)} = &-J_A \sum_{j_A} (\,|j_A\rangle\langle j_A+1| + |j_A+1\rangle\langle j_A|\,) \\ &- J_B \sum_{j_B} (\,|j_B\rangle\langle j_B+1| + |j_B+1\rangle\langle j_B|\,) \\ &+ U \sum_{j_A=j_B} |j_A, j_B\rangle\langle j_A, j_B|\,, \end{aligned} \qquad (4)$$

where j_A and j_B denote the positions of particles A and B, respectively. The eigenstates of $H^{(2)}$ can be expanded as $|\Psi\rangle = \sum_{j_A,j_B} \Psi(j_A, j_B) |j_A, j_B\rangle$, so that the Schrödinger equation $H^{(2)} |\Psi\rangle = E^{(2)} |\Psi\rangle$ leads to the difference equation

$$\begin{aligned} &-J_A\big[\Psi(j_A+1, j_B) + \Psi(j_A-1, j_B)\big] \\ &-J_B\big[\Psi(j_A, j_B+1) + \Psi(j_A, j_B-1)\big] \\ &+U\delta_{j_A,j_B}\Psi(j_A, j_B) = E^{(2)}\Psi(j_A, j_B). \end{aligned} \tag{5}$$

In order to solve the problem analytically, we need to transform the two-body difference equation (5) into a "one-body" problem. To that end, we define the center of mass $j_R = \frac{1}{2}(j_A + j_B)$ and relative $j_r = j_A - j_B$ coordinates and use for the two-particle wave function the separation ansatz[15,20]

$$\Psi(j_A, j_B) = e^{iKj_R} e^{-i\beta_K j_r} \psi_K(j_r), \tag{6}$$

where

$$\tan(\beta_K) = \frac{J_A - J_B}{J_A + J_B} \tan(K/2), \tag{7}$$

with $K \in \Omega$ the center-of-mass quasimomentum. Note that for $J_A = J_B = J$ the separation ansatz (6) reduces to that for identical particles.[12,14] The resulting recursion relation now reads

$$-J_K\big[\psi_K(j_r+1) + \psi_K(j_r-1)\big] + U\delta_{j_r,0}\psi_K(j_r) = E_K^{(2)}\psi_K(j_r), \tag{8}$$

where $J_K = \sqrt{J_A^2 + J_B^2 + 2J_AJ_B\cos(K)}$ is the collective hopping rate,[11,15] which in the case of identical particles, $J_A = J_B = J$, reduces to the standard[8,9,12,14] expression $J_K = 2J\cos(K/2)$.

Equation (8) admits two kinds of solutions, corresponding to the scattering states of asymptotically free particles and to the two-particle bound, or dimer, states.

3.1. *Scattering states*

Since the interaction between the particles is governed by a short-range — in this case a contact, $U\delta_{j_r,0}$ — potential, its action amounts to a unitary phase shift (see below), while the spectrum of such solutions is given by the sum of the spectra for two free particles A and B with momenta $q_A = K/2 + k$ and $q_B = K/2 - k$:

$$E_{K,k}^{(2)} = E_{q_A}^{(1)} + E_{q_B}^{(1)} = -2J_K\cos(k), \tag{9}$$

which spans the interval $E^{(2)}_{K,k} \in [-2J_K, 2J_K]$. The corresponding symmetric scattering wave functions, for $\sin(k) \neq 0$, are given by

$$\psi_{K,k}(j_r) = \cos(k|j_r| + \delta_{K,k}), \tag{10}$$

which, upon substitution into Eq. (8), yields for the scattering phase shift $\delta_{K,k}$,

$$\tan(\delta_{K,k}) = -\frac{U}{2J_K \sin(k)}. \tag{11}$$

For $U \to 0$ we have non-interacting particles $\psi_{K,k}(j_r) = \cos(k|j_r|)$, while for $U/J_K \to \pm\infty$ we obtain the fermionized solution $\psi_{K,k}(j_r) = \sin(k|j_r|)$, whereby the two particles never occupy the same lattice site, $\psi_{K,k}(0) = 0$.

When $\sin(k) = 0$, the scattering wave functions (10) are no longer valid, and the lattice generalization of the continuum zero-energy solution of the Schrödinger equation apply.[10,15] At the bottom and the top of the scattering band, $E^{(2)}_{K,\mp} = \mp 2J_K$ ($k = 0, \pi$), the corresponding solutions $\psi_{K,\mp}$ have the form

$$\psi_{K,-}(j_r) = 1 - \frac{|j_r|}{a_{K,-}}, \tag{12}$$

$$\psi_{K,+}(j_r) = (-1)^{j_r}\left(1 - \frac{|j_r|}{a_{K,+}}\right), \tag{13}$$

where $a_{K,\mp}$, with $a_{K,-} = -a_{K,+}$, are the scattering lengths, which are calculated by substituting Eqs. (12) into Eq. (8), resulting in $a_{K,-} = -2J_K/U$ (in units of the lattice constant). Note that, since the scattering lengths are finite for any $U \neq 0$, there exists no "zero-energy" resonance in this model.[a] Therefore, for every value of the total quasimomentum K there is only one bound state, as we shall see below.

To illustrate the foregoing discussion, in Fig. 2 we show the energy spectrum (9) for two identical bosons, $J_A = J_B = J$, and the corresponding density of states defined via

$$\rho(E, K) = \frac{L}{2\pi}\frac{\partial k}{\partial E} = \frac{L}{2\pi}\frac{1}{\sqrt{[4J\cos(K/2)]^2 - E^2}}, \tag{14}$$

with L a quantization length. As seen, the density of scattering states, i.e., the number of states per unit interval of energy, is lowest in the middle of the band, $E \simeq 0$ (and $K \simeq 0$), while for a given quasimomentum K of the center of mass motion of the two particles, $\rho(E, K)$ increases rapidly as energy $E = E^{(2)}_{K,k}$ approaches its maximal and minimal values $\pm 2J_K$.

---

[a]It contrast, however, with the extended Hubbard model in Refs. 14 and 15.

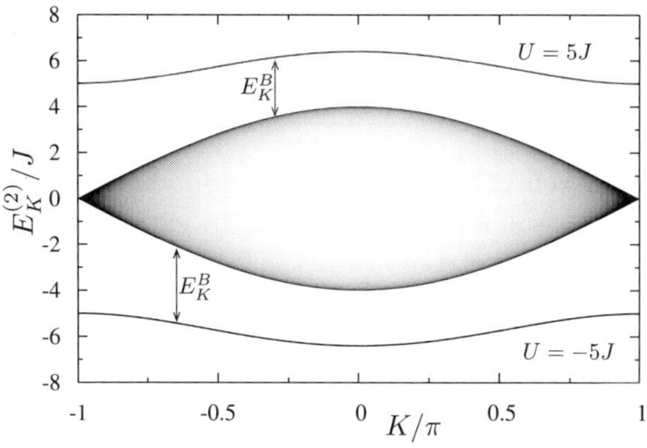

Fig. 2. Energies $E_K^{(2)}$ versus the center-of-mass quasimomentum K for a pair of identical bosons in a 1D lattice. The continuum spectrum corresponds to energies (9) of the scattering states, with the shading proportional to the density of states (14). The line below and the line above the scattering band are, respectively, the energies (15) of the attractively-bound dimer with $U = -5J$ and the repulsively-bound dimer with $U = 5J$.

3.2. *Bound states*

The two-body bound states of Hamiltonian (4) are the solutions of Eq. (8) yielding normalizable relative coordinate wave function, $\sum_{j_r} |\psi_K(j_r)|^2 < \infty$, with the corresponding energy below (for $U < 0$) or above (for $U > 0$) the scattering continuum at each value of the total quasimomentum K. Introducing into Eq. (8) the exponential ansatz $\psi_K(j_r) \propto \alpha_K^{|j_r|}$ yields the bound state energy $E_K^{(2)} = -J_K(\alpha_K + 1/\alpha_K)$, with α_K given by

$$\alpha_K = \mathcal{U}_K - \mathrm{sgn}(U)\sqrt{\mathcal{U}_K^2 + 1}, \quad \mathcal{U}_K \equiv \frac{U}{2J_K}.$$

We see that α_K is real and also $|\alpha_K| < 1$ for all $U \neq 0$. Hence, the relative coordinate wave function $\psi_K(j_r)$ is normalizable, and the energy satisfies $|E_K^{(2)}| > 2J_K$ for all K. Explicitly, the energy and the normalized wave function for the bound dimer are given by

$$E_K^{(2)} = \mathrm{sgn}(U)\sqrt{U^2 + 4J_K^2}, \tag{15}$$

$$\psi_K(j_r) = \frac{\sqrt{|\mathcal{U}_K|}}{\sqrt[4]{\mathcal{U}_K^2 + 1}} \left(\mathcal{U}_K - \mathrm{sgn}(U)\sqrt{\mathcal{U}_K^2 + 1}\right)^{|j_r|}. \tag{16}$$

In Fig. 2 we show the energies $E_K^{(2)}$ for a pair of identical bosons, $J_A = J_B = J$, interacting via on-site attractive, $U < 0$, or repulsive, $U > 0$, potential. The binding energies E_K^B are defined with respect to the edges of the scattering band (9). Note that in the case of repulsive interaction,

$U > 0$ ($\alpha_K < 0$), the sign of the wave function (16) alternates between the neighboring sites j_r. Remarkably, when $|K| = \pi$, and thereby $J_K = 0$, the relative coordinate wave function $\psi_K(j_r)$ is completely localized at $j_r = 0$ for any $U \neq 0$.

Clearly, for a given value of the dimer quasimomentum K, the stronger is the on-site interaction $|U|$, the smaller is the extent of the wave function $\psi_K(j_r)$, meaning that the constituent particles are stronger co-localized. For $|U| \gg J$, the dimer energy in Eq. (15) can be approximated as

$$E_K^{(2)} \simeq \mathcal{E}^{(2)} - 2J^{(2)} \cos(K), \tag{17}$$

where the first term $\mathcal{E}^{(2)} \equiv [U - 2J^{(2)}]$ represents the dimer "internal energy", while the second term $\epsilon^{(2)}(K) \equiv -2J^{(2)} \cos(K)$ is the kinetic energy of a dimer with quasimomentum K and an effective tunnelling rate $J^{(2)} \equiv -2J^2/U$.

4. Three Particles in a Lattice

Building on the solution of the two-body problem, in this section we consider three bosonic atoms is an optical lattice described by Hamiltonian (1).

4.1. *Spectrum of scattering states*

With the expertise gained from the previous section, we can readily deduce that in the case of three particles there are two distinct scattering continua (see Fig. 3 left panel). The first is the three-body scattering continuum of three (asymptotically) free particles, with the energy given by the sum of single-particle bands, $E_{c3} = \epsilon(k_1) + \epsilon(k_2) + \epsilon(K - k_1 - k_2) \equiv \epsilon(k_1, k_2, K - k_1 - k_2)$, where each particle quasimomentum $k_j \in \Omega$ is in the first Brillouin zone $\Omega \equiv [-\pi, \pi]$ and $K = k_1 + k_2 + k_3 \pmod{2\pi}$ is the total quasimomentum. The second is the two-body scattering continuum of a bound pair (dimer) and a free particle (monomer), with energy $E_{c2} = \mathrm{sgn}(U)\sqrt{U^2 + [4J\cos(Q/2)]^2} - 2J\cos(K - Q)$, where the first term is the energy of a dimer, Eq. (15), with quasimomentum Q.

4.2. *Bound states*

We seek the bound states $|\Psi\rangle$ of three bosons in momentum representation,

$$|\Psi\rangle = \frac{1}{(2\pi)^{3/2}} \iiint_{\Omega^3} dk_1 dk_2 dk_3 \, \Psi(k_1, k_2, k_3) \, |k_1, k_2, k_3\rangle, \tag{18}$$

where the wave function $\Psi(k_1, k_2, k_3)$ is symmetric with respect to exchange of any pair of particles. From the stationary Schrödinger equation $H|\Psi\rangle = E|\Psi\rangle$, using the conservation of total quasimomentum K, we obtain[21]

$$\Psi(k_1, k_2, k_3) = -\frac{M(k_1) + M(k_2) + M(k_3)}{\epsilon(k_1, k_2, k_3) - E}, \tag{19}$$

where functions $M(k)$ satisfy the 1D Mattis integral equation[21]

$$M(k)[1 + I_E(k)] = -\frac{U}{\pi} \int_{-\pi}^{\pi} dq \, \frac{M(q)}{\epsilon(k, q, K-k-q) - E}, \tag{20}$$

with $I_E(k)$ being a generalized Watson integral[22]

$$\begin{aligned} I_E(k) &\equiv \frac{U}{2\pi} \int_{-\pi}^{\pi} dq \, \frac{1}{\epsilon(k, q, K-k-q) - E} \\ &= -\frac{\operatorname{sgn}[E - \epsilon(k)]U}{\sqrt{[E - \epsilon(k)]^2 - 16J^2 \cos^2[(K-k)/2]}}. \end{aligned}$$

Equation (20) can be cast as a homogeneous Fredholm equation of the second kind with eigenvalue $\lambda = 1$. Hence, for a given U/J and fixed K, it is a nonlinear equation for energy E, which we solve numerically.

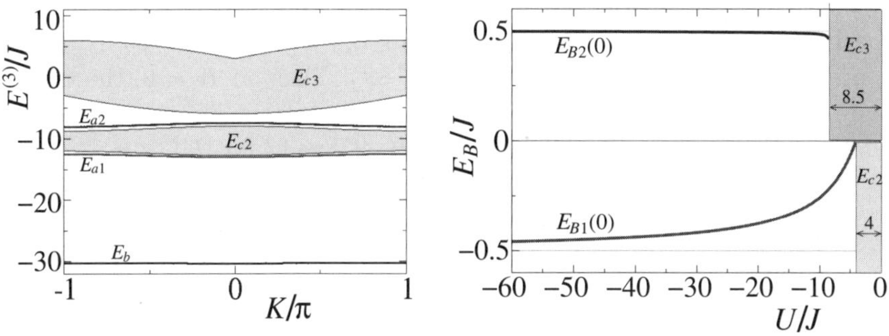

Fig. 3. Left: Full three-particle energy spectrum of Hamiltonian (1) with $U = -10J$, versus the total quasimomentum K. All bound states are obtained via exact numerical solution of Eq. (20). Right: Binding energies E_B for the off-site (weakly-bound) trimers at $K = 0$ versus the interaction strength $U < 0$.

The full three-body spectrum of Hamiltonian (1) is shown in Fig. 3 left panel, with the bound state energies denoted by E_b, E_{a1} and E_{a2}. For concreteness, here we consider attractive interaction, $U < 0$, but note that our results equally apply to the case of repulsive interaction, $U > 0$.[15,20]

In complete analogy with the two-body problem, the Bose-Hubbard Hamiltonian (1) with $|U|/J \gg 1$ has a very narrow band of on-site bound states, corresponding to three tightly bound bosons co-localized on the same lattice site, with energies $E_b \approx 3U$ far from both scattering continua.[21] But as also seen in Fig. 3, in a 1D lattice, bosons can form two new kinds of three-body bound states whose energies E_{a1} and E_{a2} lie below and above the two-body continuum E_{c2}. Some properties of these states can be deduced by energy considerations. First, these are not on-site bound states, since their energies $E_{a1(2)} \simeq U + O(J)$ are far from $3U$. Next, their binding energies, with respect to the E_{c2} band, are $E_{B1(2)} \lesssim \mp J/2$, which suggests that these are off-site weakly-bound states of a dimer and a monomer. Note that the state above the two-body continuum is bound stronger than the state below the continuum. Finally, they are not Efimov states which can exist only in 3D systems near two-boson resonances.[21,23]

As can be seen from Fig. 3 right panel, where we plot the binding energies $E_{B1(2)}$ at quasimomentum $K = 0$, there are thresholds for the existence of full bands ($K \in [-\pi, \pi]$) of the off-site bound states. For the trimer below the two-body continuum, the binding energy vanishes when $|U| \approx 4J$: at this critical value of U the trimer energy E_{a1} approaches the edge of the dimer–monomer scattering continuum $E_{c2} = -\sqrt{U^2 + 16J^2} - 2J$. On the other hand, the $K = 0$ trimer above the two-body continuum ceases to exist already for $|U| \approx 8.5J$, since then its energy E_{a2} approaches the bottom of the three-body continuum $E_{c3} = 3\epsilon(0) = -6J$ (the two continua, E_{c2} and E_{c3}, overlap for $|U| \leq 8J$). Thus, at $K = 0$, the trimer state with energy E_{a2} starts to appear well in the strong interaction regime, while for larger K the threshold is smaller: $|U| \approx 4J$ for $|K| \to \pi$.

4.3. *Effective model*

Since for strong on-site interaction $|U|/J \gg 1$ the dimer is essentially unbreakable, the off-site trimers with energies close to U must be bound due to a mechanism different from the on-site interaction alone. To identify such a mechanism, we derive an effective perturbative model, valid for $|U|/J > 8$, describing two distinguishable, hard-core particles — the dimer and the monomer. To second order in the tunneling rate J, the effective Hamiltonian reads

$$H_{\text{eff}} = H_1 + H_2 + H_{\text{int}}, \tag{21}$$

where

$$H_1 = -J \sum_j (\hat{b}_j^\dagger \hat{b}_{j+1} + \text{H.c.})$$

describes the single monomer;

$$H_2 = \mathcal{E}^{(2)} \sum_j \hat{m}_j - J^{(2)} \sum_j (\hat{c}_j^\dagger \hat{c}_{j+1} + \text{H.c.})$$

is the Hamiltonian for a dimer (*cf.* Eq. (17)), with $\hat{c}_j^\dagger$ ($\hat{c}_j$) being the dimer creation (annihilation) operator and $\hat{m}_j = \hat{c}_j^\dagger \hat{c}_j$ the number operator at site j; and finally

$$H_{\text{int}} = V^{(2)} \sum_j \hat{m}_j \hat{n}_{j\pm1} - W \sum_j (\hat{c}_{j+1}^\dagger \hat{c}_j \hat{b}_j^\dagger \hat{b}_{j+1} + \text{H.c.})$$

describes effective interactions between the dimer and the monomer, including a weak nearest-neighbor interaction $V^{(2)} = -7J^2/2U$, and an exchange interaction with the rate $W = 2J$. As we will see below, it is the exchange term that is responsible for the formation of the off-site trimers.

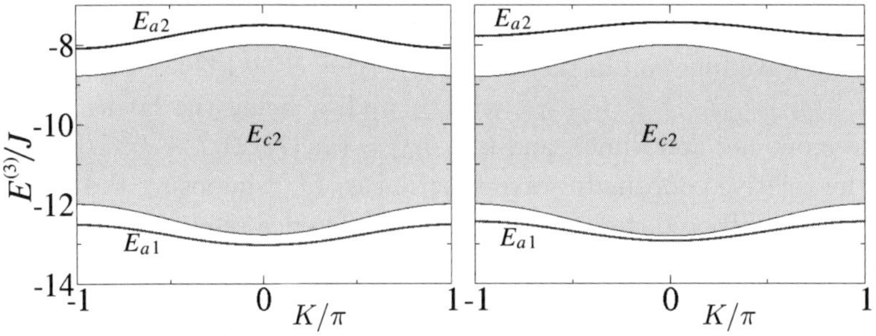

Fig. 4. Left: Magnified part of the exact three-particle spectrum of Fig. 3. Right: Dimer–monomer spectrum of the effective Hamiltonian (21) with the two bound states obtained via numerical solution of Eq. (22).

In Fig. 4 right panel, we plot the spectrum of the effective Hamiltonian (21), which contains two bound states with energies E_{a1} and E_{a2} below and above the two-body scattering continuum $E_{c2} = \mathcal{E}^{(2)} + \epsilon^{(2)}(Q) + \epsilon(K - Q)$. These dimer–monomer bound states are obtained using the Schrödinger equation for the two-body wave function $\Psi(Q, k)$ in momentum space, which leads to the integral equation

$$\Psi(Q, k) = -\frac{1}{2\pi} \int_{-\pi}^{\pi} dq \, \frac{U_{12} + V_{\cos}(Q, q) + V_{\sin}(Q, q)}{\mathcal{E}^{(2)} + \epsilon^{(2)}(q) + \epsilon(K - q) - E} \Psi(q, k), \tag{22}$$

where $K = Q + k$ is the total quasimomentum, $V_{\cos}(Q, q) = [2V^{(2)} \cos(q) - 4J \cos(K - q)] \cos(Q)$ and $V_{\sin}(Q, q) \equiv V_{\cos}(Q, q)$ with cos $\rightarrow$ sin, while $U_{12}(\rightarrow \infty)$ is an artificial dimer–monomer on-site interaction imposing the hard-core condition on Hamiltonian (21). Equation (22) reduces to a non-linear equation for the energy E solving which (numerically) we obtain E_{a1} and E_{a2}. Comparison with the exact spectrum on the left panel of Fig. 4 reveals good agreement: the continuum spectra are indistinguishable, while the small but noticeable differences in the bound-state energies are associated with the internal structure of the dimer, not accounted for by the effective model, and they gradually disappear with increasing the on-site interaction strength U.

4.3.1. *Analytic solutions*

There are two important cases, corresponding to the maximal ($K = \pi$) and minimal ($K = 0$) total quasimomentum K, for which the bound and scattering states of the effective Hamiltonian (21) can be calculated analytically employing the method of Sec. 3. To that end, we expand the two-particle eigenstates in coordinate basis $|\Psi\rangle = \sum_{j_1 \neq j_2} \Psi(j_1, j_2)\,|j_1, j_2\rangle$ with the wave function in the form $\Psi(j_1, j_2) = e^{iKj_R} e^{-i\beta_K j_r} \psi_K(j_r)$, where $j_R \equiv \frac{1}{2}(j_1 + j_2)$, $j_r \equiv j_1 - j_2$, with j_1 and j_2 being the lattice positions of the monomer and dimer, and $\tan(\beta_K) = \tan(K/2)[J - J^{(2)}]/[J + J^{(2)}]$. For the relative coordinate wave function $\psi_K(j_r)$, imposing the hard-core condition $\psi_K(0) = 0$, we then obtain the difference relations

$$\begin{aligned} J_K \psi_K(\pm 2) + W_K \psi_K(\mp 1) + [\bar{E} - V^{(2)}]\psi_K(\pm 1) = 0, \\ J_K[\psi_K(j_r + 1) + \psi_K(j_r - 1)] + \bar{E}\psi_K(j_r) = 0, \end{aligned} \tag{23}$$

with $|j_r| > 1$, $J_K \equiv \sqrt{J^2 + J^{(2)2} + 2JJ^{(2)}\cos(K)}$, $W_K \equiv W\cos(K)$, and $\bar{E} \equiv E - \mathcal{E}^{(2)}$.

Scattering solutions: Using the standard ansatz $\psi^{(+)}_{K,k}(j_r) = \cos\big(k|j_r| + \Delta^{(+)}_{K,k}\big)$ and $\psi^{(-)}_{K,k}(j_r) = \mathrm{sgn}(j_r)\cos\big(k|j_r| + \Delta^{(-)}_{K,k}\big)$ with k the relative quasi-momentum, for the corresponding phase shifts $\Delta^{(\pm)}_{K,k}$ of the symmetric (+) and antisymmetric (−) scattering wave functions we obtain

$$\tan(\delta^{(\pm)}_{K,k}) = \frac{J_K\cos(2k) + [\bar{E} \pm W_K - V^{(2)}]\cos(k)}{J_K\sin(2k) + [\bar{E} \pm W_K - V^{(2)}]\sin(k)}, \tag{24}$$

with $\bar{E} = E_{c2} - \mathcal{E}^{(2)} = -2J_K\cos(k)$. Note that in the limit of $|U| \to \infty$, as the the nearest neighbour interaction $V^{(2)}$ and the dimer hopping $J^{(2)}$ tend to zero ($J_K \to J$), Eq. (24) holds for all $K = k$. The full scattering wave function is given by a superposition $\psi_{K,k}(j_r) = A\,\psi^{(-)}_{K,k}(j_r) + B\,\psi^{(+)}_{K,k}(j_r)$, which, upon expressing through incident, reflected and transmitted waves,

$$\psi_{K,k}(j_r) = \begin{cases} e^{ikj_r} + r_K\, e^{-ikj_r} & (j_r < 0) \\ t_K\, e^{ikj_r} & (j_r > 0) \end{cases},$$

leads to $A/B = -e^{-i[\Delta^{(+)}_{K,k} - \Delta^{(-)}_{K,k}]}$. For the reflection $r_K(k)$ and transmission $t_K(k)$ amplitudes we then obtain $r_K, t_K = \frac{1}{2}\big[e^{2i\Delta^{(+)}_{K,k}} \pm e^{2i\Delta^{(-)}_{K,k}}\big]$, and the transmission and reflection probabilities are given by $T_K = |t_K|^2 = \sin^2\big(\Delta^{(+)}_{K,k} - \Delta^{(-)}_{K,k}\big)$ and $R_K = |r_K|^2 = \cos^2\big(\Delta^{(+)}_{K,k} - \Delta^{(-)}_{K,k}\big)$.

In Fig. 5 we plot $T_K(k)$ for $U/J = -10$ at total quasimomenta $K = 0$ and $|K| = \pi$. The transmission spectra for the intermediate values of K lies in between the curves for $K = 0$ and $K = \pi$. We observe the maximum transmission in the vicinity of $k = \pm\pi/2$, where $T_K(\pi/2)$ ranges from 50% to 80%. With increasing the interaction strength U, we find that the maximum transmission saturates at around 64% for all values of K, which should be contrasted with the results of Ref. 19.

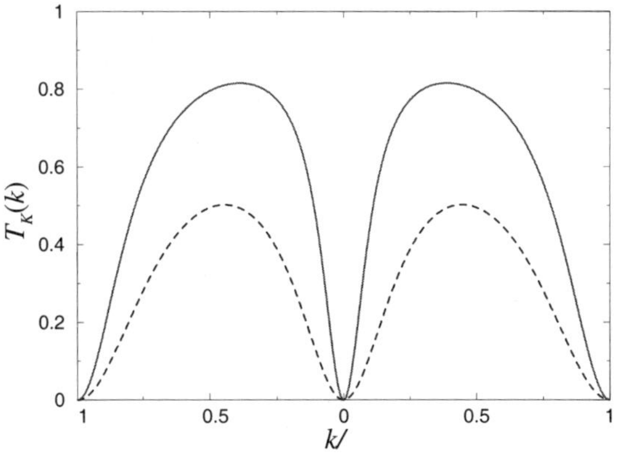

Fig. 5. Transmission probability $T_K(k)$ of a single particle with relative quasimomentum k through a bound dimer, for $U/J = -10$ and total quasimomenta $K = 0$ (red solid line) and $|K| = \pi$ (blue dashed line).

5. Bound Solutions

The dimer–monomer bound states are obtained from Eq. (23) using the exponential ansatz $\psi_K(j_r > 0) \propto \alpha_K^{j_r - 1}$ and $\psi_K(-j_r) = \pm\psi_K(j_r)$, which yields

$$\alpha_K^{(\pm)} = -\frac{J_K}{V^{(2)} \mp W_K} \tag{25}$$

for the symmetric (+) and antisymmetric (−) wave function of the bound state ($|\alpha_K| < 1$), with the corresponding energy $\bar{E}_{a1(2)} = -J_K[1 + (\alpha_K^{(\pm)})^2]/\alpha_K^{(\pm)}$.

It is now easy to see that without the exchange interaction there would be no dimer–monomer bound states (for any K). Indeed, this hypothetic ($W = 0$) problem is exactly solvable for all K, and for two hard-core bosons with nearest-neighbour interaction $V^{(2)}$ there could be only one bound state[14] when $|\alpha_K| = |J_K/V^{(2)}| < 1$, which cannot be satisfied in the range of validity ($|U|/J > 8$) of the effective model, Eq. (21). The effective nearest-neighbour interaction is, however, responsible for the asymmetry in the binding energies $E_{B1(2)} = \bar{E}_{a1(2)} \mp 2J_K$ of the exchange-bound trimers below and above the continuum E_{c2} (see Fig. 3 right panel). With increasing the on-site interaction U, the binding energies attain the universal limits $E_{B1(2)} \to \mp J/2$ which remain valid for all K, since $J^{(2)} \to 0$ as $|U/J| \to \infty$. We thus conclude that the effective particle exchange interaction binds the dimer and the monomer into the weakly bound trimer.

6. Conclusions

In this paper, we have presented complete solutions for the one-, two- and three-body problems in a 1D tight-binding lattice described by the Bose-Hubbard model.

For the case of two bosons, we have derived the scattering states and the bound states of co-localized particles, which we termed as dimers. The corresponding binding mechanism is provided by the on-site interaction between the particles. Perhaps counterintuitively, these bound dimer states exist not only for attractive, but also for repulsive interactions.[8] In free space, the repulsive interaction would inevitably lead to the pair separation, or dissociation, whereby the potential energy of the repulsion is converted into the kinetic energy of the free particles. In the lattice, however, the kinetic energy of each particle cannot take on arbitrary values, but is restricted to the values in the allowed Bloch band, which itself is bounded both from below and from above. Therefore, in the absence of energy dissipation, a pair of co-localized particles interacting even via repulsive potential is destined to stay bound together as a dimer, simply because there are no available free-particle energy states to which the dimer can dissociate.

For three bosons, we have found three families of trimers — bound states of three particles. The first strongly-bound trimer, being an analog of the dimer, corresponds to all three particles occupying the same lattice site and bound by the on-site interaction. The other two families of trimers are weakly-bound with energies just below and above the two-body scattering continuum of a single particle (monomer) and an interaction-bound dimer. Intuitively, these trimer states correspond to symmetric and antisymmetric states of a dimer and monomer at the neighboring lattice sites interacting with each other via an effective (particle) exchange interaction, which is responsible for their binding.

The phenomena discussed above are pertinent to the experiments with cold bosonic atoms in optical lattices.[7] Studying larger number of bosons in a lattice might reveal other exotic bound states, while longer range interactions, such as those between dipolar atoms,[24] or molecules,[25] will certainly play an important role in the formation of few-body bound states.[14,15,26]

Acknowledgements

This work was supported by the EC Marie-Curie Research Training Network EMALI.

References

1. N.J. Ashcroft and and N.D. Mermin, *Solid State Physics* (Thomson, New York, 1976).
2. S. Sadchev, *Quantum Phase Transitions* (Cambridge University Press, Cambridge, 1999).
3. J. Hubbard, *Proc. Roy. Soc. A* **276** 238 (1963).
4. M.P.A. Fisher, P.B. Weichman, G. Grinstein, and D.S. Fisher, *Phys. Rev. B* **40**, 546 (1989).
5. D. Jaksch, C. Bruder, J.I. Cirac, C.W. Gardiner, and P. Zoller, *Phys. Rev. Lett.* **81**, 3108 (1998).
6. M. Greiner, O. Mandel, T. Esslinger, T.W. Hänsch, and I. Bloch, *Nature* **415**, 39 (2002).
7. I. Bloch, J. Dalibard, and W. Zwerger, *Rev. Mod. Phys.* **80**, 885 (2008).
8. K. Winkler, G. Thalhammer, F. Lang, R. Grimm, J. Hecker Denschlag, A.J. Daley, A. Kantian, H.P. Büchler, and P. Zoller, *Nature* **441**, 853 (2006).
9. R. Piil and K. Mølmer, *Phys. Rev. A* **76**, 023607 (2007).
10. N. Nygaard, R. T. Piil,and K. Mølmer, *Phys. Rev. A* **78**, 023617 (2008).
11. R. T. Piil, N. Nygaard and K. Mølmer, *Phys. Rev. A* **78**, 033611 (2008).
12. M. Valiente and D. Petrosyan, *J. Phys. B* **41**, 161002 (2008).
13. M. Valiente and D. Petrosyan, *Europhys. Lett.* **83**, 30007 (2008).
14. M. Valiente and D. Petrosyan, *J. Phys. B* **42**, 121001 (2009).
15. M. Valiente, arXiv:1001.3805 (2010).
16. D. Petrosyan, B. Schmidt, J.R. Anglin, and M. Fleischhauer, *Phys. Rev. A* **76**, 033606 (2007).
17. B. Schmidt, M. Bortz, S. Eggert, M. Fleischhauer and D. Petrosyan, *Phys. Rev. A* **79**, 063634 (2009).

18. L. Wang, Y. Hao and S. Chen , *Eur. Phys. J. D* 48, 229 (2008).
19. L. Jin, B. Chen, and Z. Song, *Phys. Rev. A* **79**, 032108 (2009).
20. M. Valiente, D. Petrosyan and A. Saenz, *Phys. Rev. A* **81**, 011601(R) (2010).
21. D.C. Mattis, *Rev. Mod. Phys.* **58**, 361 (1986).
22. G.N. Watson, *Q. J. Math. Oxford* **10**, 266 (1939).
23. V.N. Efimov, *Sov. J. Nucl. Phys.* **12**, 589 (1971).
24. A. Griesmaier, J. Werner, S. Hensler, J. Stuhler, and T. Pfau, Phys. Rev. Lett. **94**, 160401 (2005).
25. K.-K. Ni, S. Ospelkaus, M.H.G. de Miranda, A. Pe'er, B. Neyenhuis, J.J. Zirbel, S. Kotochigova, P.S. Julienne, D.S. Jin, and J. Ye, Science **322**, 231 (2008).
26. M. Valiente, Ph.D. Thesis (unpublished).

SLOW LIGHT AND PHASE TRANSITION IN THE ARRAY OF ATOMIC POLARITONS

I. O. BARINOV, E. S. SEDOV, A. P. ALODJANTS*, S. M. ARAKELIAN

Department of Physics and Applied Mathematics, Vladimir State University, Vladimir, 600000, Russia
**E-mail: alodjants@vpti.vladimir.ru*

We study a critical phenomenon for atomic polaritons in a cavity array for the interaction of two-level atoms with quantized optical radiation in the strong coupling regime. We have shown that the phase transition to the superfluid Bardeen-Cooper-Schrieffer (BCS)-type state of low branch polaritons occurs. Such a transition results in the appearance of a macroscopic polarization of atomic medium at non-zero frequency. The principal result is that the group velocity of polaritons essentially depends on the order parameter of the system, i.e. on the average photon number in the cavity array. We have shown that atomic band-gap structure under consideration admits a complete polariton localization within the Brillouin zone which can be used for observing slow light phenomena.

Keywords: Atomic polaritons; band-gap structures; slow light; tight-binding approximation; phase transition.

1. Introduction

Present remarkable achievements with trapped atomic gases evoke great interest in investigations of phase transitions in such a systems – see, e.g., Refs. 1 and 2. Although Bose-Einstein condensation (BEC) of the atoms has been observed in many labs, the requirement to use extremely low (up to microKelvins) temperatures strictly limits the utilization of such an effect for practical purposes. This represents an important reason for studying relatively high temperature phase transitions. Usually, such transitions take place in the coupled matter-field systems for which the effective mass of the particles is many order times smaller than free mass of atoms (or electrons). In particular, evidence for a BEC of polaritons in semiconductor nanostructures has been reported recently – see, e.g., Refs. 3–5. In such a medium low branch polaritons represent a linear superposition of excitons (two level oscillators) localized in quantum wells and photons of quantum optical field. The mass of polaritons in such a structure can be only two times larger than the photon mass in the cavity.

Another intriguing problem that has been discussed long time is photonic phase transition (or BEC of photons) – see, e.g., Ref. 6. The first attempts to consider some analogy between lasing under the threshold region and second-order phase transition in ferromagnets have been made in Ref. 7. Simple Dicke model that describe an ensemble of two level atoms interacting with quantized electromagnetic field has been proposed to observe the phase transition of the photons Ref. 8–13. In quantum optics the phase transition in such a system has been interpreted as a transition to some superradiant (coherent or condensed) photonic state with zero chemical potential.[8] Simultaneously in Ref. 9 it is pointed out that in this case the building up of spontaneous static field (the field with zero frequency) in the medium is takes place. In fact such a phase transition creates some ferroelectric state in the medium what is not so interesting from the practical point of view. Moreover in Ref. 10 it has firstly been declared that Thomas-Reiche-Kuhn sum rule formally contradicts nontrivial solution of phase transition for simple Dicke model (a more recent discussion on this problem is established in Ref. 11).

From our point of view even for the simple Dicke model we need to consider coupled matter-field states – polaritons for which chemical potential μ is non-zero as a rule. In fact fixing the total number of excitations it is possible to obtain an additional parameter representing chemical potential that is not limited by sum rule in this case – cf. Ref. 13. Apparently the photonic condensate in such a case can be considered as a limiting state of low branch polaritons achieved due to large values of atom-field detuning.

In our previous papers[14,15] we propose to use some coherent properties of the cavity polaritons arising due to atom-field interaction under the strong coupling condition for cloning and spatial storing quantum optical information. At the same time we have shown that polaritonic model of atom-field interaction is valid under the low excitation density limit when atoms predominantly populate their ground state.

In the present paper we continue the investigation of thermodynamic and critical properties of polaritons in two level atomic system. Modern nanofabrication and nanophotonic technologies make it possible to study slow light and phase transitions of polaritons by using array of photonic cavities (or coupled resonator waveguides) doped by two level atoms – see e.g. Refs. 16–18. We propose a spatially-periodical atomic structure called it polaritonic crystal (PolC), that represents a lattice of weakly coupled trapped two-component atomic ensembles interacting with optical field in the cavity – c.f. Ref. 19. In the present paper we will show that the group velocity of polaritons essentially depends on the order parameter of the system, i.e. on the photonic field amplitude in the cavity array.

The paper is arranged as follows. In Sec. 2 we establish the quantum field theory approach for the model of polaritonic band-gap structure in the frameworks of all available current facilities for trapping and manipulating atoms by laser field. In Sec. 3 we focus our attention on the dispersion properties of polaritons in a spatially-periodic atomic structure of PolC. In particular we consider a thermodynamic approach to study the polariton properties for practically used (non-zero) temperatures and finite excitation densities. In Sec. 4, we have shown that the phase transition in atomic system under consideration is very close to normal metal-superconductor transition typically studied in the BCS theory – see e.g. Ref. 20. Finally, in Sec. 5 we examine polariton group velocity dependence on the temperature in the presence of normal state-superfluid phase transition. In conclusion we summarize the results obtained.

2. The Model Of Polaritonic Crystal (PolC)

General model of 2D-polaritonic crystal under consideration is shown in Fig. 1. We propose array of the M single-mode microcavities with nearest-neighbor interactions in XY-plane – see Fig. 1. Each of the cavity contains the ensemble of ultracold two-level atoms with $|a\rangle$ and $|b\rangle$ internal levels and interacting with quantized electromagnetic fields in z-direction. There exist different ways to produce such a system experimentally. One of them is based on the confinement of two-level atoms in a special photonic band-gap structure, i.e. photonic crystal or CROW.[21] Notably, in contrast to papers,[17,18] where coupled resonators with doped atoms are studied, in our configuration (see Fig. 1), overlapping of optical fields and atomic wave functions happen in the XY-plane being *orthogonal* to the cavities axis z. Another promising candidate for experimental realization of such a system is connected with trapping atoms in hollow core (photonic crystal or microstructure) optical fibers – see e.g. Ref. 22. In this case we have to describe a polaritonic waveguide array representing some extension of tunnel coupled optical fiber or waveguide array well known in nonlinear optics – cf. Ref. 23.

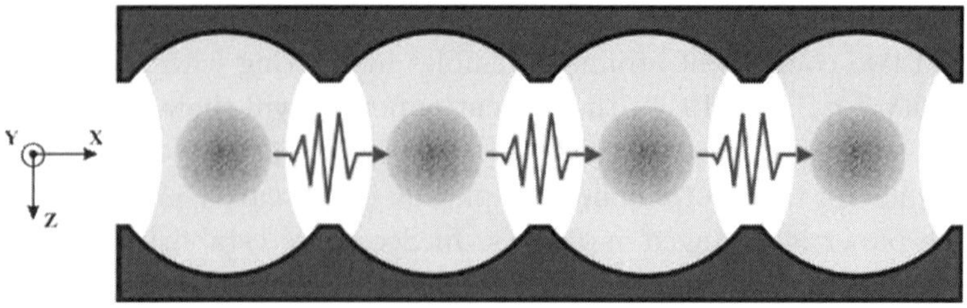

Fig. 1. Model of polaritonic crystal: the array of the microcavities with trapped macroscopic portions of two level atoms.

The total Hamiltonian H for the atom-light coupled system in Fig. 1 can be represented as:

$$H = H_{\rm at} + H_{\rm ph} + H_{\rm int} \tag{1}$$

where $H_{\rm at}$ is the part of Hamiltonian that describes two level atoms in periodical structure, part $H_{\rm ph}$ is responsible for photonic field distribution and term $H_{\rm int}$ characterizes the atom-light interaction in each cavity. In some approximations it is possible to show that the named parts of total Hamiltonian H can be written down as (for more details see also Ref. 19)):

$$H_{\rm at} = \hbar \sum_{n=1}^{M} \Big(\omega_{n,\rm at}^{(a)} a_n^\dagger a_n + \omega_{n,\rm at}^{(b)} b_n^\dagger b_n - \frac{\gamma_a}{2} \left(a_n^\dagger a_{n-1} + a_n^\dagger a_{n+1} + H.C. \right) - \frac{\gamma_b}{2} \left(b_n^\dagger b_{n-1} + b_n^\dagger b_{n+1} + H.C. \right) \Big) \tag{2}$$

$$H_{\rm ph} = \hbar \sum_{n=1}^{M} \Big(\omega_{n,\rm ph} \psi_n^\dagger \psi_n - \frac{\alpha}{2} \left(\psi_n^\dagger \psi_{n-1} + \psi_n^\dagger \psi_{n+1} + H.C. \right) \Big), \tag{3}$$

$$H_{\rm int} = \hbar \sum_{n=1}^{M} \frac{g_n}{\sqrt{N}} \left(\psi_n^\dagger a_n^\dagger b_n + b_n^\dagger a_n \psi_n \right), \tag{4}$$

where annihilation (creation) operators $a_n(a^\dagger)$, $b_n(b^\dagger)$ in (2) characterize dynamical properties of atomic ensemble (atomic quantum modes) at lower ($|a\rangle$) and upper ($|b\rangle$) levels in the n-site array ($n = 1, 2, ..., M$) respectively; operator $\psi_n(t)$ describes the time behavior of cavity field in the n-site array respectively. Coupling coefficients $\gamma_{a,b}$ are the nearest-neighbor hopping constants that depend on the atomic cloud wave function (Wannier functions) overlapping integrals under the so-called tight binding approximation

– cf. Ref. 24. The $\hbar\omega_{n,\mathrm{at}}^{(a)}$ and $\hbar\omega_{n,\mathrm{at}}^{(b)}$ characterize the energy of the atoms at $|a\rangle$ and $|b\rangle$ levels respectively.

Parameter α in (3) characterizes a spatial field overlapping between the neighbor cavities; $\omega_{n,\mathrm{ph}}$ is angular frequency of single mode light field for n-th cavity.

The interaction of two-level atoms with quantized electromagnetic field in (4) is considered under the rotating wave approximation and determined by constant g_n. For simplicity we assume all cavities identical to each other and contain the same atom number $N_n = N$. At the same time we suppose that atom-light coupling coefficients in (4) are equal to each other at all sites, i.e. $g \equiv g_1 = ... = g_M$.

In addition, we consider the following approximations for the problem.

First, we consider ideal atomic gas to consist of non-interacting particles.

Second, we assume that spatial degrees of freedom are "frozen" for both atomic and optical systems in the cavities. In this case characteristic angular frequencies $\omega_{n,\mathrm{at}}^{(a,b)}$ ($\omega_{n,\mathrm{ph}}$) depend on the trapping potential of the atoms (photons) for each site as well. This approach is valid for a relatively small number of atoms, i.e. when $N \leq 10^4$ – cf. Ref. 25.

Third, we are working under the so-called strong atom-light coupling regime for which the coupling parameter g for each site of the lattice is much larger than the inverse coherence time τ_{coh} of combined atom-optical system, i.e. (cf. Refs. 3 and 14)

$$g \gg 2\pi / \tau_{\mathrm{coh}} \, . \tag{5}$$

Physically, τ_{coh} is connected with the time necessary to achieve thermal equilibrium for atomic system under the interaction with quantized optical field in PolC structure. We also neglect the spectral lines broadening mechanisms due to preserving the condition (5) at low temperatures (mK's). We consider pure (or thermodynamical equilibrium) states for coupled atom-light system as a result.

Let us proceed to $\vec{k}$ - representation in expressions (1)–(4). Taking into account periodical properties of the system we can represent operators ψ_n, a_n, b_n in the form:

$$a_n = \frac{1}{\sqrt{M}} \sum_{\vec{k}} a_{\vec{k}} e^{i\vec{k}\vec{n}}, \quad b_n = \frac{1}{\sqrt{M}} \sum_{\vec{k}} b_{\vec{k}} e^{i\vec{k}\vec{n}}, \quad \psi_n = \frac{1}{\sqrt{M}} \sum_{\vec{k}} \psi_{\vec{k}} e^{i\vec{k}\vec{n}}, \tag{6}$$

where $\vec{n}$ is a lattice vector.

To be more specific we consider *one dimensional* spatially-periodical structure for PolC, i.e. $\vec{k}\vec{n} = nk_x l$, $n = 1, 2, ..., M$, l is the lattice constant.

Inserting (6) into (2)–(4) we arrive to the expression for Hamiltonian (1) represented in k - space, i.e.

$$H = \hbar \sum_{\vec{k}} \left(\omega_{ph}(k) \psi_{\vec{k}}^{\dagger} \psi_{\vec{k}} + \frac{\omega_{at}(k)}{2} \left(b_{\vec{k}}^{\dagger} b_{\vec{k}} - a_{\vec{k}}^{\dagger} a_{\vec{k}} \right) \right.$$

$$\left. + \frac{g}{\sqrt{N_{tot}}} \sum_{\vec{q}} \left(\psi_{\vec{k}}^{\dagger} a_{\vec{q}}^{\dagger} b_{\vec{k}+\vec{q}} + b_{\vec{k}+\vec{q}}^{\dagger} a_{\vec{q}} \psi_{\vec{k}} \right) \right), \tag{7}$$

where $N_{\rm tot} = NM$ is a total number of atoms for all sites.

In (7) $\omega_{\rm ph}(k)$ and $\omega_{\rm at}(k)$ characterize the dispersion properties of photonic and atomic system in PolC-structure respectively, and are determined by:

$$\omega_{\rm ph}(k) = \omega_{n,\rm ph} - 2\alpha cos(kl), \quad \omega_{\rm at}(k) = \omega_{n,\rm at}^{(b)} - \omega_{n,at}^{(a)} - 2\gamma cos(kl), \tag{8}$$

where we introduce effective coupling coefficient $\gamma = \gamma_b - \gamma_a$ for atoms in lattice.

The operator of excitations $N_{{\rm ex},\vec{k}}$ with quasi-momentum $\vec{k}$ for band-gap system in Fig. 1 can be defined as

$$N_{{\rm ex},\vec{k}} = \psi_{\vec{k}}^{\dagger} \psi_{\vec{k}} + \frac{1}{2} \sum_{q} \left(b_{\vec{k}+\vec{q}}^{\dagger} b_{\vec{k}+\vec{q}} - a_{\vec{q}}^{\dagger} a_{\vec{q}} \right). \tag{9}$$

It is easy to see that operator $N_{{\rm ex},\vec{k}}$ commutes with Hamiltonian $H_{\vec{k}}$ defined in (7). Therefore, the excitation density $\rho_{\vec{k}}$, defined as

$$\rho_{\rm ex} \equiv \frac{\left\langle N_{{\rm ex},\vec{k}} \right\rangle}{N_{\rm tot}} = \lambda_{\vec{k}}^2 + \frac{1}{2N_{\rm tot}} \sum_{\vec{q}} \left\langle b_{\vec{k}+\vec{q}}^{\dagger} b_{\vec{k}+\vec{q}} - a_{\vec{q}}^{\dagger} a_{\vec{q}} \right\rangle, \tag{10}$$

is the conserving quantity for the considered coupled atom-field system, where $\lambda_{\vec{k}} = \left(\left\langle \psi_{\vec{k}}^{\dagger} \psi_{\vec{k}} \right\rangle / N_{\rm tot} \right)^{1/2}$ is normalized field amplitude with momentum $\vec{k}$. The first term in (10) describes a contribution of optical field to the density of excitations. The second one is $S_{\vec{k}} = \frac{1}{N_{\rm tot}} \sum_{\vec{q}} \left\langle b_{\vec{k}+\vec{q}}^{\dagger} b_{\vec{k}+\vec{q}} - a_{\vec{q}}^{\dagger} a_{\vec{q}} \right\rangle$, and is responsible for collective atomic population inversion.

Physically $\lambda_{\vec{k}}$ represents order parameter in the problem of phase transition. The excitation density $\rho_{\vec{k}}$ is more atom-like when inequality

$$\lambda_{\vec{k}}^2 \ll 1 \tag{11}$$

takes place for order parameter. Physically relation (11) implies small average photon number in comparison with the total number of atoms in PolC, i.e. $\left\langle \psi_{\vec{k}}^{\dagger} \psi_{\vec{k}} \right\rangle \ll N_{\rm tot}$.

A low density limit is obtained from (10), (11) and looks like

$$\rho_{\vec{k}} \approx -0.5, \quad S_{\vec{k}} \simeq -1. \tag{12}$$

Roughly speaking the polaritons can be treated as bosonic quasi-particles in this limit only.

Second, the saturation limit for atomic ensemble is achieved taking into account (11) for excitation density $\rho_{\vec{k}} \approx 0, (S_{\vec{k}} \simeq 0)$.

Third, for $\rho_{\vec{k}} \approx +0.5, (S_{\vec{k}} \simeq 1)$ the inversion regime in two-level atomic system coupled with optical field occurs.

For large values of field amplitude, e.g. for $\lambda_{\vec{k}} > 1$, the excitation density $\rho_{\vec{k}}$ becomes more photon-like. In this case the polaritonic model under discussion breaks down. At the same time, nonlinear effects for atom-field interaction become more important. But it is not the case of our consideration and we are going to hold relation (11) in the paper.

3. Polariton Dispersion in the Cavity Array

Under the tight-binding approximation expression (7) represents *many body* Hamiltonian for atom-field interaction in *momentum space* and describes periodical structure in Fig. 1. Neglecting inhomogeneous broadening of atomic system and introducing collective polarization operator $P_{\vec{k}} = \frac{1}{\sqrt{N_{\text{tot}}}} \sum_{\vec{q}} a^{\dagger}_{\vec{q}} b_{\vec{k}+\vec{q}}$ it is possible to represent Hamiltonian (7) in the form:

$$H = \hbar \sum_{\vec{k}} \left(\omega_{\text{ph}}(k) \psi^{\dagger}_{\vec{k}} \psi_{\vec{k}} + \omega_{\text{at}}(k) P^{\dagger}_{\vec{k}} P_{\vec{k}} + g(\psi^{\dagger}_{\vec{k}} P_{\vec{k}} + P^{\dagger}_{\vec{k}} \psi_{\vec{k}}) \right). \tag{13}$$

The Hamiltonian (13) can be diagonalized using linear Bogoliubov transformation describing dark and bright state polaritons – cf. Ref. 15.

To determine the order parameter $\lambda_{\vec{k}}$ we use a variational (thermodynamic) approach, see e.g. Ref. 8. In the case the grand canonical ensemble with finite (nonzero) chemical potential μ with partition function $Z_{\vec{k}}(N_{\text{tot}}, T) = Tr(e^{-\beta H'_{\vec{k}}})$ should be explored; $H'_{\vec{k}} = H_{\vec{k}} - \mu N_{\text{ex},\vec{k}}$ being a modified Hamiltonian, $\beta \equiv (k_B T)^{-1}$. Notably, thermodynamic approach under consideration is valid for large total number of atoms in the PolC structure, i.e. for $N_{tot} = NM \gg 1$. Since the number of atoms N at each site is not so large we need to have a large number of cavities ($M \gg 1$) as well.

In semiclassical limit, while neglecting fluctuations of optical field and also atom-field correlations, one can obtain

$$\tilde{\omega}_{\rm ph}\lambda_{\vec{k}} = \frac{g^2\lambda_{\vec{k}}\tanh\left(\frac{\hbar\beta}{2}\sqrt{\tilde{\omega}_{\rm at}^2+4g^2\lambda_{\vec{k}}^2}\right)}{\sqrt{\tilde{\omega}_{\rm at}^2+4g^2\lambda_{\vec{k}}^2}}, \tag{14}$$

where we made denotations $\tilde{\omega}_{\rm ph} \equiv \omega_{\rm ph}(k) - \mu$, $\tilde{\omega}_{\rm at} \equiv \omega_{\rm at}(k) - \mu$ and used the coherent basis representation for partition function $Z_{\vec{k}}(N_{\rm tot}, T)$ for order parameter (field amplitude) $\lambda_{\vec{k}}$ with quasi-momentum $\vec{k}$.

By its form the Eq. (14) is similar to BCS gap equation applied for studying phase transition to superfluid (or condensed) state for various systems in solid state physics – see e.g. Refs. 13, 20 and 26.

We examine Eq. (14) in the framework of phase transition of atomic polaritons to superfluid state for atomic polaritons. In particular, parameter combination $2g\lambda_{\vec{k}}$ plays the role of the frequency gap, and normalized photon number $\lambda_{\vec{k}}$ in the cavity array represents the order parameter. The normal state of polaritons is obtained from (14) for $\lambda_{\vec{k}} = 0$ and occurs for temperature $T \geq T_{\rm c}$ ($T_{\rm c}$ is the critical temperature of phase transition). On the other hand, nontrivial solutions of Eq. (14) with $\lambda_{\vec{k}} \neq 0$ characterize the superfluid states of polaritons for temperatures $T < T_{\rm c}$.

By using (10) and partition function $Z_{\vec{k}}(N_{\rm tot}, T)$ we can get an expression for polariton excitation density $\rho_{\vec{k}}$ at finite (non-zero) temperatures

$$\rho_{\vec{k}} = \lambda_{\vec{k}}^2 - \frac{1}{2}\frac{|\tilde{\omega}_{\rm at}|\tanh\left(\frac{\hbar\beta}{2}\sqrt{\tilde{\omega}_{\rm at}^2+4g^2\lambda_{\vec{k}}^2}\right)}{\sqrt{\tilde{\omega}_{\rm at}^2+4g^2\lambda_{\vec{k}}^2}} \tag{15}$$

The phase transition to superfluid state occurs for density $\rho_{\vec{k},{\rm c}} = -\frac{1}{2}\tanh\left(\frac{\hbar|\tilde{\omega}_{\rm at}|}{2k_{\rm B}T_{\rm c}}\right)$ obtained from (15) for critical temperature of phase transition $T = T_{\rm c}$ (when $\lambda_{\vec{k}} = 0$).

Putting (15) into (14) we obtain for chemical potential μ of the polaritonic system:

$$\mu_{1,2} = \frac{1}{2}\left[\omega_{\rm at}(k) + \omega_{\rm ph}(k) \pm \omega_{\rm R,eff}(k)\right], \tag{16}$$

where $\omega_{\rm R,eff}(k) = \sqrt{\delta^2 + 4g_{\rm eff}^2}$ is the effective Rabi splitting frequency; $g_{\rm eff} = g\sqrt{2(\lambda_{\vec{k}}^2 - \rho_{\vec{k}})}$ defines the effective atom-field coupling coefficient depending on the temperature, excitation density $\rho_{\vec{k}}$ and normalized field amplitude $\lambda_{\vec{k}}$; $\delta = \omega_{\rm ph}(k) - \omega_{\rm at}(k) = \tilde{\delta} - 2(\alpha - \gamma)cos(kl)$ is the phase mismatch depending on quasi-momentum k, $\tilde{\delta} \equiv \omega_{n,{\rm ph}} - (\omega_{n,{\rm at}}^{(b)} - \omega_{n,{\rm at}}^{(a)})$ is the detuning for $kl = \pi/2 + \pi n$, $n = 0, 1, \ldots$.

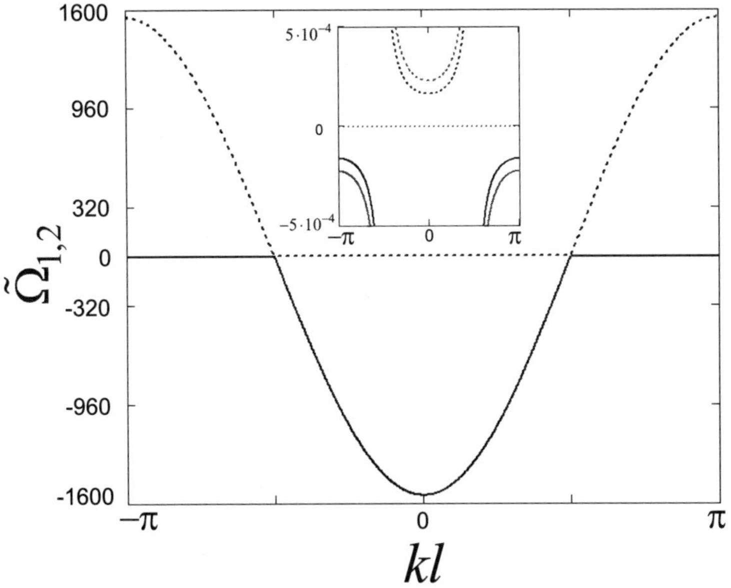

Fig. 2. Dependences of characteristic angular frequencies for upper ($\tilde{\Omega}_1(k)$) and lower ($\tilde{\Omega}_2(k)$) branch of polaritons versus reduced quasi-momentum (i.e. Bloch vector) kl in the first Brillouin zone. The angular frequency detuning is $\tilde{\delta} = 0$, lattice constant is $l = 2\mu m$. In the inset the "fine structure" of Rabi splitting is presented. The solid curves correspond to polariton dispersion taking into account finite (non-zero) cavity field amplitudes $\lambda_{\vec{k}} = 0.5$ and excitation density $\rho_{\vec{k}} = -0.45$. The dotted line corresponds to dispersion (frequency spacing) for two level atoms in the lattice.

For Rabi splitting frequency we have $\omega_{\mathrm{R,eff}}(k) = 0$ in resonance condition $\delta = 0$ for atom-field interaction and when $\rho_{\vec{k}} = \lambda_{\vec{k}}^2$. The latter equality requires a positive density $\rho_{\vec{k}} > 0$ that corresponds to inversion in two-level atomic system.

We are now focusing our attention on polariton properties of atomic system without inversion, i.e. for only negative (low enough) excitation densities ($\rho_{\vec{k}} < 0$). In this limit the Eqs.(16) define the upper (μ_1) and lower (μ_2) branch polariton frequencies versus both temperature and excitation densities.

In Fig. 2 the reduced dispersion relation $\tilde{\Omega}_{1,2}(k) \equiv [\mu_{1,2}(k) - (\omega_{n,\mathrm{at}}^{(b)} - \omega_{n,\mathrm{at}}^{(a)})]/2g$ for upper and lower branch of polaritons is presented in the first Brillouin zone of the considered periodical structure. We consider, as an example, the interaction of quantized optical field with ensembles of two-level rubidium atoms having a mean resonance frequency of transition $382THz$; suppose that the wave vector of the resonant light field $k_z \simeq 0.8 \times 10^7$ m^{-1} ($\lambda \simeq 785$nm), corresponding to the weighted mean of the rubidium D-lines. In the case the effective photon mass can be estimated as $m_{\rm ph} \simeq 2.8 \times 10^{-36} kg$ which implies low branch polariton mass $m_{\rm ph} \simeq 5.6 \times 10^{-36} kg$. We consider also a zero detuning Rabi splitting frequency (the atom-field coupling parameter) $g/2\pi = 500MHz$ for all numerical simulations. The spontaneous emission lifetime for rubidium D-lines is $\tau_{\rm spont} \approx 27ns$ which corresponds to spontaneous emission rate equal to about $37MHz$. Thereby, the condition (5) is still fulfilled in the case under discussion.

Fig. 2 describes polariton dispersion in low density limit (12). In the inset to Fig. 2 solid curves demonstrate polariton dispersion for non-zero order parameter $\lambda_{\vec{k}}$ and excitation value $\rho_{\vec{k}} = -0.45$.

For relatively small photon number in the cavity array, i.e. under the condition (11), the effective Rabi splitting angular frequency $\omega_{\rm R,eff}(k)$ can be represented as

$$\omega_{\rm R,eff}(k) \approx \omega_R(k) + \frac{4g^2\lambda_{\vec{k}}^2}{\sqrt{\delta^2 - 8g^2\rho_{\vec{k}}}} \tag{17}$$

where $\omega_R(k) = \sqrt{\delta^2 - 8g^2\rho_{\vec{k}}}$ is the Rabi splitting angular frequency for the normal state of polaritons at $\lambda_{\vec{k}} = 0$. The last term in (17) describes the temperature dependent additional frequency gap between polaritonic branches occurring due to the phase transition at $T < T_{\rm c}$ – see Fig. 2.

The principal feature of dispersion curves in Fig. 2 is the presence of minimum at $k = 0$ for low branch polaritons. For small quasi-momentum values, such as $kl \ll 1$, from Eqs.(16) one can obtain:

$$\mu_{1,2}(k) \approx \frac{\hbar k^2}{2m_{1,2}}, \tag{18}$$

where

$$m_{1,2} = \frac{2m_{\rm at}m_{\rm ph}\sqrt{\tilde{\Delta}^2 + 4g_{\rm eff}^2}}{(m_{\rm at} + m_{\rm ph})\sqrt{\tilde{\Delta}^2 + 4g_{\rm eff}^2} \pm (m_{\rm at} - m_{\rm ph})\tilde{\Delta}} \tag{19}$$

are masses of upper (m_1) and lower (m_2) polariton branch. In (19) we introduce maximal detuning $\tilde{\Delta} = \tilde{\delta} - 2(\alpha - \gamma)$ accessible at $k = 0$; $m_{\rm at} = \frac{\hbar}{2\gamma l^2}$ and $m_{\rm ph} = \frac{\hbar}{2\alpha l^2}$ are the effective masses of atoms and photons in the lattice. In a real experimental situation (when relation $m_{\rm ph}/m_{\rm at} \ll 1$ (i.e. $\alpha \gg \gamma$) is held) the masses of polaritons $m_{1,2} = \frac{2m_{\rm ph}}{1 \pm m_{\rm ph}/m_{\rm at}} \simeq 2m_{\rm ph}$ are small enough for resonant atom-optical interaction at $\tilde{\Delta} = 0$.

Thus, the relation (18) for $\mu_2(k)$ describes the free quasi-particles (polaritons) at the bottom of the lower curve in Fig. 2. Such a peculiarity of the low branch polariton dispersion can be used to achieve the BEC-state with quasi-momentum $k = 0$ cf. Refs. 4 and 14.

4. Phase Transition for Low Branch Polaritons

By using (15) we can bring Eq. (14) to the form:

$$\Omega_{\mathrm{ph}} = \frac{\tanh\left(\frac{\hbar\beta g}{2}\sqrt{\Omega_{\mathrm{at}}^2 + 4\lambda_{\vec{k}}^2}\right)}{\sqrt{\Omega_{\mathrm{at}}^2 + 4\lambda_{\vec{k}}^2}}, \tag{20}$$

where $\Omega_{\mathrm{at}} = -\delta/2g + \sqrt{(\delta/2g)^2 + 2(\lambda_{\vec{k}}^2 - \rho_{\vec{k}})}$, $\Omega_{\mathrm{ph}} = \delta/2g + \sqrt{(\delta/2g)^2 + 2(\lambda_{\vec{k}}^2 - \rho_{\vec{k}})}$.

The critical temperature T_{c} of the phase transition between normal and superfluid (BEC) state can be easily found out from (20) putting $\lambda_{\vec{k}} = 0$. Indeed one can obtain:

$$T_{\mathrm{c}} \simeq \frac{\hbar g \Omega_{\mathrm{at}}^{(0)}}{2k_{\mathrm{B}}\mathrm{atanh}\left(-2\rho_{\vec{k}}\right)} \approx \frac{\hbar g\sqrt{-2\rho_{\vec{k}}}}{k_{\mathrm{B}}\ln\left(2/(1+2\rho_{\vec{k}})\right)}, \tag{21}$$

where $\Omega_{0,\mathrm{at}}^{(0)} \equiv \Omega_{\mathrm{at}}|_{\lambda_{\vec{k}}=0}$. The last expression in (21) is true for $\delta = 0$ at low temperature limit, when the condition

$$k_{\mathrm{B}}T \ll \hbar g \tag{22}$$

is satisfied.

In Fig. 3 the phase boundary for T_{c} according to Eq. (21) is presented. It is clearly seen that the critical temperature vanishes for positive detuning δ. Physically the fact is connected with "heavy" (i.e. atom-like) polaritons for lower branch – see (19). On the other hand for negative detuning ($\delta < 0$) the polaritons become more photon-like, and temperature T_{c} increases. Notably a very low excitation density requires a large value of coupling constant g for high enough temperatures of phase transition – cf. (21). On the other hand, for a given value of g the phase transition temperature T_{c} decreases when the density of excitations approaches the value corresponding to the low density limit described by condition (12).

In Fig. 4 we plotted dependence of the order parameter $\lambda_{\vec{k}}$ as a function of T/T_{c} in accordance with numerical solution of Eq. (20) for various detunings δ. Fig. 4 demonstrates a second-order continuous phase transition for photon number in the cavity array that occurs at low (mK) temperatures.

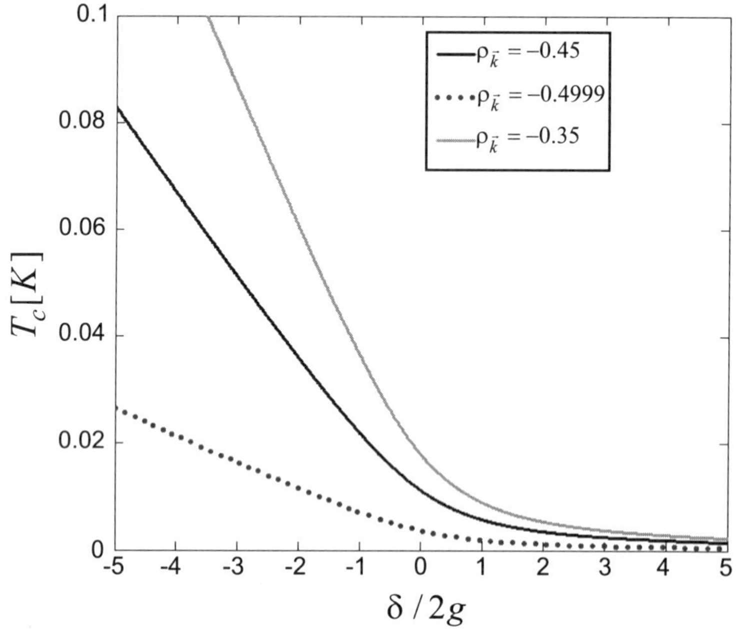

Fig. 3. Dependence of critical temperature T_c on normalized atom-filed frequency detuning $\delta/2g$.

The zero temperature order parameter $\lambda_{\vec{k},0} \equiv \lambda_{\vec{k}}\big|_{T=0}$ decreases for positive detuning δ when the low branch polaritons become more atom-like. In this region for the order parameter we have

$$\lambda_{\vec{k},0} \simeq \sqrt{\frac{1-4\rho_{\vec{k}}^2}{8\Omega_{\mathrm{ph}}^{(0)}\sqrt{\left(\frac{\delta}{2g}\right)^2 - 2\rho_{\vec{k}}}}}, \tag{23}$$

where $\Omega_{\mathrm{ph}}^{(0)} \equiv \Omega_{\mathrm{ph}}\big|_{\lambda_{\vec{k}}=0}$.

In the low temperature limit (22) the behavior of the order parameter, presented in Fig. 4, is described by the expression:

$$\lambda_{\vec{k}} \simeq \lambda_{\vec{k},0}\sqrt{1-e^{x_c - x}}, \tag{24}$$

where we introduce the dimension-less parameter $x = \hbar g|\Omega_{\mathrm{at}}^{(0)}|/k_{\mathrm{B}}T$, $x_c = x|_{T=T_c}$

For high enough temperatures when $k_{\mathrm{B}}T \gg \hbar g$ and for $T \to T_c$ from (20) under the condition (11) for the order parameter $\lambda_{\vec{k}}$ it is possible to

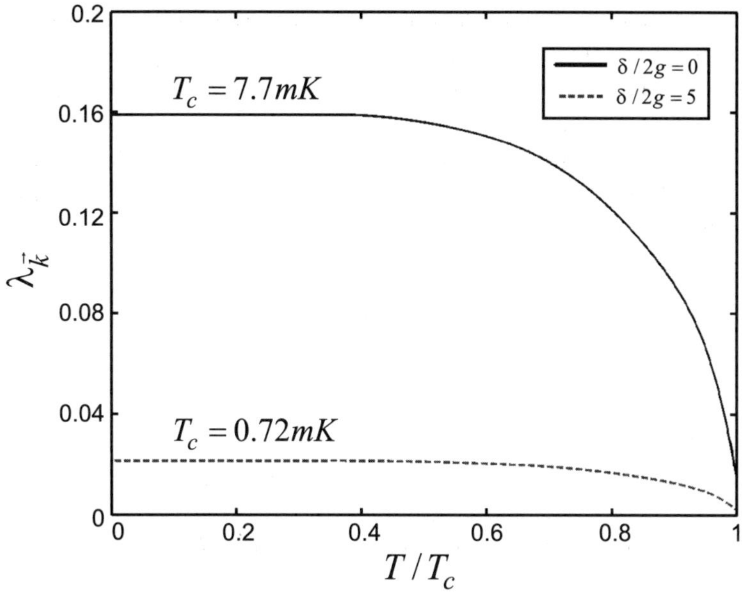

Fig. 4. Dependence of the order parameter $\lambda_{\vec{k}}$ versus reduced temperature $T/T_{\rm c}$. The excitation density $\rho_{\vec{k}} = -0.45$.

find out:

$$\lambda_{\vec{k}} \simeq \sqrt{\frac{\left(\Omega_{\rm at}^{(0)}\right)^2 \Omega_{\rm ph}^{(0)} x_c}{8\sinh(x_c)\sqrt{\left(\frac{\delta}{2g}\right)^2 - 2\rho_{\vec{k}}}}\left(1 - \frac{T}{T_{\rm c}}\right)} \quad . \tag{25}$$

Experimentally it is easier to study the dependence of the order parameter $\lambda_{\vec{k}}$ vs atom-field detuning δ or vs angular frequency of optical field $\omega_{ph}(k)$ when the temperature of atomic system array is fixed. Near the critical point the behavior for dependences in Fig. 4 can be described as [cf.(25)]

$$\lambda_{\vec{k}} \simeq \sqrt{\frac{\hbar\omega_{\rm ph,\mathit{c}}\left(\Omega_{\rm at,\mathit{c}}\right)^3 \Omega_{\rm ph,\mathit{c}}}{16k_{\rm B}T\sinh(x_c)\left[\left(\frac{\delta_c}{2g}\right)^2 - 2\rho_{\vec{k}}\right]}\left(1 - \frac{\omega_{\rm ph}}{\omega_{\rm ph,\mathit{c}}}\right)}, \tag{26}$$

where $x_c = \hbar g|\Omega_{\rm at,\mathit{c}}|/k_{\rm B}T$, and $\Omega_{\rm at,\mathit{c}} = -\delta_c/2g + \sqrt{(\delta_c/2g)^2 - 2\rho_{\vec{k}}}$, $\Omega_{\rm ph,\mathit{c}} = \delta_c/2g + \sqrt{(\delta_c/2g)^2 - 2\rho_{\vec{k}}}$; $\delta_c = \omega_{\rm ph,\mathit{c}} - \omega_{\rm at}$ is the critical detuning, $\omega_{\rm ph,\mathit{c}}$

is a critical frequency of optical field for which the phase transition to the state with $\lambda_{\vec{k}} = 0$ occurs. Such a critical frequency $\omega_{\mathrm{ph},c}$ depends on the field distribution in a spatially-periodic system in general (see Fig. 1).

Now let us pay attention to the optical properties of atomic system.

An important feature of a two-level atomic system is the existence of the average collective stationary polarization of $P_{\vec{k}}(t) = P_{\vec{k}} e^{-i\mu t}$ and inversion $S_{\vec{k}}$ of the atomic level occupation that looks like:

$$P_{\vec{k}} = -\frac{g\lambda_{\vec{k}}\tilde{\omega}_{\mathrm{at}} \tanh\left(\frac{\hbar\beta}{2}\sqrt{\tilde{\omega}_{\mathrm{at}}^2 + 4g^2\lambda_{\vec{k}}^2}\right)}{|\tilde{\omega}_{\mathrm{at}}|\sqrt{\tilde{\omega}_{\mathrm{at}}^2 + 4g^2\lambda_{\vec{k}}^2}}, \tag{27}$$

$$S_{\vec{k}} = -\frac{|\tilde{\omega}_{\mathrm{at}}| \tanh\left(\frac{\hbar\beta}{2}\sqrt{\tilde{\omega}_{\mathrm{at}}^2 + 4g^2\lambda_{\vec{k}}^2}\right)}{\sqrt{\tilde{\omega}_{\mathrm{at}}^2 + 4g^2\lambda_{\vec{k}}^2}}. \tag{28}$$

For the normal state of polaritons ($\lambda_{\vec{k}} = 0$) at $T \geq T_{\mathrm{c}}$ from (27), (28) we obtain

$$P_{\vec{k}} = 0, \quad S_{\vec{k}} = -\tanh\left(\frac{\hbar|\tilde{\omega}_{\mathrm{at}}^2|}{2k_{\mathrm{B}}T}\right) \tag{29}$$

Thus, the existence of condensed solution of Eq. (14) with $\lambda_{\vec{k}} \neq 0$ leads to the appearance of stationary polarization $P_{\vec{k}}(t)$ in the cavity array of two level atoms on non-zero frequency that is equal to chemical potential μ. Physical meaning of angular frequencies $\tilde{\omega}_{\mathrm{ph}}$ and $\tilde{\omega}_{\mathrm{at}}$ in Eq. (14) becomes clear: they determine the detunings of both optical field frequency and the frequency $\tilde{\omega}_{\mathrm{ph}}(k)$ of two-level atomic transition $\tilde{\omega}_{\mathrm{at}}(k)$ from the frequency μ of polarization in atomic ensemble.

With the help of routine definition $\chi = \chi^{(1)} + 3\chi^{(3)}|E|^2$ it is easy to find out the expressions for linear ($\chi^{(1)}$) and cubic nonlinear ($\chi^{(3)}$) susceptibilities of atomic system:

$$\chi^{(1)} \simeq -\frac{N_{\mathrm{tot}}|\mu_{ab}|^2}{V\hbar g\Omega_{\mathrm{at}}^{(0)}} \tanh\left(\frac{x}{2}\right), \tag{30}$$

$$\chi^{(3)} \simeq \frac{2N_{\mathrm{tot}}|\mu_{ab}|^4\xi}{3V(\hbar g)^3{\Omega_{\mathrm{at}}^{(0)}}^3}\left[\tanh\left(\frac{x}{2}\right) - \frac{x}{2\cosh^2\left(\frac{x}{2}\right)}\right], \tag{31}$$

where $\xi = 1 + \Omega_{\mathrm{at}}^{(0)} \Big/ 2[(\delta/2g)^2 - 2\rho_{\mathrm{ex}}]^{1/2}$, μ_{ab} is dipole matrix element.

The expressions (30), (31) represent a finite temperature generalization of well known result for susceptibilities of two level atomic system under the thermal equilibrium far from resonance ($\Omega_{\rm at}^{(0)^2} \gg \lambda_{\vec{k}}^2$) and in the absence of inhomogeneous broadening – see Ref. 27. The maximal (absolute) value of χ corresponds to zero temperature. The strong nonlinearity of atomic medium in (31) is characterized by a large value of order parameter $\lambda_{\vec{k}}$ when polaritonic model of atom-field interaction breaks down. Near the critical point of phase transition the nonlinearity is negligibly small ($\lambda_{\vec{k}} \to 0$) and we have a linear macroscopic polarization of atomic medium – cf. (27).

5. Polariton Group Velocity

Now let us consider the group velocity $v_{1,2} = \frac{\partial \mu_{1,2}(k)}{\partial k}$ behavior for polaritons. With the help of Eqs. (18), (19) it is possible to find out:

$$v_{1,2}(k) = \frac{\hbar \sin(kl)}{2lm_{ph}} \left(1 + \frac{m_{\rm ph}}{m_{at}} \pm \left(1 - \frac{m_{\rm ph}}{m_{\rm at}} \right) \frac{\delta}{\sqrt{\delta^2 + 4g_{\rm eff}^2}} \right). \tag{32}$$

In particular, for $kl \ll 1$ the dependence of group velocities $v_{1,2}$ for polaritons is approximately linear, i.e. $v_{1,2}(k) \simeq \frac{\hbar k}{m_{1,2}}$. In this limit the magnitude of polaritons group velocity is restricted by the inequality

$$2\gamma l^2 k \le v_{1,2}(k) \le 2\alpha l^2 k, \tag{33}$$

where upper bound corresponds to photon-like polaritons. The lower bound in (33) describes atom-like polaritons and can be achieved under the condition $\frac{g_{\rm eff}^2}{\Delta^2} \ll \frac{m_{\rm ph}}{m_{\rm at}} \ll 1$. Thereby, the manipulation of group velocity of light in PolC structure can be realized by varying polaritonic effective masses $m_{1,2}$ with the help of atom-field detuning $\tilde{\Delta}$.

Notably, the group velocities $v_{1,2}$ vanish on the boundaries of Brillouin zone, i.e. $v_{1,2} = 0$ for $kl = n\pi$, $n = 0, \pm 1, ...$, and so, the PolC structure admits a complete polariton localization within the zone.

The ability to reduce the group velocity of polaritons can be used for observing "slow" light phenomenon in PolC. At the same time we can explore such a possibility for spatial storing and retrieving quantum optical information - cf. Ref. 18. For this case we can control the group velocity of propagating optical field by means of the atom-field detuning Δ (or $\tilde{\Delta}$).

In Fig. 5 we present the phase boundaries for the group velocities of low branch polaritons versus reduced temperature $T/T_{\rm c}$ of atomic system. We assume that the polaritons have the quasi-momentum k values which are different but close to each other. The chosen parameters in Fig. 5 correspond to atom-like polaritons with small k in the bottom region of dispersion curve (see Fig. 2) under the condition $|\tilde{\delta}| \gg |\alpha - \gamma| k^2 l^2$ (only in this limit we can

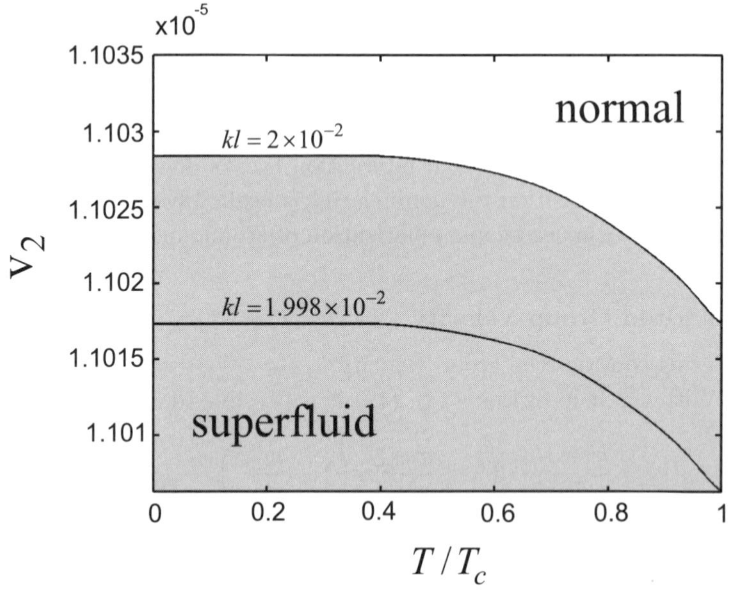

Fig. 5. Reduced group velocity $v_2 \equiv v_2(k)$ for the lower branch polaritons as a function of reduced temperature T/T_c (the detuning $\tilde{\delta}/2g = 5$); v_2 represented in the units of the speed of light in the vacuum c.

assume that polaritons with different k have approximately the same critical temperature).

Thereby, in the limit of $\tilde{\Delta} \gg 2|g_{eff}|$ and $\frac{m_{ph}}{m_{at}} = \frac{\gamma}{\alpha} \ll \frac{g_{eff}^2}{\tilde{\Delta}^2} \ll 1$ we can obtain for polariton group velocities

$$v_2(k) \approx \frac{2\hbar k g^2(\lambda_{\vec{k}}^2 - \rho_{\vec{k}})}{m_{\rm ph}\tilde{\Delta}^2} = \frac{4\alpha l^2 k g^2(\lambda_{\vec{k}}^2 - \rho_{\vec{k}})}{\tilde{\Delta}^2}. \tag{34}$$

Eq. (34) demonstrates a significant reduction of group velocity value for atom-like polaritons. In fact under conditions of Fig. 5 $v_2(k)/c \simeq 10^{-5}$. The expression $v_{2,c}(k) = \frac{4\alpha l^2 k g^2 |\rho_{\vec{k},c}|}{\tilde{\Delta}^2}$ describes critical velocity of low branch polaritons (for given k) that can be easily found out for the critical temperature of the phase transition $T = T_c$. On the other hand, the maximal group velocity of polaritons $v_{2,0}(k) = \frac{4\alpha l^2 k g^2 |\rho_{\vec{k}}|}{\tilde{\Delta}^2}$ is achieved for the order parameter $\lambda_{\vec{k}} = \lambda_{\vec{k},0}$ at $T = 0$.

Thereby, there exists some temperature domain and the order parameter value region (see Fig. 4) where the polaritons (with given quasi-momentum k) exist in superfluid state, and have the velocity value such as $v_{2,c}(k) \leq$

$v_2(k) \leq v_{2,B}(k)$, where $v_{2,B}(k)$ is the velocity of polaritons at the phase boundary under the transition to superfluidity. Fig. 5 illustrates the results of the analysis: shaded region corresponds to superfluid state for atom-like polaritons with various quasi-momentum k starting from $kl = 2 \cdot 10^{-2}$.

Thus, the phase transition results in superfluid low branch polariton state with group velocities $v_2(k)$ for which the order parameter (light field amplitude) $\lambda_{\vec{k}} \neq 0$. Otherwise polaritons are in normal state with velocities $v_2(k) \geq v_{2,B}(k)$ and the order parameter $\lambda_{\vec{k}} = 0$. The discussed effects result in the possibility to strong control the optical pulse propagation in the periodical structure (Fig. 1) under the condition of coupled atom-light states existence.

6. Conclusion

Let us briefly summarize the results obtained in the paper. We propose a new type of spatially periodical structure, – polaritonic crystal (PolC), to observe a slow light phenomenon due to coupled atom-field states (polaritons) in the lattice. Under the tight-binding approximation such a system realizes an array of weakly coupled trapped two-component atomic ensembles interacting with optical field in a tunnel-coupled one dimensional cavity array. We have considered the problem of formation of BCS state for polaritons arising due to interaction of two-level atoms with optical field in the system under discussion. We have shown that the second order continuous phase transition results in formation of superfluid state of polaritons under certain conditions. We have shown that the macroscopic polarization of atomic medium on non-zero frequency occurs in this case. It is important that the observable group velocity for propagating optical wave packet can be essentially reduced due to localization of atom-field excitations (polaritons) in the polaritonic crystal structure. It can be smoothly controled by using some parameters (frequency detuning) of atom-field system.

Acknowledgments

This work is partially supported by Russian Foundation for Basic Research (Grants Nos. 08-02-99011, 09-02-91350) and by some other grants from Agency of Education and Science of Russian Federation.

References

1. M. Greiner, O. Mandel, T. Esslinger, T. W. Hansch, and I. Bloch, Nature **415**, 39 (2002).
2. F. Dimer, B. Estienne, A. S. Parkins, and H. J. Carmichael, Phys. Rev. A **75**, 013804 (2007).
3. H. Deng, G. Weihs, C. Santori, J. Bloch, and Y. Yamamoto, Science **298**, 199 (2002).
4. J. Kasprzak, M. Richard, S. Kunderman, et al, Nature **443**, 409 (2006).
5. H. Deng, D. Press, S. Gotzinger, et al, Phys. Rev. Lett. **97**, 146402 (2006).
6. R. Y. Chiao, J. Boyce, Phys. Rev. A. **60**, 4114 (1999).
7. V. De Giorgio, M. O. Scully, Phys. Rev. A. **2**, 1170 (1970).
8. K. Hepp and E. H. Lieb, Ann. Phys. (N.Y.) **76**, 360 (1973).
9. V. I. Emeljanov and Yu. L. Klimontovich, Phys. Lett. A **59**, 366 (1976).
10. K. Rzazewski, K. Wodkiewicz, and W. Zakowicz, Phys. Rev. Lett. **35**, 432 (1975).
11. G. Liberti and R. L. Zaffino, Phys. Rev. A **70**, 033808 (2004).
12. J. Larson, M. Lewenstein, New J. of Phys. **11**, 063027 (2009).
13. P. R. Eastham and P. B. Littlewood, Phys. Rev. B **64**, 235101 (2001).
14. V. A. Averchenko, A. P. Alodjants, S. M. Arakelian, et al., Rus. J. Quantum Electronics **36**, 1 (2006).
15. A. P. Alodjants, S. M. Arakelian, S. N. Bagaev, I. A. Chekhonin, V. S. Egorov, J. Rus. Laser Res. **27**, 400 (2006); A. P. Alodjants, S. M. Arakelian, and A. Yu. Leksin, Laser Physics **17**, 1 (2007).
16. T. Aoki, B. Dayan, E. Wilcut, W.P.Bowen, A.S. Parkins, T.Y. Kippenberg, K.Y. Vahala, H.J. Kimble, Nature, **443**, 671, (2006).
17. M.J. Hartmann, F. Brandao, M.B. Plenio, Nature, **2**, 849, (2006).
18. L. Zhou, J. Lu, C. P. Sun, Phys. Rev.A, **76**, 012313 (2007).
19. I.O. Barinov, A. P. Alodjants, S. M. Arakelian, Rus., Quantum Electronics **39**, 685 (2009).
20. A. A. Abrikosov, L. P. Gor'kov, and I. E. Dzyaloshinskii, *Methods of quantum field theory in statistical mechanics*, (Prentice-Hall, Inc., Englewood Cliffs, New Jersey, 1963).
21. Z.S. Yang, N.H.Kwong, R.Binder, A.L.Smirl, JOSA B., **22**, 2144 (2005); M.F. Yanik, S. Fan, Phys. Rev. Lett., **92**, 083901-1 (2004).
22. F. Benabid, F. Couny, J. C. Knight, T. A. Birks, P. St J. Russell, Nature **434**, 488 (2005).
23. R. Morandotti, U. Peschel, J. S. Aitchison, H. S. Eisenberg, Y. Silberberg, Phys. Rev. Lett. **83**, 4756 (1999).
24. A. Smerzi, A. Trombettoni, P.G Kevrekidis, A.R. Bishop, Phys. Rev.Lett **89**, 170402-1 (2002).
25. J. R. Anglin, A. Vardi, Phys.Rev.A **64**, 013605 (2001).
26. L.V. Keldysh, Yu. V. Kopaev, Fizika Tverdogo Tela (Sov. Solid State Phys.) **6**, 2791 (1964).
27. Robert W. Boyd. Nonlinear Optics, Academic Press, Amsterdam (2003).

PART C

Nonlinear Optical Effects and Spectroscopy

FORMATION OF NARROW OPTICAL RESONANCES USING SUBMICRON-THIN ATOMIC VAPOR LAYERS

D. SARKISYAN* and A. PAPOYAN

Institute for Physical Researches, National Academy of Sciences,
Ashtarak-2, 0203, Ashtarak, Armenia
**E-mail: david@ipr.sci.am*

We address the problem of formation of spectrally narrow optical resonances based on resonant absorption and fluorescence processes in an extremely thin vapor cells with smoothly variable thickness L comparable with resonant wavelength ($\lambda = 794$ nm for D_1 line of ^{85}Rb, ^{87}Rb atoms). Studying and exploiting unusual and spectacular behavior of resonant processes strongly dependent on L/λ ratio, we demonstrate several experimental techniques capable of quantitative atomic spectroscopic studies and related applications. Particularly, we have shown realization of advantageous atomic frequency reference schems and methods for determination of frequency shifts of split Zeeman components of hyperfine transitions and modification of their transition probabilities in external magnetic field B =1 − 4000 G. We discuss and compare usefulness of the proposed techniques for practical applications, such as measurement and mapping of strongly inhomogeneous magnetic field with sub-micron spatial resolution.

Keywords: Atomic spectroscopy; wavelength-thin media; narrow resonance; frequency reference; magneto-optics.

1. Introduction

In original experiments[1,2] it was revealed that the resonant interaction of laser radiation with atoms in a thin film of alkaline vapor with thickness $L = 10 - 1000$ μm may underlie a novel method for sub-Doppler spectroscopy, as was earlier predicted theoretically in Refs. 3–5. Distinguishing peculiarity of thin vapor films is caused by the fact that the duration of the atom-laser radiation interaction, governed by wall-to-wall trajectories, is anisotropic. As a particular consequence, under special experimental conditions in thin atomic vapor column, velocity selective optical pumping/saturation (VSOP) sub-Doppler resonance peak of reduced absorption takes place in transmission spectra at the exact position of atomic resonance.[1–3] The VSOP peak appears because an atom (as an example, let us consider energy levels of ^{87}Rb shown in Fig. 6) in the ground level $F_g = 1$ absorbs a laser photon, goes to the excited level $F_e = 2$, and then gets de-excited spontaneously to either the ground state $F_g = 1$ or to $F_g = 2$. This phenomenon is termed optical pumping (OP). As a result of OP, a part of atoms resides in the state $F_g = 2$ thus reducing the number of atoms absorbing from the level $F_g = 1$; therefore, also the absorption from this level is reduced. The efficiency of OP process is determined by the expression:[1–3,10]

$$\eta \sim \frac{\Omega^2 \gamma_N t}{\left(\Delta + \vec{k}\vec{v}\right)^2 + \Gamma^2}, \tag{1}$$

where t is the radiation-atom interaction time, Δ is the frequency detuning from resonance, and Γ is the sum of homogeneous and inhomogeneous broadenings. It is seen from Eq. (1) that the longer the time of interaction t, the higher the efficiency of OP. For atoms flying perpendicularly to the laser beam, the interaction time is $t_D = D/v$ (D being the diameter of the laser beam), whereas for atoms flying along the beam we have $t_L = L/v$. Since the beam diameter is $\simeq 1$ mm, while the distance between the windows L is ~ 10 μm,[1,2] t_D is two orders of magnitude longer than t_L. But since for atoms flying perpendicularly to the laser beam $\vec{k}\vec{v} = 0$, the expression Eq. (1) takes its maximum at $\Delta = 0$. Therefore, the VSOP peak is located exactly at the atomic transition.

Huge progress in this field was stimulated by recently developed extremely thin cells (ETC) of submicron thickness L of atomic vapor of alkali metals.[6] These cells allow one to study even more pronounced peculiarities of resonant absorption, reflection, fluorescence as well as magneto-optical processes for the case when L is of the order or less than resonant optical radiation wavelength λ.[7–19] Particularly, very good VSOP contrast is obtained for $L = \lambda$, which is favorable for a straightforward recording of a narrow resonance in a single-passage transmission spectrum. Also several new striking features, which are absent for micrometric-thin cells, were revealed in ETC spectra. As is was shown by experiments on absorption and resonance fluorescence spectra in ETC, the main parameter determining the nature of interaction is not the thickness L itself, but the ratio L/λ. In particular, it was found that for low pump intensity (< 1 mW/cm^2), the spectral width of absorption on a hyperfine transition exhibits oscillating behavior, reaching its minimum value when $L = (2m + 1)\lambda/2$ (m is an integer). The most narrow spectral width, which is 3-4 times less than the Doppler-broadened absorption width in ordinary cm-size cells, is achieved when $L = \lambda/2$. In contrast, the maximum value of the absorption spectral width, which is close to the Doppler linewidth in ordinary cm-size cells, is achieved when $L = m\lambda$. This oscillating linewidth effect is called "collapse and revival of Dicke-type coherent narrowing" effect (CRDN).[8–14] It was demonstrated that also the absorption magnitude exhibits oscillating behavior depending on L/λ ratio.

For relatively high pump intensity (> 1 mW/cm^2), a VSOP resonance with a spectral width close to the natural one, and centered exactly at the atomic resonance frequency, occurs when $L = m\lambda$.[10–12,14,15] The most narrow and high contrast VSOP resonance is observable when $L = \lambda$.

Important to note that in contrast to the transmission spectra, the spectral width of the fluorescence spectra increases monotonically, without oscillations with L, remaining below the Doppler broadening even when the thickness of vapor column is increased up to ~ 4 μm.[7,11] The narrowest spectral width of the fluorescence spectra is obtained when L is in the range $\lambda/4 < L \leq \lambda/2$ (1.5 - 1.8 times narrower than the absorption spectral width for the same thickness). The experimentally observed CRDN effect, as well as appearance of VSOP resonance in the absorption and fluorescence spectra of ETC are rather well described by developed theoretical models.[9,10,15,16,18]

Furthermore, the use of ETC with thickness $L < 100$ nm (for this case it is more suitable to use "nanocell" rather than "ETC" term) is shown to be efficient for studies of atom – dielectric surface (cell's window) van der Waals interaction. As is expected, this interaction is manifested by a strong red frequency shift observed in transmission and fluorescence spectra of Cs and Rb atomic vapor (in some cases also the blue shift is possible).[20]

Narrow optical resonances in transmission and fluorescence spectra of ETC give a possibility to directly study the behavior of hyperfine optical atomic transitions of Rb and Cs in external magnetic fields.[21–25] The VSOP resonance in the transmission spectrum of an ETC with $L = \lambda$, as well as the narrow fluorescence peak with $L = \lambda/2$, are split in an external magnetic field B into several components with frequency positions and probabilities depending on B-field. The latter can be straightforwardly used for investigation of optical transitions between individual Zeeman sublevels of hyperfine states in a wide range of magnetic field between 1 G and 5000 G.

The recent interest to the electromagnetically induced transparency (EIT) phenomenon is caused by a possibility to form narrow optical resonances.[26] As the size of the vapor cell is reduced, the lifetime of the ground-state coherence underlaying the EIT becomes shorter because of quenching collisions of atoms with the cell windows. Therefore, one could expect that the EIT effect should be completely washed out already at $L \simeq 1$ μm. However, in Refs. 19 and 27 the EIT resonance, along with new features and distinctions from the case of ordinary cells, was successfully observed in the ETC under special conditions. The latter is caused again by the contribution of atoms flying nearly parallel to the cell windows, which is enhanced thanks to their longer interaction time with a laser field. Thus, the ETC containing atomic vapor of metal is a promising and efficient tool for studies of atom-light, atom-atom, atom-surface, and atom-external magnetic filed interactions. Besides pure fundamental interest, these phenomena have definite applied potential. The scope of this work covers several practical schemes for formation of narrow optical resonances in extremely thin cells; the paper also addresses and explores possible applications, in particular for optical magnetometry and high-gradient B-field mapping, and frequency referencing.

2. Experimental Method

2.1. *Design of extremely thin cells with controllable thickness L*

The first design of the ETC, which consists of windows and a vertical side arm (a metal reservoir), was presented in Ref. 6. Later, this design has been somewhat modified; the typical example of recent version is presented in Fig. 1. We use 2.5 mm-thick window wafers, 25 – 30 mm in diameter or 15 $\times$ 30 mm rectangular, fabricated either from commercial monocrystalline sapphire or garnet ($Y_3Al_5O_{12}$), which are both chemically resistant to the chemically aggressive vapors of alkali metals. In order to minimize the windows birefringence, the sapphire plates are cut across the c-axis (the garnet is free of birefringence). The wafers have to be perfectly polished (better than $\lambda/10$). In order to form the wedged gap in the vertical direction, which is needed to get a variable thickness of atomic vapor column, see the inset in Fig. 1, an Al_2O_3 strip, $\simeq$ 1000 nm-thick, 5 mm-long, and 1 mm-wide, is deposited on the inner surface of one of the windows in its lower part (in some cases, a platinum strip is used as a spacer: Pt is also immune against highly corrosive hot Rb atomic vapor). Prior to deposition, a 2 mm hole is drilled in the bottom of the windows, into which a 50 mm long tube of the same diameter, made of commercial sapphire, is inserted. Then, the entire construction is assembled and glued in a vacuum furnace. After the gluing,

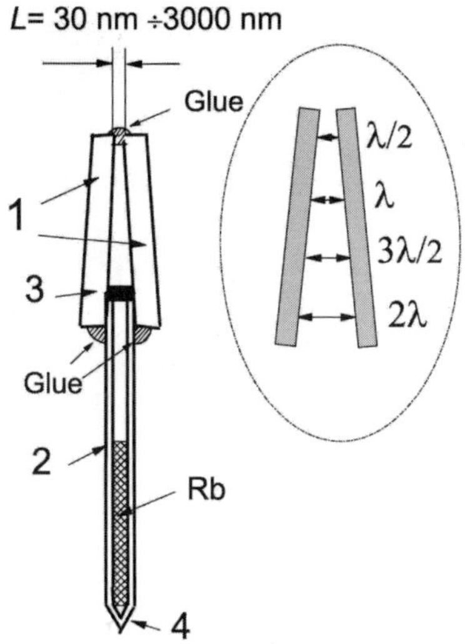

Fig. 1. Design of ETC. *1*- windows of the cell; *2*- sapphire side arm; *3*- spacer; *4*- glass extension.

a glass extension is sealed to the sapphire tube, and the ETC is filled with a natural mixture of the ^{85}Rb and ^{87}Rb (or Cs, or Cs-Rb mixture, etc.), as is done for an ordinary glass cell. The amount of rubidium is chosen so that the sapphire tube, with an inner diameter of about 1 mm, is almost completely filled. After sealing off of the ETC, the glass tube has a length of $5 - 6$ mm.

Note that the vapor pressure in the cell is determined by the temperature of the upper surface of the metallic column (for more detailed information, see Ref. 19). A photograph of the ETC with smoothly variable thickness wedged in the vertical direction is shown in Fig. 2. The wedged gap in this case was formed using a platinum spacer strip of 2 μm thickness. The presented ETC has garnet windows of 0.8 mm thickness (the use of thin wafers in some cases is more convenient), and in order to increase the wafers thickness at the bottom to fit the ∅ 2 mm side arm, 2 additional garnets plates are glued to the main wafers.

Adjusting the heater construction, one can achieve the following limiting thermal conditions: the temperature attainable at the upper surface of metallic Rb column is approximately 500°C; the temperature of the glass extension (which is outside of the heater) is about 80°C. At such a low temperature, metallic Rb practically does not react with glass. As for the

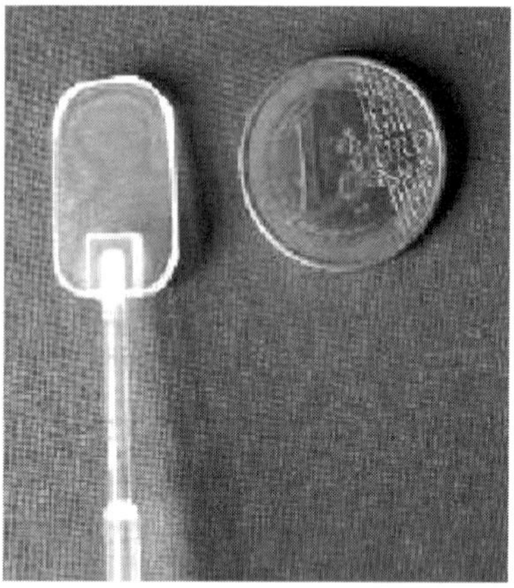

Fig. 2. ETC with thickness smoothly variable from 50 nm at the top to 2 μm near the bottom. A clearly visible interference pattern is caused by the gap thickness which is of the order of light wavelength.

interaction of hot Rb vapor with commercial sapphire or garnet, no chemical interaction occurs up to temperature of about 1000°C.[19]

2.2. *Measurement of the thickness L of the ETC*

Interferometric method described in Ref. 8 is used to determine the wedge-shaped thickness L of the ETC vapor gap. The following procedure is implemented. Since the thickness of the vapor gap varies just by a few micrometers or less across the whole window aperture of $\simeq$ 3 cm, the two inner surfaces of the ETC windows are highly parallel to each other within the laser beam aperture of $\simeq$ 1 mm. Hence, behavior of the transmission and reflection of the ETC is similar to that of a low finesse Fabry-Pérot etalon, and well-known interferometric formulas describing the thickness dependence of reflected or transmitted power can be used:

$$\frac{R_2}{R_1} = \frac{(1 - R_1^2)}{R_1} \times \frac{K \sin^2(\varphi/2)}{1 + K \sin^2(\varphi/2)}, \tag{2}$$

where R_1 is the reflection from the first surface of the ETC (see Fig. 3), R_2 is the reflection by the gap (i.e., the joint reflection by the two inner surfaces of the ETC),

$$K = \frac{4R_1}{(1 - R_1^2)}, \quad \varphi = \frac{4\pi L}{\lambda},$$

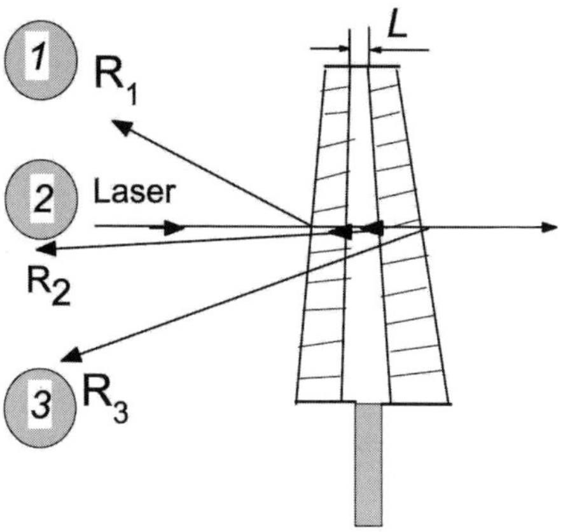

Fig. 3. Three reflected beams R_1, R_2 and R_3 originated from four surfaces of the ETC.

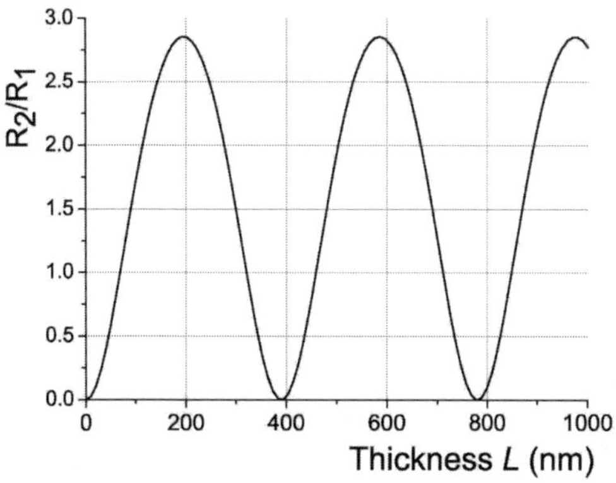

Fig. 4. The ratio R_2/R_1 as a function of gap thickness L. The wavelength is $\lambda = 780$ nm; ETC windows refractive index is $n = 1.8$ (garnet wafers).

and L is the distance between inner surfaces of the ETC. In order to spatially separate *1*, *2* beams the window wafers are made slightly wedged. R_3 is the reflection from the outer surface of the rear window (beam *3*).

In Fig. 4 we present the ratio R_2/R_1 as a function of the gap thickness L for the case of $\lambda = 780$ nm laser wavelength and garnet windows with $n = 1.80$ (for sapphire $n = 1.76$). The curve for the thickness measurement has an oscillating behavior, which imposes ambiguity for the determination of L. As it is seen, for $L = m\lambda/2$ (m is integer), the reflection $R_2 = 0$

(particularly, for $L = 390$ nm, 780 nm, 1170 nm, etc.), which is very convenient for the experimental adjustment. This inconvenience was overcome by performing the measurements presented in Fig. 3 for several wavelengths (thus, 3 different laser wavelengths have been used in Ref. 8). However, as we will show below, the determination of $L = \lambda/2$, the thickness of particular interest, is easy enough for a wedged cell simply from the transmission spectrum, and having this reference, the ambiguity can be lifted for larger L, even for single wavelength interference case.

2.3. *Experimental arrangement for spectroscopic measurements*

The experimental arrangement for transmission and fluorescence measurements is presented in Fig. 5. Extended cavity diode laser (ECDL) with $\lambda = 794$ nm, $\simeq 30$ mW power and $\simeq 1$ MHz spectral linewidth is used. Faraday isolator (FI) helps to avoid the optical feedback. Diode laser radiation wis focused to a spot size of $\varnothing \simeq 0.5$ mm on the ETC nearly perpendicular to its window surface (the latter is very important[1,7]). The ETC is placed in the oven made from non-magnetic materials, which has 3 openings: 2 for the laser beam transmission, and one for registration of resonance fluorescence in perpendicular direction to the laser beam. The temperature of the

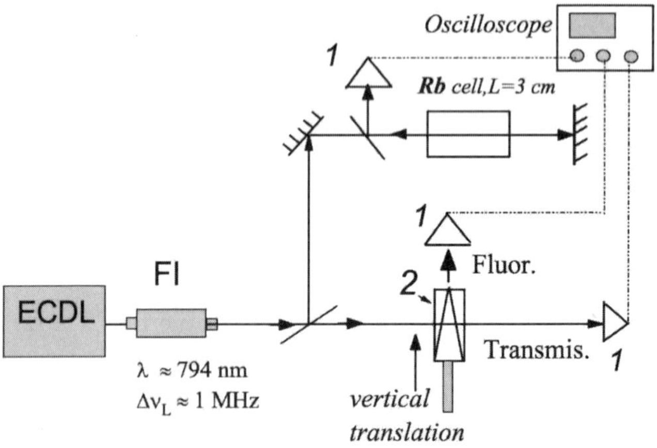

Fig. 5. Sketch of the experimental setup. *ECDL* – diode laser with $\lambda = 780$ nm; *FI* – Faraday isolator; *1* – photodiodes; *2* – extremely thin cell inside the oven; the upper branching is an auxilliary frequency reference formation unit.

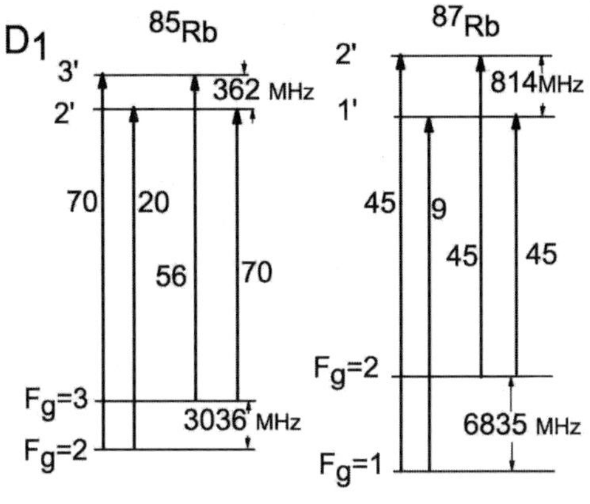

Fig. 6. Energy levels with hyperfine structure of ^{85}Rb, ^{87}Rb D_1 line.

ETC is kept at $\simeq 110-120°C$ on the side arm (this defines the Rb atomic vapor pressure), and $20-30°C$ higher at the windows in order to prevent Rb vapor condensation. This regime corresponds to the number density of Rb atoms $N \simeq 10^{13}$ cm^{-3}. In order to measure and exploit the transmission and fluorescence spectra at different ETC thicknesses, the oven with the ETC is mounted on a translation stage for smooth vertical displacement as indicated by an arrow in Fig. 5.

To obtain transmission and/or fluorescence spectra, the laser radiation frequency is linearly scanned in appropriate spectral region around the studied transition group(s) of ^{85}Rb or ^{87}Rb D_1 line. A fraction of radiation ($\simeq 50\%$) is branched to an auxiliary reference setup (upper part in Fig. 6) providing well-identified frequency marks in the measurement region. Different reference schemes were used depending on the task: i) resonant transmission or fluorescence spectra of 3 cm-long room temperature Rb cell; ii) saturated absorption (SA) spectra obtained in a room temperature 3 cm-long Rb cell; iii) transmission spectra of hyperfine atomic transitions in an auxiliary ETC, which is preferable in most cases. The spectra were recorded by photodetectors with operation amplifiers (*1*) followed by Tektronix TDS 2014B digital four-channel storage oscilloscope. The basic experimental arrangement shown in Fig. 6 was used for investigations presented in Section III. For the results presented in the Section IV, there are some modifications of the experimental setup, which are described in the corresponding section.

3. Resonant Absorption and Fluorescence in ETC

3.1. *Key features of resonant absorption in ETC*

In seminal papers of Romer and Dicke it was shown that the microwave absorption of a gas in a vapor cell with a size smaller than the wavelength ($\lambda = 2.5$ cm) can have a sub-Doppler spectrum.[28] This unexpected effect (one could anticipate a linewidth increase due to frequent collisions of atoms with the cell walls as happens for the electromagnetically induced transparency[26]) can be treated by predominant contribution of slow-longitudinal-velocity atoms to the absorption signal. As a result, the Doppler broadening is strongly reduced for the radiation propagating perpendicular to the ETC's windows. In order to explain the fact that minimal spectral width of the absorption is achieving for $L = \lambda/2$, also coherent effect has to be taken into account.[8,10,16] The development of an ETC with atomic vapor of alkali metals made it possible to study Dicke-narrowing effect in optical domain. To avoid misunderstanding, we should note that there is also another, well-known process of Dicke-narrowing caused by buffer gas collisions (DNB),[29] which differs from the one discussed here.

There are several factors, which could influence the spectral profile (lineshape) and magnitude in transmission, fluorescence and reflection spectra of atomic vapor confined in an ETC, and make them sensitive to L/λ ratio: i) the transient behavior of atomic excitation (i.e. build-up of the absorption coefficient) leads to a Dicke-type sub-Doppler narrowing with enhancement of slow atom contribution, and the lineshape exhibits the most narrow width when $L = (2n+1)\lambda/2$ (i.e. with λ pseudo-periodicity) This narrowing is practically absent (is collapsed) when $L = 2m(\lambda/2)$: this is so called CRDN, or Dicke coherent narrowing and its collapse;[8,9,12] ii) due to Fabry-Pérot nature of the ETC, both the pump and the re-emitted fields form a partially standing wave inside the ETC, the intensities of transmitted and reflected radiations depend on L/λ (with $\lambda/2$ periodicity), which affects the transmission, fluorescence and reflection spectra. In addition, the pump intensity can strictly influence the shape and magnitude of the transmission, fluorescence and reflection spectra; also there is a strong the van der Waals (vdW) interaction between an atom and dielectric windows of an ETC, however vdW influent only when $L/\lambda \ll 1$.[20]

One can conclude that because of CRDN and VSOP formation conditions, the most striking features of resonant processes in the ETC occur at thicknesses $L = (2m+1)\lambda/2$ and $L = m\lambda$. We present below the experimental results for these two cases of particular interest and importance.

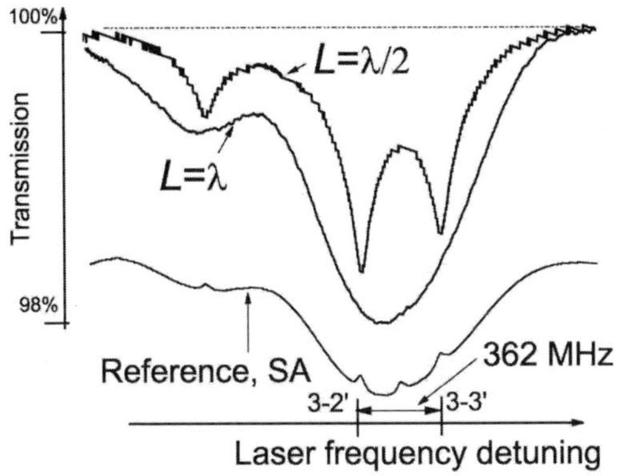

Fig. 7. Transmission spectra of ^{85}Rb $F_g = 3 \longrightarrow F_e = 2, 3$ and ^{87}Rb $F_g = 2 \longrightarrow F_e = 2$ for $L = \lambda/2$ and $L = \lambda$. The laser intensity is < 0.1 mW/cm^2. Strong narrowing is seen for $L = \lambda/2$. Lower curve is a reference SA spectrum.

3.2. *Low and high laser intensity cases*

Within last several years, the CRDN effect was studied on D_1 and D_2 lines of Cs and D_2 line of Rb ($\lambda = 852$, 894 and 780 nm), and in all cases the expected prominent L/λ-dependent features have been observed.[8,11,12] Here we mainly present new original results, which have not been published elsewhere, although the observed peculiarities of recorded spectra are similar to that obtained earlier.

It is convenient to observe CRDN effect on D_1 line of either Rb (Fig. 6) or Cs due to simplicity of atomic energy levels. Resonance transmission spectrum of ^{85}Rb $F_g = 3 \longrightarrow F_e = 2, 3$ and ^{87}Rb $F_g = 2 \longrightarrow F_e = 2$ transitions of D_1 line (the subscripts g and e refer to lower and upper states, respectively) for the ETC with $L = \lambda/2$ and $L = \lambda$ are shown in Fig. 7. The laser beam (ECDL, $\lambda = 794$ nm, $\gamma_L = 1$ MHz) is focused by $F =$ 35 cm lens into the ETC at nearly normal incidence. The laser intensity is < 0.1 mW/cm^2. As it is seen, there is a strong narrowing of spectral width in the case of $L = \lambda/2$. The full width at half maximum (FWHM) for this case is 150 MHz (well fitted for each transition by "pseudo-Voigt2" function), while for $L = \lambda$ the FWHM is $\simeq 460$ MHz (somewhat smaller than FWHM $\simeq 500$ MHz for the transmission profile of an ordinary cm-size cell, which is well fitted by a Gaussian function).

In Fig. 8 the transmission spectrum on the same transitions is presented for $L = \lambda$ thickness and laser intensity $\simeq 10$ mW/cm^2 (upper curve).

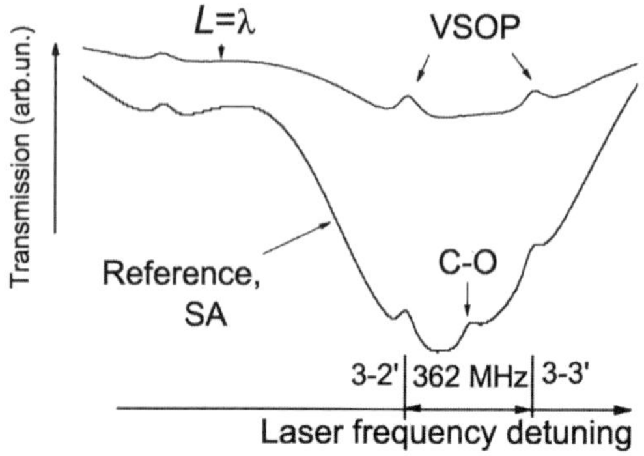

Fig. 8. Transmission spectrum of ^{85}Rb $F_g = 3 \longrightarrow F_e = 2, 3$ and ^{87}Rb, $F_g = 2 \longrightarrow F_e = 2$ for $L = \lambda$. The laser intensity is $\simeq 10$ mW/cm^2. Narrow VSOP features are clearly visible. Lower curve is a reference SA spectrum.

Though the spectral width here is much larger than for $L = \lambda/2$, one can clearly see the VSOP resonances appearing exactly at the frequency positions of atomic transitions (these resonances are absent when the laser intensity is less than 0.1 mW/cm^2). The lower curve shows simultaneously recorded SA spectrum. The complicating crossover (CO) resonance recorded in the SA spectrum is absent in the $L = \lambda$ ETC spectrum, which is a very important advantage: although there is a beam reflected from the outer rear surface of the ETC propagating backwards with respect to the main beam, it does not cause formation of CO resonance, since the interaction time t_L for atoms with a longitudinal velocity $v_z = 2\pi\varepsilon/k$ (where ε is half of the frequency separation of corresponding excited levels) is too short at $L = \lambda$ for developing an efficient optical pumping needed for buildup of CO resonance. The FWHM of the peak of reduced absorption in the reference SA spectrum presented in Fig. 8 is $\simeq 30$ MHz, and is $\simeq 50$ MHz for the case of $L = \lambda$ (under the optimal experimental parameters the minimal spectral widths could be 6 MHz and 10 MHz, correspondingly). A theoretical model, which adequately takes into account the peculiarities of real ETC at low intensity, is presented in Ref. 9; this model is generalized in Refs. 10, 15, 16, 18 to include also dependence on radiation intensity.

It is important to note that for the case of $L = \lambda/2$, the atomic vapor density can be risen up to $10^{16} - 10^{17}$ cm^{-3} (this depends on atomic transition probability), and the absorption will be still below 100% due to small value of L ("optically thin" absorption regime). The absorption is simply governed by $N \times L$ product (N being the atomic number density), which is a determinative parameter. The latter peculiarity enables to study the broadening and shift of atomic transitions at very high atomic densities from transmission spectra recorded with the help of ETC with $L = \lambda/2$: the self-broadening and frequency shifts become pronounced at atomic vapor densities $> 10^{14}$ cm^{-3}, when on-resonance D line absorption of atomic vapor in ordinary cm-size cells is in a deep "optically thick" regime. These studies have been implemented in Ref. 17. The diode laser radiation frequency (spectral width $\simeq 1$ MHz) was scanned in the vicinity of $5^2S_{1/2} \longrightarrow 6^2P_{3/2}$ blue transitions ($\lambda = 420.2$ nm) of ^{85}Rb and ^{87}Rb. A strong red shift of all the atomic transitions was detected at the temperature $T >$ 300°C corresponding to $N > 10^{16}$ cm^{-3} (the experiment was carried out jointly with G. Pikhler's group, Institute of Physics, Zagreb).

3.3. *Resonant fluorescence by ETC*

In ordinary cm-size cells, the behaviors of the resonant absorption and fluorescence spectra are very similar to each other for low absorption and low laser intensity regime.[29] In the case of an ETC, the following five distinctions have been revealed: i) the fluorescence spectrum spectral width is always narrower ($1.5 - 1.8$ times for $L = \lambda/2$, and $\simeq 3$ times for $L = \lambda$) than the absorption width for the same transition, a general result which is related with longer buildup time for the fluorescence, i.e. absorption and emission cycle, thus imposing atomic velocity restrictions (the atom has to emit a photon before being quenched on the cell window); ii) surprisingly, the CRDN effect, which is present in absorption spectra, is not observable in fluorescence: as opposed to absorption, the increase of L results in monotonic increase of fluorescence spectral width, without oscillating behavior versus L; iii) unlike the absorption case, the fluorescence spectral width remains below the Doppler broadening for cm-size cells, even when the thickness of vapour column is increased to $L \simeq 4$ μm; iv) the VSOP resonance peak in ETC transmission at $L = m\lambda$ is observable for radiation intensities of several mW/cm^2, while the intensity required to observe VSOP resonance dip in the fluorescence spectrum is more than by one order of magnitude larger; v) the VSOP resonance peaks in transmission spectrum for $L = m\lambda$ appear for $I_L \simeq$ a few mW/cm^2 simultaneously on "cycling" and "open" transitions, while the VSOP resonance dips in fluorescence first appear on open transitions, and then, with further increase of L (or I_L), also on a cycling transition (only for $F_g < F_e$).[16]

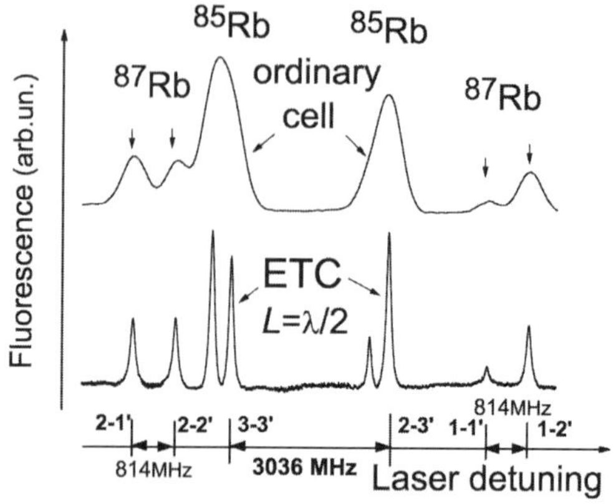

Fig. 9. Fluorescence spectrum on Rb D_1 line from an ETC with $L = \lambda/2$ (lower curve). All the 8 atomic transitions of D_1 line (see Fig. 6) are fully resolved. Upper curve is spectrum of fluorescence emitted by an ordinary cm-size cell.

Although sensitive detection and high laser intensity are needed to record resonant fluorescence emitted by an ETC, there are two advantages one can gain in this case: a narrower sub-Doppler spectral width, and a nearly zero background of the fluorescence signal.

In particular experiment, the laser beam (ECDL, $\lambda = 794$ nm, $\gamma_L = 1$ MHz, intensity is $\simeq 10$ mW/cm^2) is sent at nearly normal incidence onto the ETC operating at $T_{SA} \simeq 110°$C, and $T_W \simeq 140°$C temperature regime. The fluorescence spectrum of Rb D_1 line for $L = \lambda/2$ is shown in Fig. 9 (lower trace). The upper trace is the fluorescence emitted by an ordinary cm-size cell. All the hyperfine transitions labeled correspondingly are resolved in the lower trace. Under optimal conditions (i.e. low I_L and sensitive detection), the FWHM of the fluorescence spectrum for an individual transition can be reduced down to $\simeq 70$ MHz, in contrast to $\simeq 500$ MHz for the case of ordinary cm-size cell. Moreover, the relative amplitudes of fluorescence peaks are proportional to the transition probabilities presented in Fig. 6.

The lineshape and magnitude of the resonant fluorescence on $F_g = 3 \rightarrow F_e = 2, 3$ transitions of ^{85}Rb D_1 line, obtained for thicknesses $L = \lambda/2$ for different values of laser intensity: 4, 8, 20, 40, 80, 120, 160, 208, and 240 mW/cm^2 (curves labeled from *1* to *9*, respectively) can bee seen in Fig. 10. The striking point here is that even for $I_L = 240$ mW/cm^2, there is no sign of any dip in the fluorescence spectrum, meanwhile as it is shown below, dips caused by velocity selective optical pumping process appear for $L = \lambda$

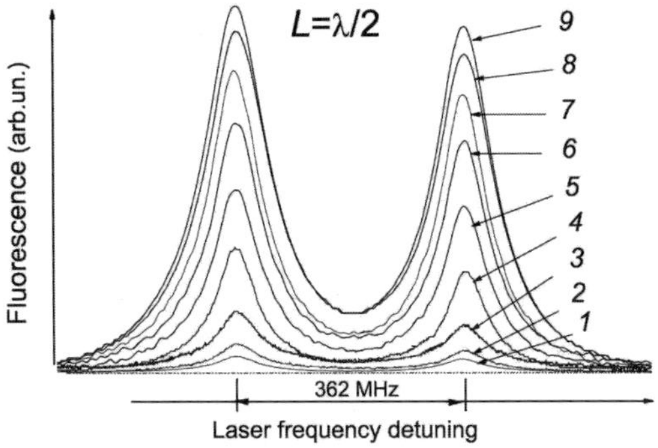

Fig. 10. Spectrum of fluorescence emitted by ETC with $L = \lambda/2$ on Rb D_1 line for different values of laser intensity. The laser intensities corresponding to curves labeled *1* to *9* are: 4, 8, 20, 40, 80, 120, 160, 208, and 240 mW/cm^2.

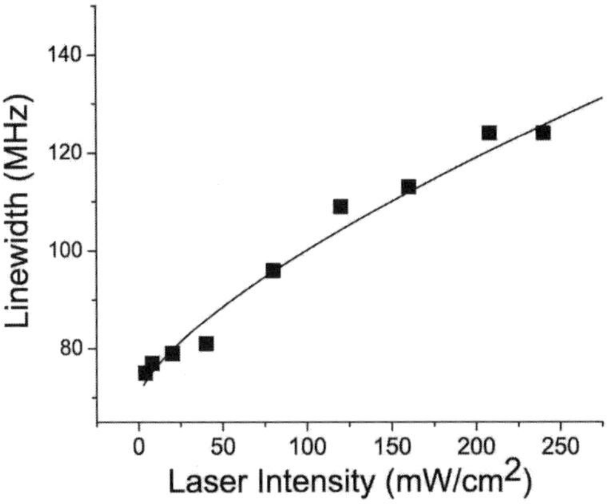

Fig. 11. Spectral width of fluorescence on an individual transition versus I_L. Squares: experiment, solid line: calculated by empiric formula (see text).

when the intensity is $\simeq 30$ mW/cm^2 (see Fig. 13). Similar behavior exhibits also transmission spectrum for $L = \lambda/2$: the experimental results obtained at $I_L \simeq 1$ W/cm^2 show no dips of reduced absorption in the transmission spectrum; in contrast, $I_L \simeq$ several mW/cm^2 is sufficient to form dips at $L = \lambda$ (see Fig. 8). These effects are caused by a strong Dicke-type coherent narrowing for a thickness $L = \lambda/2$.

The spectral width of fluorescence for an individual transition (after fitting with 2 Lorentzian curves, see below) as a function of the laser intensity is presented in Fig. 11. The solid curve is plotted using empiric formula γ_{FL}

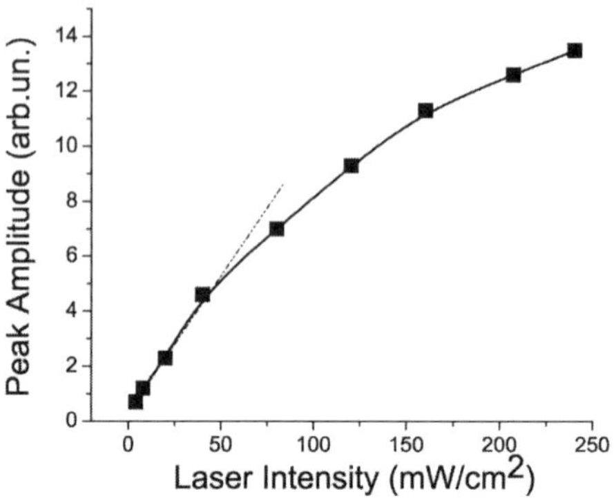

Fig. 12. Peak amplitude of the fluorescence of an individual transition versus I_L. Squares: experiment, solid line is drawn to guide the eye, dotted line: linear dependence.

[MHz] = 70 [MHz] $+1.2 \times I^{0.7}$ [mW/cm^2]. According to this formula, the γ_{FL} value at $L = \lambda/2$ would reach the Doppler width of 500 MHz for laser intensity $\simeq$ 4.5 W/cm^2. Thus, the fluorescence spectral width measured at $L = \lambda/2$ can be used as a direct laser intensity gauge.

The peak amplitude of the fluorescence for an individual transition as a function of the laser intensity is presented in Fig. 12. It is interesting to note that linear dependence is observable up to $I_L = 50 \div 60$ mW/cm^2, while for an ordinary cm-size cell the saturation appears already for the laser intensity of several mW/cm^2. This is caused by the fact that the lifetime of the ground-states becomes shorter as the cell thickness is reduced because of spin-exchange collisions of atoms with the cell windows. Thus, the value of $\Gamma = (2\pi t)^{-1}$, the denominator of the formula Eq. (1), where $t = L/u$ (u is the thermal velocity) is large enough ($\Gamma > 100$ MHz) for the case of $L = \lambda/2 \simeq 0.4$ μm, and one could expect reduction of optical pumping efficiency. Further decrease of the thickness would cause additional increase of fluorescence saturation intensity.

We present in Fig. 13 the fluorescence spectra for $L = \lambda$ (upper curves) for the laser intensity $\simeq$ 30 mW/cm^2. As it is seen, the velocity selective optical pumping dips of reduced fluorescence appear at all the transitions of ^{87}Rb, $F_g = 2 \rightarrow F_e = 1, 2$ and ^{85}Rb, $F_g = 3 \rightarrow F_e = 2, 3$ (a) and ^{85}Rb, $F_g = 2 \rightarrow F_e = 2, 3$ (b). The spectral width of the reduced fluorescence dip is $\simeq$ 20 MHz. Optimization of experimental parameters, in particular, use of lower laser intensity and uniform area of the thickness $L = \lambda$ may result in further narrowing (to less than 10 MHz). Note, that the dips of

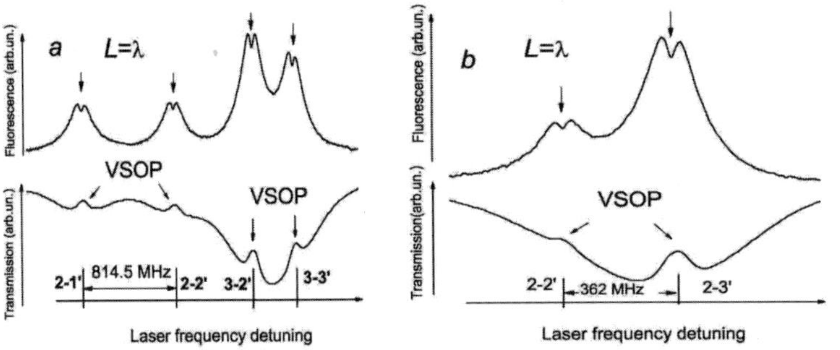

Fig. 13. Fluorescence spectra (upper curves) for $L = \lambda$, $I_L \simeq 30$ mW/cm^2 recorded at ^{87}Rb, $F_g = 2 \rightarrow F_e = 1, 2$ and ^{85}Rb, $F_g = 3 \rightarrow F_e = 2, 3$ (a) and ^{85}Rb, $F_g = 2 \rightarrow F_e = 2, 3$ (b) transitions. Dips of reduced fluorescence with the spectral width $\simeq 20$ MHz marked by arrows appear at all transitions. Lower curves show the VSOP resonances in the transmission spectra for $L = \lambda$.

reduced fluorescence appear exactly at the frequency of atomic transitions. The lower curves show VSOP resonances in the transmission spectra for $I_L = 30$ mW/cm^2. We should mention that for VSOP formation in the transmission spectrum when $L = \lambda$, laser intensity is of several mW/cm^2 and spectral width in the range of $1 \div 50$ MHz is sufficient, while for dip formation in the fluorescence spectrum the laser intensity has to be > 20 mW/cm^2, and the spectral width should not exceed a few MHz. Thus, the VSOP formation in transmission spectra when $L = \lambda$ is a more robust process.

The lineshape and magnitude of the resonant fluorescence on $F_g = 3 \rightarrow F_e = 2, 3$ transitions for the laser intensity $\simeq 10$ mW/cm^2 as a function of ETC thickness L varying in the range $\lambda/2 \div 4\lambda$ with $\lambda/2$ step is shown in Fig. 14. As was mentioned above, the narrowest spectral width is achieved when L is in the range $\lambda/4 < L < \lambda/2$, particularly for $L = \lambda/2$ ($\simeq 70$ MHz in Fig. 14). For this value of laser intensity there are no dips of reduced fluorescence up to $L = 2\lambda$. The dips determined by $I \times L$ product appear for $L > 2\lambda$. It is possible to form narrow dips of reduced fluorescence in the range of $\lambda - 10\lambda$ either keeping invariable I_L and increasing L, or keeping invariable L and increasing I_L. Note that even at the largest thickness $L = 4\lambda$ (≈ 3.2 μm) presented in Fig. 14 the spectral width of fluorescence is still sub-Doppler (< 500 MHz). Spectral width of fluorescence for an individual transition as a function of the ETC thickness L is presented in Fig. 15. For correct determination of spectral width we have done a fitting with three

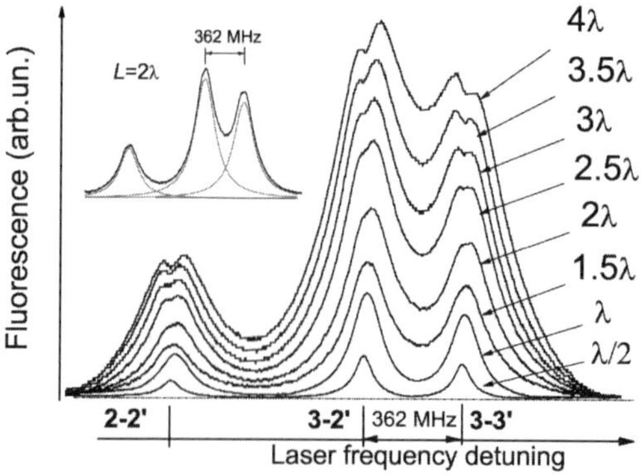

Fig. 14. Fluorescence spectrum on ^{87}Rb $F_g = 2 \longrightarrow F_e = 2$, ^{85}Rb $F_g = 3 \longrightarrow F_e = 2, 3$ transitions emitted for $I_L \simeq 5$ mW/cm^2 by an ETC with variable thickness L (from $\lambda/2$ to 4λ, with $\lambda/2$ step).

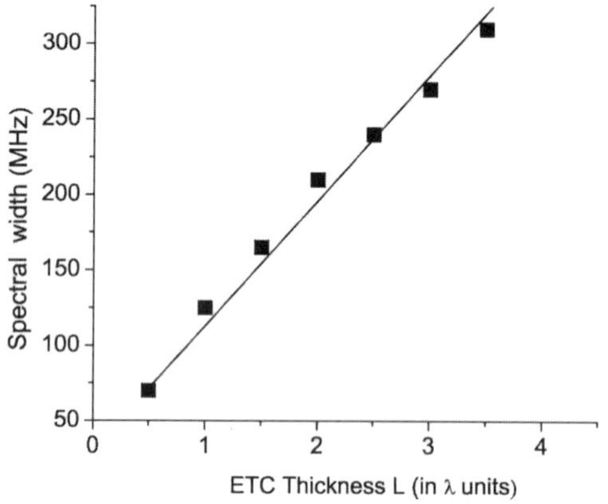

Fig. 15. Spectral width of the fluorescence of an individual transition versus thickness L. Squares: experiment, solid line: calculated by empiric formula (see text).

Lorentzian curves. An example for the case of $L = 2\lambda$ is shown in the inset (three grey lines). The solid curve is plotted by empiric formula: γ_{FL} [MHz] $=$ 30 [MHz] + 83 $\times L$ (in λ units). The estimate predicts that up to $L \simeq$ 4.5 μm the spectral width of the fluorescence is less than Doppler width for a cm-size cell. Relatively weak dependence of γ_{FL} on thickness L means that in the case of $\pm$ 15 nm accuracy of determination of the thickness $L \approx \lambda/2$ (which is realistic with the use of the curve presented in Fig. 6), we get only several MHz of additional uncertainity in spectral broadening. This relative insensitivity is important for the application presented in Section IV.

3.4. *Frequency reference for atomic transitions based on ETC with $L = \lambda$ and $L = \lambda/2$*

In Ref. 14 it was demonstrated that the atomic frequency reference based on ETC with $L = \lambda$ has several advantages as compared with a widely used saturated absorption (SA) technique, namely: i) the absence of complicating crossover resonances; ii) the ratio of VSOP peaks amplitudes for a ETC with $L = \lambda$ is close to the ratio of corresponding transition probabilities, unlike the case of SA; iii) a single-beam transmission spectrum; iv) the laser power required for a ETC spectral reference is as low as $\simeq 0.1$ mW. Recently, an atomic frequency reference based on $L = 0.5 - 1$ mm Rb atomic vapor cell was proposed by Mikata et al.[30] They exploited the advantage of a thin cell discussed here to form a narrow optical resonance: for atoms flying parallel to the window of thin cell, the time of interaction is $t_D = D/v$, whereas for atoms flying along the beam one has $t_L = L/v$, and since the beam diameter is a few mm, while the distance between the windows is less than 1 mm, thus $t_D > t_L$.

As we have mentioned above (see expression Eq. (1)), the longer the interaction time the higher the optical pumping efficiency. However, in the case of an ETC with thickness $L \simeq \lambda$ the interaction time anisotropy is significant since $t_D/t_L \simeq 1000$. For the same reason, a pump and probe technique combined with a time delay of the probe beam was implemented in Ref. 30 in order to increase the optical pumping efficiency for atoms flying parallel to the cell windows. In the case of an ETC, just a single-beam transmission is sufficient to obtain strong resonance narrowing. In Fig. 16 we present comparison of the results from Ref. 30 with the ones obtained with an ETC for the same transitions $F_g = 3 \rightarrow F_e = 2, 3$ of ^{85}Rb D_1. The grey line profile in both *a*) and *b*) shows the resonances with spectral width $\simeq 50$ MHz formed by a pump and probe technique. The black line in Fig. 16*a* shows the resonance fluorescence signal obtained by ETC with $L = \lambda/2$ (spectral width $\simeq 70$ MHz). Although the black profile demonstrates wider spectral width, however it gives more correct ratio of the peak amplitudes $A(3 - 2')$ of $F_g = 3 \rightarrow F_e = 2$ and $A(3 - 3')$ of $F_g = 3 \rightarrow F_e = 3$ transitions: $A(3 - 2')/A(3 - 3') = 1.2$ (the theoretical value of transitions probabilities ratio is 1.25, see Fig. 6). For the case of single-beam transmission for $L = \lambda$ (Fig. 16*b*) the VSOP resonances have a spectral width $\simeq 20$ MHz, i.e. that more than twice times narrower than for

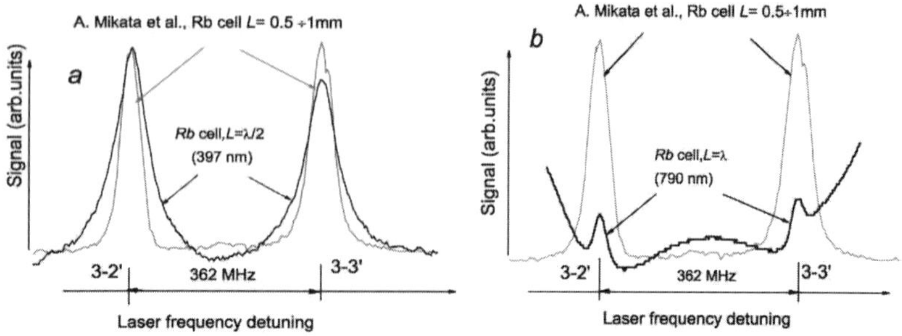

Fig. 16. The grey line curves repeated in *a*) and *b*) show the resonances with $\simeq$ 50 MHz spectral width formed by a technique presented in Ref. 30. Black line: the fluorescence signal for ETC with $L = \lambda/2$ ($\simeq$ 70 MHz width) (*a*), and the VSOP resonances in transmission spectrum for ETC with $L = \lambda$ ($\simeq$ 20 MHz width) (*b*).

the grey profile, and the ratio of VSOP resonances is equal to 1.2. Hence, an ETC with thickness either $L = \lambda/2$ (fluorescence) or $L = \lambda$ (transmission) can be employed as a convenient and advantageous frequency reference for atomic resonance transitions.

4. Magneto-Optical Processes in ETC

4.1. *Fluorescence of an ETC with* $L = \lambda/2$ *in external magnetic field: HLZT method*

It is well known that energy levels of atoms placed in an external magnetic field undergo resonance frequency shifts and variation of transition probabilities. Below we demonstrate that using $L = \lambda/2$ fluorescence spectra it is possible to efficiently study these changes employing the fluorescence emitted by an ETC with $L = \lambda/2$: in spite of certain drawbacks (signal weakness, relatively high required excitation intensity), it has advantages of narrow sub-Doppler spectral width ($\simeq$ 70 MHz) and zero detection off-resonance background.[23] We have called this method "half-λ Zeeman technique" (HLZT).

The sketch of an experimental setup is presented in Fig. 17*a*. The circularly polarized beam of extended cavity diode laser (ECDL, $\lambda = 794$ nm, $P_L \simeq 30$ mW, $\gamma_L < 1$ MHz) resonant with ^{87}Rb D_1 transition frequency, was directed onto the Rb ETC (*2*) with the vapor column thickness $L = \lambda/2$ at nearly normal incidence. The cell located in a special oven with 3 openings (2 inlets for laser beam passage, and an orthogonal

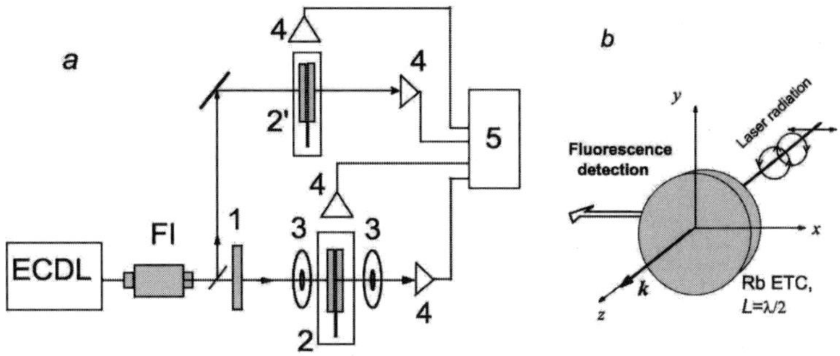

Fig. 17. *a*) Sketch of the experimental setup. *ECDL* – diode laser, *FI*- Faraday isolator, *1* – $\lambda/4$ plate, *2* – ETC in the oven, *2'* – an auxiliary ETC and the oven, *3* – ring magnets, *4* – photodetectors, *5* – digital storage oscilloscope. *b*) Mutual orientation of laser beam, *B*-field, and detection of fluorescence.

inlet for side fluorescence detection) operated at the following temperature regime: $T_{SA} \simeq 110 - 120°\text{C}$, $T_W \simeq 140 - 150°\text{C}$, which corresponds to $N \simeq 6 \times 10^{12} - 1.5 \times 10^{13}$ cm^{-3}. This geometry allows simultaneous recording of fluorescence and transmission spectra. The fluorescence signal was detected by a photodiode *4* with an aperture of 1 cm^2, which was placed at 90° angle to the laser propagation direction. The photodiode signal was recorded by Tektronix TDS 2014B digital four-channel storage oscilloscope (*5*). A Glan prism was used to purify initial linear radiation polarization of the laser radiation; to produce a circular polarization, a $\lambda/4$ plate (*1*) was utilized. Configuration of the experiment is shown in Fig. 17*b*. Magnetic field was directed along the laser radiation propagation direction $\vec{k}$ ($\vec{B}//\vec{k}$).

About 50% of the pump power was branched by a beam splitter to an auxiliary Rb ETC (*2'*), which was used as a frequency and amplitude reference for $B = 0$. Moderate longitudinal magnetic field ($B < 250$ G) was applied to the ETC by a system of Helmholtz coils (not shown in Fig. 17*a*). The *B*-field strength was measured by a calibrated Hall gauge. It is important to note that HLZM allows to apply much stronger magnetic field using widely available strong permanent ring magnets (PRM): in spite of strong inhomogeneity of magnetic field (in our case it can reach 150 G/mm), the variation of *B*-field inside atomic vapor column is < 0.1 G, i.e. by several orders less than the applied B value, thanks to small thickness of the ETC ($L = 397$ nm).

The allowed transitions between magnetic sublevels of hf states for ^{87}Rb D$_1$ line in the case of σ^+ (left circular) polarized excitation and selection

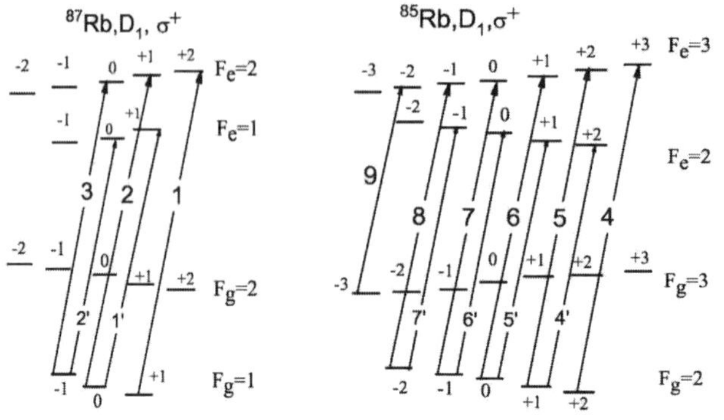

Fig. 18. The hfs energy level diagram of D_1 line of ^{87}Rb (left) and ^{85}Rb (right) in magnetic field and possible atomic Zeeman transitions for σ^+-polarized exciting laser radiation.

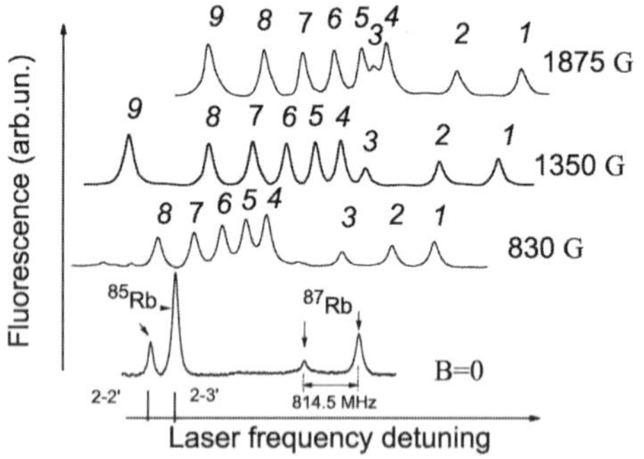

Fig. 19. Fluorescence spectra of ^{87}Rb $5S_{1/2}$, $F_g = 1 \longrightarrow 5P_{1/2}$, $F_e = 1, 2$ and ^{85}Rb $5S_{1/2}$, $F_g = 2 \longrightarrow 5P_{1/2}$, $F_e = 2, 3$ transitions for different values of magnetic field B, with σ^+ radiation excitation. The lower curve is the fluorescence spectrum for $B = 0$.

rules $\Delta m_F = +1$ are depicted in Fig. 18 (HLZM works well also for σ^- excitation).

The fluorescence spectra on ^{87}Rb $5S_{1/2}$, $F_g = 1 \longrightarrow 5P_{1/2}$, $F_e = 1, 2$ and ^{85}Rb $5S_{1/2}$, $F_g = 2 \longrightarrow 5P_{1/2}$, $F_e = 2, 3$ transitions for the case of different values of longitudinal magnetic field B with σ^+ laser radiation excitation (power 2 mW and beam diameter $\varnothing = 2$ mm) are presented in Fig. 19. The spectra are shifted vertically for convenience. The lower curve is the fluorescence spectrum for $B = 0$. It is easy to identify all the nine transitions. The labels denote corresponding transitions between magnetic

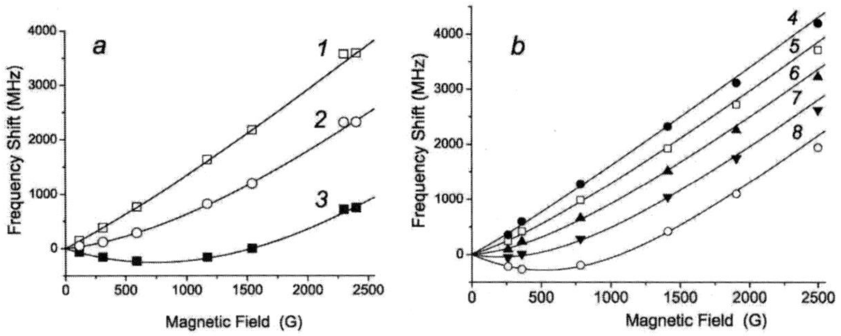

Fig. 20. Frequency shift of the fluorescence peaks *1, 2, 3* (*a*) and *4, 5, 6, 7, 8* (*b*) relative to the initial position for $B = 0$. Theoretical curves are shown by solid lines.

sublevels shown in Fig. 18. A valuable feature is that the fluorescence amplitude of an individual transition is proportional to its probability, which is of importance for studying the dynamics of variation of transition probabilities with B. Thus, the probability of the fluorescence peak labeled *8* increases rapidly with B, and for $B \simeq 1000$ G the ratio of fluorescence peak amplitudes for *4* and *8* equals to $A(4)/A(8) \approx 1$, meanwhile the initial ($B = 0$) ratio is ≈ 15. As is clearly visible in Fig. 19, the most convenient peak for magnetic field measurement is the one labeled *1* (^{87}Rb $F_g = 1, m_F = +1 \longrightarrow F_e = 2, m_F = +2$ transition), since it has the largest amplitude among the transitions *1,2,3*. Also, this peak does not overlap with any other transition at any value of magnetic field strength.

Fig. 20*a* shows the frequency shift of the fluorescence peaks *1, 2, 3* relative to the initial frequency position (for $B = 0$). A linear Zeeeman shift for the peaks *1* and *2* is observed up to $B \simeq 500$ G, with shift rates 1.16 and 0.93 MHz/G correspondingly. Similar shifts for peaks *4, 5, 6, 7, 8* are shown in Fig 20*b*. Corresponding theoretical curves showing a good agreement with the experiment are depicted by solid lines (the calculations were done by the group of M. Auzinsh, University of Latvia). For $B > 1000$ G the shift rate of the most preferable marker peak *1* is 1.59 MHz/G.

For low magnetic field the probabilities for the fluorescence peaks labeled *4, 5, 6, 7, 8* are strongly different (the most strong is *4*), however they are increasing with B, and eventually become practically equal to each other. In Fig. 21 we present the ratio of amplitudes (probabilities) for the fluorescence peaks *5, 6, 7, 8* relative to the peak number *4*. The observed result is confirmed by theoretical curves shown by solid lines.

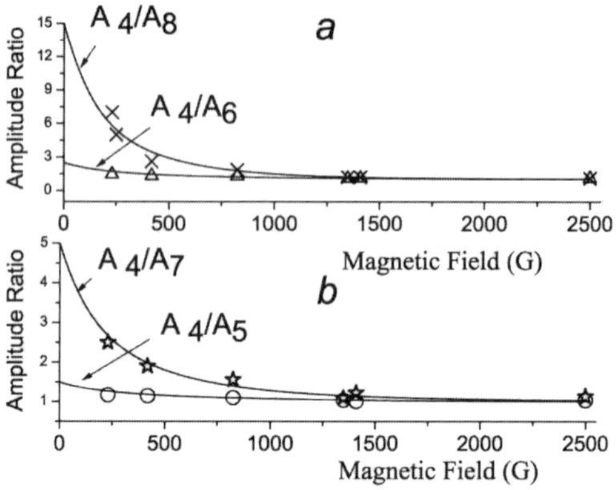

Fig. 21. Magnetic field dependence of ratio of the fluorescence peak *4* amplitude to amplitudes of *5, 6, 7, 8*. Theoretical curves are shown by solid lines.

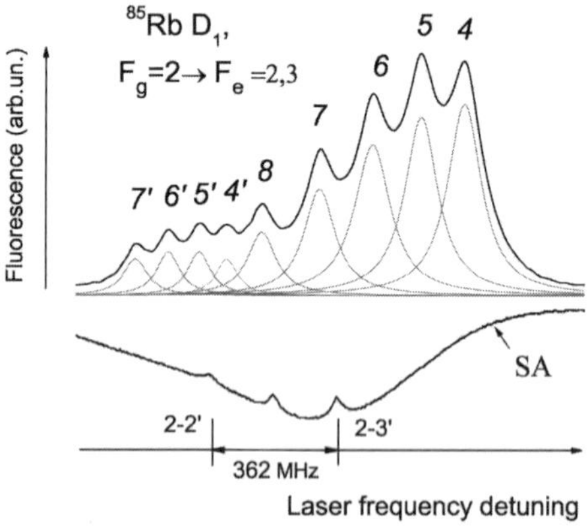

Fig. 22. Fluorescence spectrum covering all the nine components of $5S_{1/2}$, $F_g = 2 \longrightarrow 5P_{1/2}$, $F_e = 2,3$ transitions for B =250 G. Fitting by Lorentzian profiles is done for any individual component (grey lines), showing a good overall profile reproduction. Lower curve is the reference SA spectrum.

It is interesting to note, that even at moderate magnetic field B = 250 G all the nine components of $5S_{1/2}$, $F_g = 2 \longrightarrow 5P_{1/2}$, $F_e = 2,3$ transitions shown in diagram Fig. 18 are well resolved (fitting is done by nine Lorentzian profiles — grey lines) as presented in Fig. 22. For $B = 0$ the probabilities ratio for the σ^+ transitions *4:5:6:7:8* is 15:10:6:3:1, and the probabilities ratio for *4':5':6':7'* is 2:3:3:2. Thus, the advantage of HLZT is that it allows one to monitor and study not only the frequency shift of an individual Zeeman transition, but also the modification in its transition probability in the region of 10 G $\div$ 2500 G. Moreover, as is seen from Fig. 10, a correct ratio of the peak amplitudes of $F_g = 3 \rightarrow F_e = 2$ and $F_g = 3 \rightarrow F_e = 3$ transitions (equal 1.2) is observed up to laser intensity $\simeq$ 100 mW/cm^2. This means that the fluorescence peak amplitudes ratio is proportional to probabilities ratio up to laser intensity $\simeq$ 100 mW/cm^2, contrary to the case of saturated absorption technique, see Ref. 14. Finally, as we have noted, a $\pm$ 15 nm thickness deviation from $L = \lambda/2$ causes a spectral width broadening just by several MHz, which is favorable for HLZT application.

4.2. *Fluorescence of an ETC with $L = \lambda$ in external magnetic field: LFZT method*

As it was shown in Fig. 13, narrow dips of reduced fluorescence with spectral width 10–20 MHz appear at all transitions of Rb D_1 line in the fluorescence spectrum for the thickness $L = \lambda$ and relatively high laser intensity ($\simeq$ 30 mW/cm^2). It is natural to expect splitting of the fluorescence spectrum of a $L = \lambda$ ETC in an external magnetic field containing dips of a reduced fluorescence.

Fluorescence spectra of $5S_{1/2}$, $F_g = 2 \longrightarrow 5P_{1/2}$, $F_e = 2, 3$ transitions (eight components) for B = 1430 G are shown in Fig. 23. Upper curve represents the case of $L = \lambda$ and laser intensity $\simeq$ 30 mW/cm^2, with dips on fluorescence peaks marked by the arrows (we call this method "$L = \lambda$-Fluorescence Zeeman technique" (LFZT)). The middle curve shows the $L = \lambda/2$ spectrum for laser intensity $\simeq$ 10 mW/cm^2. The lower curve is a frequency reference realized by SA technique. The use of LFZT is favorable for weak transitions since it yields larger peak amplitude of the fluorescence. Moreover, increasing the thickness L up to several λ will enable one to increase the fluorescence amplitude as it is shown in Fig. 14. The dips of reduced fluorescence, which appear exactly at the atomic transitions (see Fig. 13), are good "markers" of frequency position of atomic transitions in external magnetic field. However, the main benefit of HLZT is that it provides narrower spectral width of the signal, which causes better spectral resolution, and thus better accuracy of the amplitude measurement needed for the measurement of modification of atomic transition probability. Additionally, the spatial resolution of HLZT (397 nm) is better than that of LFZT method (794 nm or more when the thickness $L > \lambda$) which is important when it is required to measure strongly non-homogeneous magnetic

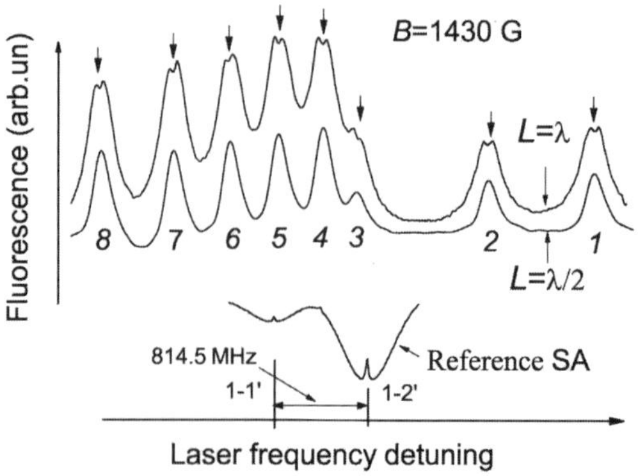

Fig. 23. Fluorescence spectrum covering eight components of $5S_{1/2}$, $F_g = 2 \longrightarrow 5P_{1/2}$, $F_e = 2, 3$ transitions for $B = 1430$ G. Upper curve: $L = \lambda$ (with dips in fluorescence marked by arrows); middle curve: $L = \lambda/2$. Lower curve: reference SA spectrum.

field. Finally, the contrast of the VSOP resonances in LZT (see Figs. 24, 25) is much better than the contrast of dips of reduced fluorescence.

4.3. *ETC transmission for $L = \lambda$ in external magnetic field: LZT method and comparison with HLZT and LFZT*

It was revealed in Refs. 19, 25 that the use of VSOP resonances formed in transmission spectra of ETC with thickness $L = \lambda$ (this technique is called "$L = \lambda$ Zeeman technique" (LZT)), could have several advantages as compared with the HLZT: in particular, the spectral width of VSOP resonances used in LZT is at least $\simeq 4$ times narrower, resulting in much higher resolution; also the laser power required for LZT is $\simeq 10$ times lower than that needed for HLZT; and finally, recording of resonant transmission spectra does not require sensitive detectors with a large aperture as in the case of fluorescence spectra. Nevertheless, in some cases HLZT could be preferable, in particular when transitions probability is small enough. Moreover, the fluorescence signal appears on a flat (zero) background, which is different from LZT where the VSOP resonances appear inside absorption spectrum, possibly affecting correct determination of frequency positions and amplitudes of peaks.

For comparison of these two methods at the same value of B-field, we present in Fig. 24 a fluorescence spectrum for $L = \lambda/2$ (upper curve-HLZT, $B = 2480$ G) and transmission spectrum spectrum $L = \lambda$ (middle

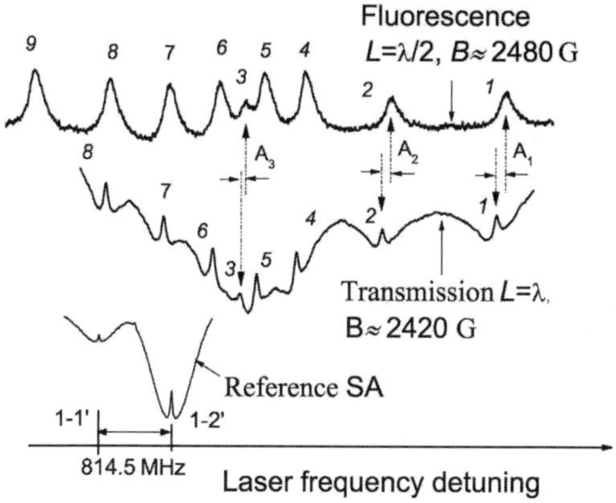

Fig. 24. Spectra of of ^{87}Rb $F_g = 1 \longrightarrow F_e = 1, 2$ and ^{85}Rb $F_g = 2 \longrightarrow F_e = 2, 3$ transitions: HLZT fluorescence spectrum (upper curve, $B = 2480$ G); LZT transmission spectrum (middle curve, $B = 2420$ G). Lower curve is a reference SA spectrum.

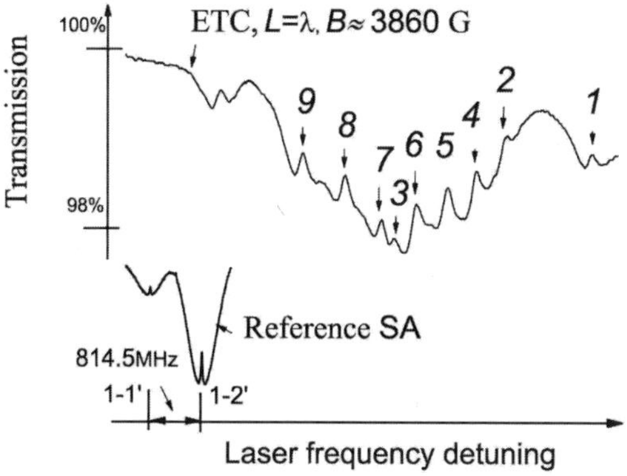

Fig. 25. Transmission spectrum of ^{87}Rb, $F_g = 1 \longrightarrow F_e = 1, 2$ and ^{85}Rb, $F_g = 2 \longrightarrow F_e = 2, 3$ transitions for $L = \lambda$ (LZT method) at $B = 3860$ G. VSOP peaks labeled *1* $\div$ *9* denote corresponding transitions shown in Fig. 18. Lower curve is a reference SA spectrum.

curve- LZT, $B = 2420$ G); the lower curve is a reference SA spectrum. This figure helps to illustrate the disscussed advantages and drawbacks of HLZT and LZT. In particular, due to narrower spectral width of the VSOP resonances, an atomic transition labeled *3* is fully resolved in the case of LZT method, while it partially overlapped with peaks *5* and *6* in HLZT.

Somewhat different values of B for the upper and middle curves provide additional information: as it is seen the largest frequency shift value among transitions *1, 2, 3* versus B is recorded for *1* ($A1 \simeq 1.7$ MHz/G). $A2 \simeq 1.5$ MHz/G and $A3 \simeq 1.1$ MHz/G for peaks *2* and *3*.

The contrast of VSOP resonances of reduced absorption in LZT (middle curve) is much higher compared with the contrast of dips of reduced fluorescence presented in the upper curve of Fig. 23. The HLZT method based on the fluorescence for $L = \lambda/2$ (397 nm) provides better spatial resolution than the LZT based on transmission spectrum for $L = \lambda$ (794 nm), which can be of importance for measurement and mapping of strongly non-homogeneous magnetic field. On the other hand, the LZT signal registration is simpler, also the photodetector recording the transmission signal can be placed far from the oven, unlike the fluorescence signal registration.

In order to increase the attainable strength of external magnetic field applied to an atomic vapor contained in ETC, the permanent ring magnets embracing the cell have to be as close to each other as possible. The main limitation for the distance between PRMs is imposed by a longitudinal dimension of the cell oven (see Fig. 17a). A new oven was developed specially for high B-field applications,[25] with longitudinal dimension less than 3 cm. Transmission spectrum of ETC with $L = \lambda$ for $B = 3860$ G is presented in Fig. 25. The labels *1 - 9* denote corresponding transitions between magnetic sublevels shown in Fig. 18. We can see in Fig. 25 that the VSOP peak labeled *1* (^{87}Rb, $F_g = 1, m_F = +1 \longrightarrow F_e = 2, m_F = +2$) remains the most convenient frequency marker also for higher B-fields measurements: in the whole range of $1 \div 4000$ G, it has the largest peak amplitude among transitions *1, 2, 3,* and does not overlap with any other transition.

In recent papers by M. Momeen et al.[31] and G. Školnik et al.,[32] the Zeeman splitting of the D_1/D_2 lines of the Rb atomic vapor has been studied in the range $B = 10 - 100$ G using widely used "saturated absorption" (SA) method. Compared with our HLZT, LFZT and LZT methods, SA technique allows to study an overall behavior being unable to distinguish individual transition contributions. Moreover, SA becomes absolutely unusable for $B > 100$ G.

The experimental results presented in this Section can underlie the development of simple and efficient magnetometers based on three elaborated methods: HLZT, LFZT, or LZT, which are based on an ETC filled with Rb. Having sub-micrometer spatial resolution, these magnetometers would be specially efficient for determination of strongly non-homogeneous magnetic field B in the range of $1 \div 4000$ G.

5. Conclusions

Extremely thin cells (ETC) containing atomic vapor of alkali metals are developed for studies of resonant absorption, fluorescence, reflection as well as magneto-optical processes in the case when the atomic vapor thickness L is comparable with the laser radiation wavelength λ, namely for L from 30 nm up to several μm. The design of ETC allows smooth variation of thickness by lateral translation of a wedge-shaped gap, which is needed for realization of narrow resonance regime; the measurement of ETC thickness L is done by an interference technique. The operation temperature limit for ETC is 500°C.

Resonant absorption and fluorescence processes on ^{85}Rb, ^{87}Rb D_1 line are studied in the ETC with thickness of atomic vapor column L variable in the range from 100 to 3000 nm using a 1 MHz-linewidth ECDL with $\lambda =$ 794 nm and output power 30 mW. Unusual peculiarities of ETC spectra, which were earlier revealed on Rb D_2 and Cs D_1, D_2 lines (jointly with the group of M. Ducloy, D. Bloch from Paris-Nord University and with S. Cartaleva's group, Institute of Electronics, Sofia) are exploited to form spectrally narrow resonances. Particularly, it is experimentally confirmed that the main parameter determining the spectral profile of resonant absorption is the ratio L/λ (λ is the atomic transition wavelength of Rb D_1 line). The phenomena of Dicke-type coherent narrowing (DCN) and collapse of DCN are more pronounced for the Rb D_1 line because of larger frequency separation between the hyperfine levels. The minima of resonant absorption spectral width take place for $L = (2m+1)\lambda/2$. The smallest width, 3 - 4 times narrower than the Doppler broadening in a "usual" cm-size cell, occurs at $L = \lambda/2$.

The following distinctions between absorption and fluorescence spectra are observed also for Rb D_1 line: i) the spectral width of fluorescence is always narrower ($1.5 \div 1.8$ times for $L = \lambda/2$, and $\simeq 3$ times for $L = \lambda$) than the transmission width for the same transition; ii) in contrast to absorption, the increase of L results in monotonic increase of the fluorescence spectral width; iii) the fluorescence spectral width remains below the Doppler width up to $L \simeq 4$ μm, while the transmission width becomes equal to Doppler one even for smaller L values ($\simeq 3$ μm); iv) the VSOP resonances in ETC transmission spectra appear at $L = m\lambda$ for radiation intensities $\simeq$ mW/cm^2, while the intensity required to observe VSOP resonances in fluorescence spectra is an order of magnitude higher. The advantages of frequency reference method based on ETC compared with one recently reported and based on thin cell $L = 0.5 - 1$ mm are demonstrated.

The "$\lambda/2$-Zeeman technique" (HLZT) based on resonant fluorescence spectrum emitted by a ETC with $L = \lambda/2$ is shown to be efficient for studies of individual transitions between the Zeeman sublevels of hyperfine levels in an external magnetic field in the range of 10 – 2500 G. Possible applications, such as magnetometers with nanometric spatial resolution are described. Implementation of HLZT allowed us to reveal strong modification (more than 15 times) of atomic transition probabilities, as well as merging of ^{85}Rb and ^{87}Rb D_1 line atomic transitions, which are separated by a few GHz at $B = 0$, in a strong magnetic field ($B > 2000$ G). Another method, "λ-Fluorescence Zeeman technique" (LFZT), exploits narrow dips in fluorescence spectrum for an ETC with $L = \lambda$ observed at relatively high laser intensity is $\simeq 30$ mW/cm^2. Splitting of these dips occurring already at $B = 1$ G makes justified the use of LFZT to get precise "markers" of the exact atomic transition frequency in an external magnetic field. The 3-rd proposed technique, "λ-Zeeman technique" (LZT), is based on formatioin of narrow VSOP resonances in transmission spectra of ETC with thickness $L = \lambda$ and their splitting in a B-field. LZT is efficient for magnetic field in the range of 1 G $\div$ 4000 G. Yielding correct frequency positions and probabilities of split transition components, the LZT can serve as a powerful tool for atomic spectroscopy and magneto-optics. On the other hand, LZT is the most simple and convenient method of optical magnetometry as compared with HLZT and LFZT. The benefits of such a magnetometer are high spatial resolution and wide range of measured B values, which allow to use it for measurement and mapping of strongly inhomogeneous magnetic field.

Acknowledgments

The authors acknowledge the valuable participation of A.S. Sarkisyan in fabrication of ETCs. We are grateful to M. Ducloy, A. Weis, Yu. Malakyan, S. Cartaleva, M. Auzinsh, D. Bloch, Y. Pashayan-Leroy, C. Leroy, P. Moroshkin, and G. Pikhler for useful and stimulating discussions. We thank A. Sargsyan and H. Hakhumyan for valuable experimental assistance. The work was done with a support provided by INTAS Grant 06-1000017-9001. Research partly conducted in the scope of the International Associated Laboratory IRMAS.

References

1. S. Briaudeau, D. Bloch, M. Ducloy, Europhys. Lett. **35**(5), 337 (1996).
2. S. Briaudeau, D. Bloch, M. Ducloy, Phys. Rev. A **59**, 3723 (1999).

3. A.Ch. Izmailov, Opt. and Spectr. 74(1), 25 (1993); A.Ch. Izmailov, Laser Phys. Lett. **3**(3), 132 (2006).
4. T.A. Vartanyan, D.L. Lin, Phys. Rev. A **38**, 5197 (1995); T. Vartanyan, Rus. Physics-Uspekhi **44**, 1203 (2001).
5. B. Zambon, G. Nienhuis, Opt. Commun. **143**, 308 (1997).
6. D. Sarkisyan, D. Bloch, A. Papoyan, M. Ducloy, Opt. Commun. **200**, 201 (2001).
7. D. Sarkisyan, T. Becker, A. Papoyan, P. Thoumany, H. Walther, Appl. Phys. B: Las. Opt. **76**, 625 (2003).
8. G. Dutier, A. Yarovitski, S. Saltiel, A. Papoyan, D. Sarkisyan, D. Bloch, M. Ducloy, Europhys. Lett. **63**, 35 (2003).
9. G. Dutier, S. Saltiel, D. Bloch, M. Ducloy, J. Opt. Soc. Am. B **20**, 793 (2003).
10. G. Nikogosyan, D. Sarkisyan, Yu. Malakyan, J. Opt. Technol. **71**, 602 (2004).
11. D. Sarkisyan, T. Varzhapetyan, A. Sarkisyan, Yu. Malakyan, A. Papoyan, A. Lezama, D. Bloch, M. Ducloy, Phys. Rev. A **69**, 065802 (2004).
12. D. Sarkisyan, T. Varzhapetyan, A. Papoyan, D. Bloch, M. Ducloy, Proc. SPIE **6257**, 625701 (2006); D. Sarkisyan, A. Sargsyan, A. Papoyan, Y. Pashayan-Leroy, Proc. SPIE **6604**, 660405 (2007).
13. I. Hamdi, P. Todorov, A. Yarovitski, et al., Laser Phys. **15**, 987 (2005).
14. A. Sargsyan, D. Sarkisyan, A. Papoyan, Y. Pashayan-Leroy, P. Moroshkin, A. Weis, A. Khanbekyan, E. Mariotti, L. Moi, Laser Phys. **18**, 1 (2008).
15. A.D. Sargsyan, D.H. Sarkisyan, Y. Pashayan-Leroy, C. Leroy, P. Moroshkin, A. Weis, J. Contemp. Phys. **43**, 7 (2008).
16. C. Andreeva, S. Cartaleva, L. Petrov, S. M. Saltiel, D. Sarkisyan, T. Varzhapetyan, D. Bloch, M. Ducloy, Phys. Rev. A **76**, 013837 (2007).
17. T. Varzhapetyan, A. Nersisyan, V. Babushkin, D. Sarkisyan, S. Vdović, G. Pichler, J. Phys. B: At. Mol. Opt. Phys. **41**, 185004 (2008).
18. S. Cartaleva S. Saltier, A. Sargsyan, D. Sarkisyan, D. Slavov, P. Todorov, K. Vaseva, J. Opt. Soc. Am. B **26**, 1999 (2009).
19. D. Sarkisyan, A. Papoyan, "Optical processes in micro- and nanometric thin cells containing atomic vapor", New Trends in Quantum Coherence and Nonlinear Optics (Horizons in World Physics, Vol.263), Ed.: R.Drampyan, Nova Science Publishers, New York, Chapter 3, pp.85-124 (2009).
20. M. Fichet, G. Dutier, A. Yarovitsky, P. Todorov, I. Hamdi, I. Maurin, S. Saltiel, D. Sarkisyan, M.P. Gorza, D. Bloch, M. Ducloy, Europhys. Lett. **77**, 540001 (2007).
21. A. Papoyan, D. Sarkisyan, K. Blush et al., Laser Phys. **13**, 1467 (2003).
22. D. Sarkisyan, A. Papoyan, T. Varzhapetyan, J. Alnis, K. Blush, M. Auzinsh, J. Opt. A: Pure Appl. Opt. **6**, S142 (2004).
23. D. Sarkisyan, A. Papoyan, T. Varzhapetyan, K. Blush, M. Auzinsh, J. Opt. Soc. Am. B **22**, 88 (2005).
24. T. Varzhapetyan, H. Hakhumyan, D. Sarkisyan, V. Babushkin, A. Atvars, M. Auzinsh, J. Contemp. Phys. **42**, 223 (2007).
25. A. Sargsyan, G. Hakhumyan, A. Papoyan, A. Atvars, M. Auzinsh, Appl. Phys. Lett. **93**, 021119 (2008).

26. R. Wynands, A. Nagel, Appl. Phys. B: Lasers Opt. **68**, 1 (1999) and references therein.
27. A. Sargsyan, D. Sarkisyan, A. Papoyan, Phys. Rev. A **73**, 033803 (2006).
28. R.H. Romer, R.H. Dicke, Phys. Rev. **99**, 532 (1955).
29. W. Demtroder, Laser Spectroscopy: Basic Concepts & Instruments, Springer, Heidelberg, 1996.
30. A. Mikata, U. Tanaka, Sh. Urabe, Appl. Opt. **47**, 640 (2008).
31. M.U. Momeen, G. Rangarajan, P.C. Deshmukh, J. Phys. B: At. Mol. Opt. Phys. **40**, 3163 (2007).
32. G. Školnik, N.Vujičić , T. Ban, Opt. Commun. **282**, 1326 (2009).

MODELLING MAGNETO-OPTICAL RESONANCES IN ATOMIC RUBIDIUM AT D_1 EXCITATION IN EXTREMELY THIN CELLS WHILE MAINTAINING A SELF-CONSISTENT SET OF THEORETICAL PARAMETERS

L. KALVANS*, M. AUZINSH, R. FERBER, F. GAHBAUER and A. JARMOLA

Laser Centre, The University of Latvia, 19 Rainis Boulevard, LV-1586 Riga, Latvia
**E-mail: linards.kalvans@lu.lv*

A. PAPOYAN and D. SARKISYAN

Institute for Physical Research, NAS of Armenia, Ashtarak-0203, Armenia

Nonlinear magneto-optical resonances of alkali atoms have been studied both experimentally and theoretically with high accuracy in ordinary vapor cells. By using an extremely thin cell (ETC), Doppler broadening is reduced so that all hyperfine transitions can be resolved. To study the magneto-optical resonances, an ETC containing a natural mixture of atomic rubidium was placed at the center of a three-axis Helmholtz coil system, which canceled the ambient magnetic field. The magnetic field was controlled and scanned from −60 G to +60 G along the axis that was perpendicular to both the polarization direction and propagation direction of the laser radiation. The theoretical model was based on the optical Bloch equations; careful comparison of experimental and theoretical results allowed us to find the right model to account for transit relaxation, laser power density, and Doppler broadening, the theoretical results presented here were obtained by a self-consistent set of parameters.

Keywords: Magneto-optical resonances.

1. Introduction

The introduction of atomic vapor cells of size of few hundred nanometers has provided a number of new research and application opportunities in magnetometry and optoelectronics.[1,2] Such cells are called extremely thin cells (ETCs). In the ETCs sub-Doppler spectroscopy becomes possible when the exciting laser light is propagating normally with respect to the cell windows. In such a setup only the atoms flying nearly parallel to the windows are able to interact with the exciting laser, as the rest of the atoms suffer from frequent collisions with the cell walls, which prevents them from reaching the excited state and fluorescing. In this paper the above properties of ETCs are exploited for research of magneto-optical resonances, which occur in the fluorescence plots versus magnetic field at zero field values. The resonance is expected to be bright[3] when the quantum number of the total angular momentum of the ground state hyperfine level F_g is less than the corresponding quantum number for the excited state F_e. Otherwise ($F_g \geq F_e$), the resonance is expected to be dark.[4,5] The resonances are formed as a result of the fact that at zero magnetic field the magnetic sublevels of the ground state form a quantum superposition state, which is coupled to the exciting laser field (bright resonance) or not coupled (dark resonance). When the magnetic field is applied, the superposition state is destroyed, which leads to changes in the fluorescence signal. Despite the simple principles of the theoretical description, the theoretical calculations have produced a number of controversial results. One of the reasons for these calculations have produced controversies might be lack of ability to resolve distinct hyperfine transitions under the Doppler profile. This problem could be overcome by using ETCs. The first results of this approach[6] produced some unexpected results – resonances observed bright in an ordinary vapor cell appeared to be dark in an ETC. Our goal for this study was to clarify the situation by recording experimental data of magneto-optical resonances and applying a theoretical model proven to be successful in cm-sized cells.[7,8] In order to do so we had to find the best adjustment for the parameters in the model to fit the specific conditions of ETC.

The D_1 line of alkali metals consists of four hyperfine components. The hyperfine level splittings and relative strengths of transistions are shown in Fig. 1. The FWHM of Doppler profile at room temperature (in an

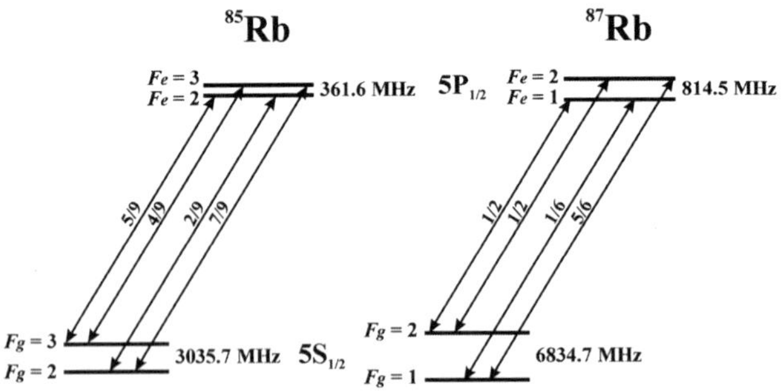

Fig. 1. Hyperfine level structure and transitions of the D_1 line of rubidium. The fractions on the arrows indicate the relative transition strengths.[9,10]

ordinary vapor cell) is about 500 MHz, which fully covers the excited state splitting for the ^{85}Rb isotope and partially for ^{87}Rb. However, in the ETC the Doppler profile is reduced, as mostly the atoms flying nearly parallel to the cell windows determine the fluorescence signal, and, thus, the hyperfine transitions become completely resolved.

2. Experimental Setup

The ETC was produced at the Institute of Physical Research in Ashtarak, Armenia, while the experiments described here were performed in Riga. The principle of the ETC is described in Refs. 1, 11. The cell windows, are two perfectly polished wafers made from YAG crystal that are glued in such a way that the distance between windows changes from 50 nm to roughly 2 μm, thus allowing the experimenter to choose the desired windows separation Rubidium is stored in a side arm bonded to the main cell. Appropriate vapor pressure was maintained by heating the side arm to 150°C and windows to 200°C in a two-chambered oven, thus preventing a formation of metal coating on the cell windows. A Toptica DL 100 external-cavity, single-mode diode laser with a wavelength of 794.3 nm and a typical linewidth of a few megahertz was used for the experiment. The laser wavelength was monitored by a High Finesse WS-7 wavemeter without using any active stabilization. To maintain the necessary linear polarization and control the power density, the laser beam was passed through a Glan-Thomson polarizer. The beam profile was checked by a Thorlabs BP 104-VIS beam profiler. A typical beam covered an area of 0.15 cm^2 (although it was different for a number of experiments), and the power density was altered between 10 mW/cm^2 and 2000 mW/cm^2.

In order to apply the magnetic field, three pairs of Helmholtz coils were used, two of which compensated the ambient laboratory magnetic field, while the third generated the scan magnetic field in the same direction along which the fluorescence was detected. The excitation, observation and magnetic field geometry can be seen in Fig. 2. In most cases the magnetic field was scanned from negative to positive values at fixed laser frequency, but for some transitions the double-scan technique[12] was used – the laser was scanned slowly across the transition while rapidly changing the magnetic field. In a typical scan the laser frequency would change by 2–5 MHz while the magnetic field was changed over the whole range (up to $\pm$60 G).

The fluorescence that emerged via the side walls of the ETC was detected by means of a Thorlabs FDS-100 photodiode without singling out any particular polarization direction. Results were recorded for both ru-

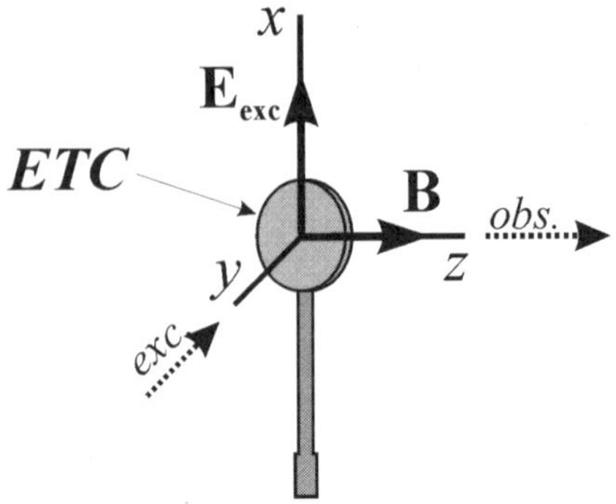

Fig. 2. Experimental geometry. The relative orientation of the laser beam (exc), laser light polarization ($\mathbf{E}_{exc}$), magnetic field ($\mathbf{B}$), observation direction (obs), and ETC walls are shown.

bidium isotopes and for all four hyperfine transitions of each isotope. To tune laser to a particular hyperfine transition, peaks in the fluorescence versus laser frequency were detected at zero magnetic field. For each transition and power density several scans were made and averaged. If there was significant deviation in results of some scans they were excluded from the averaged results.

3. Theoretical Model

A theoretical model had been developed previously in order to describe bright and dark resonances in ordinary vapor cells[7,8] where a detailed description can be found. Here we present just the final equations with brief comments. The equations are based on optical Bloch equations, which describe the time evolution of the semi-classical atomic density matrix ρ:[13]

$$i\hbar\frac{\partial\rho}{\partial t} = \left[\hat{H},\rho\right] + i\hbar\hat{R}\rho. \tag{1}$$

The Hamiltonian $\hat{H}$ is considered to consist of the unperturbed atomic Hamiltonian – H_0, an operator H_B, which describes the atomic interaction with the external magnetic field, and the dipole interaction operator $\hat{V}$. The relaxation operator $\hat{R}$ includes the spontaneous decay rate Γ_N and transit relaxation rate γ. Applying the rotating wave approximation and considering the Doppler shits due to the classical motion of atoms, stochastic differential equations are obtained. The latter are further simplified by the decorrelation approach.[14] Finally the optical coherences are eliminated and we obtain rate equations for Zeeman coherences:

$$\frac{\partial \rho_{g_i g_j}}{\partial t} = \left(\Gamma_{p,g_i e_m} + \Gamma^*_{p,e_k g_j}\right) \sum_{e_k,e_m} \left(d_1^{g_i e_k}\right)^* d_1^{e_m g_j} \rho_{e_k e_m}$$

$$- \sum_{e_k,g_m} \left[\Gamma^*_{p,e_k g_j} \left(d_1^{g_i e_k}\right)^* d_1^{e_k g_m} \rho_{g_m g_j} + \Gamma_{p,g_i e_k} \left(d_1^{g_m e_k}\right)^* d_1^{e_k g_j} \rho_{g_i g_m}\right]$$

$$-i\omega_{g_i g_j} \rho_{g_i g_j} + \sum_{e_i,e_j} \Gamma^{e_i e_j}_{g_i g_j} \rho_{e_i e_j} - \gamma \rho_{g_i g_j} + \lambda \delta(g_i, g_j) \qquad (2)$$

and

$$\frac{\partial \rho_{e_i e_j}}{\partial t} = \left(\Gamma^*_{p,e_i g_m} + \Gamma_{p,g_k e_j}\right) \sum_{g_k,g_m} d_1^{e_i g_k} \left(d_1^{g_m e_j}\right)^* \rho_{g_k g_m}$$

$$- \sum_{g_k,e_m} \left[\Gamma_{p,g_k e_j} d_1^{e_i g_k} \left(d_1^{g_k e_m}\right)^* \rho_{e_m e_j} + \Gamma^*_{p,e_i g_k} d_1^{e_m g_k} \left(d_1^{g_k e_j}\right)^* \rho_{e_i e_m}\right]$$

$$-i\omega_{e_i e_j} \rho_{e_i e_j} - \Gamma \rho_{e_i e_j}. \qquad (3)$$

These equations are pretty easy to understand qualitatively as each term describes some logical subprocess in atom-laser interaction. The density matrix elements for ground and excited state Zeeman coherences are denoted by $\rho_{g_i g_j}$ and $\rho_{e_i e_j}$. In equation 2 the first term describes induced transfers that repopulate the ground state, $\Gamma_{pg_i e_m}$ and Γ^*_{p,ekg_i} stand for the coupling induced by the laser between ground and excited states and are dependent on the Rabi frequency Ω_R, the natural linewidth Γ_N, the laser linewidth, the laser detuning from the frequency of a particular transition, and the central velocity of the atomic velocity group while performing averaging over the Doppler profile; $d_1^{e_i g_j}$ is the dipole transition matrix element between the ground state i and the excited state j. The second term (whole sum in the square brackets) describes the population loss in the ground state due to laser absorption processes. The third term describes the splitting of the Zeeman sublevels in the magnetic field and can be calculated by means of the Breit-Rabi formula.[15,16] The fourth term stands for the repopulation of the ground state via the spontaneous decay of the excited state, the fifth – relaxation of the ground state due to transit relaxation (or collisions with with the cell walls in the case of ETC). And finally, the sixth term is there for the repopulation of the ground state due to depolarized atoms arriving in the interaction region (either flying in or being depolarized after collision with the cell wall).

In the equation 3 the first term describes how population of the excited state increases due to light absorption while the second denotes the population lost via induced transitions. The third term describes level splitting

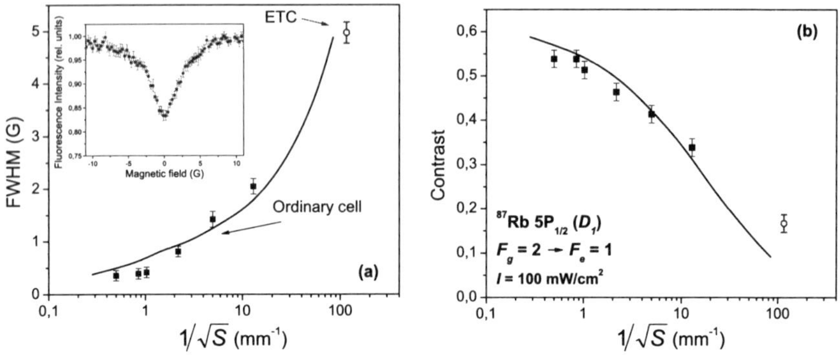

Fig. 3. Resonance (a) width and (b) contrast as a function of the inverse size of the interaction region with the beam: the square root of the cross-sectional area of the beam in the case of the ordinary cell[8] and the characteristic length before collision in case of the ETC. Filled squares experiment in an ordinary cell; open circles: experiment in an ETC; solid curve: theory. The inset in (a) shows a typical dark resonance in the ETC. Data are for the $F_g = 2 \rightarrow F_e = 1$ transition of ^{87}Rb at a laser power density of 100 mW/cm^2.

in the magnetic field and the fourth, the relaxation of the ground state via spontaneous emission (rate – Γ_N) and transit/collisions (rate – γ).

The equations 2 and 3 are considered for stationary conditions $\partial\rho/\partial t = 0$, where they form a system of linear equations, which can be solved numerically to obtain the density matrices for the ground and excited states. The assumption of stationary conditions means that the characteristic relaxation and coupling times of atomic system are much lower than the interaction times at the same conditions (magnetic field values), which holds true for our experimental conditions. Once the density matrices are known, the fluorescence signal can be calculated to a constant factor for particular observation geometry as follows:

$$I_f(\tilde{\mathbf{e}}) = \tilde{I}_0 \sum_{g_i, e_i, e_j} d^{(ob)*}_{g_i e_j} d^{(ob)}_{g_i e_i} \rho_{e_i e_j}, \tag{4}$$

where $\tilde{I}_0$ is a constant of proportionality.

One of the parameters that determined the width and contrast of the resonances in the theoretical description of ordinary vapor cells was the transit relaxation rate γ, which is the rate at which atoms fly out of the region of interaction with the laser radiation. This transit relaxation rate thus is related to the laser beam diameter. The solid squares in Fig. 3 mark the dependence of the resonance width and contrast on the inverse square root of the cross-sectional area S of the beam, which is related to the inverse of the laser beam diameter in the ordinary cell. In the ETC, resonances are significantly broader than in ordinary cells, and their contrast is much smaller. The question arises: are the resonance width and contrast in the ETC still controlled by a transit relaxation rate γ, and, if so, what is the length corresponding to this transit relaxation rate? In our experiments the laser beam diameter was much larger than the ETC wall separation; thus, the characteristic length in the ETC should be related to the wall separation rather than to the laser beam diameter. However, a Voigt profile fit to the fluorescence peaks in an LIF excitation spectrum of the ETC revealed that the Doppler width in the ETC is approximately 60 MHz (FWHM) when the wall separation L is equal to the wavelength λ of the laser radiation. The Doppler width corresponding to that temperature in an ordinary cell would be about ten times greater. One could therefore say that a typical atom that fluoresces and travels at the average thermal speed flies on a trajectory that makes an angle with respect to the ETC wall such that the flight path between the walls is about ten times the wall separation. Indeed, the abscissa of the open circles in Fig. 3, which correspond to the resonance width and contrast in the ETC, have been chosen to correspond to a length dimension that is about ten times the wall separation, and the circles lie rather close to the extrapolation of the theoretical model developed for ordinary cells. Strictly speaking, it is not correct to extend the theoretical calculation with the parameter values chosen for ordinary cells into the regime of the ETC. However, Fig. 3 suggested that a single theoretical approach might be able to describe qualitatively the width and contrast of nonlinear magneto-optical resonances in alkali vapors over several orders of magnitude of a single dimension, which is related to a transit relaxation rate γ. Hence, we formulated the hypothesis that the shapes of nonlinear magneto-optical resonances in an ETC can be described by the same model that describes these signals in ordinary cells when the particular characteristics of the ETC are reflected in appropriately chosen values of the model parameters.

4. Analysis of Results

According to the rules stated in section 1 one would expect to observe three dark and one bright resonance for each isotope. The experimental and theoretical results in Fig. 4 and Fig. 5 clearly indicate the three dark resonances for transitions $F_g = 3 \rightarrow F_e = 2$, $F_g = 3 \rightarrow F_e = 3$, and $F_g = 2 \rightarrow F_e = 2$ for ^{85}Rb, and $F_g = 2 \rightarrow F_e = 1$, $F_g = 2 \rightarrow F_e = 2$, and $F_g = 1 \rightarrow F_e = 1$ for ^{87}Rb. On the other hand the small peaks of

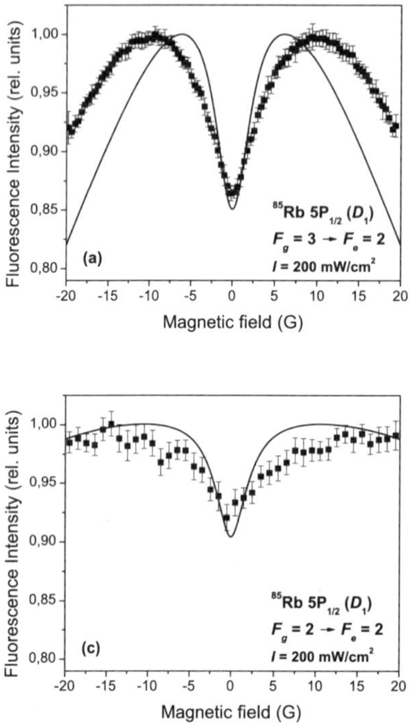

Fig. 4. Fluorescence intensity versus magnetic field for ^{85}Rb at D_1 excitation. Filled squares, experiment; solid line, theory. The excited state, total angular momentum of the ground F_g and excited states F_e of the transition, and laser power density I are given in each panel. Wall separation $L = \lambda$, the wavelength of the light.

the bright resonance observed in the ordinary cell[7,8] have disappeared both from the theoretical and experimental results. This might be explained by the increased transit/collision relaxation rate γ in the ETC conditions, which is the main parameter determining the shape of the resonance – the smaller is its value the more narrow is the peak of the resonance. In the case of the ETC it has most likely become so wide that it is not resolvable under the wide shape of the decreasing fluorescence as a result of the magnetic field shifting the Zeeman sublevels out of resonance with the laser field.

A lot of effort was applied to fit the experimental and theoretical results as the theoretical model does not allow to predict exact Rabi frequencies for particular laser power density. In addition, the most effective value of the transit/collision rate γ had to be established. Of no less importance was the estimation of the residual Doppler profile in the ETC and the value of the background signal present in the experimental data. In order

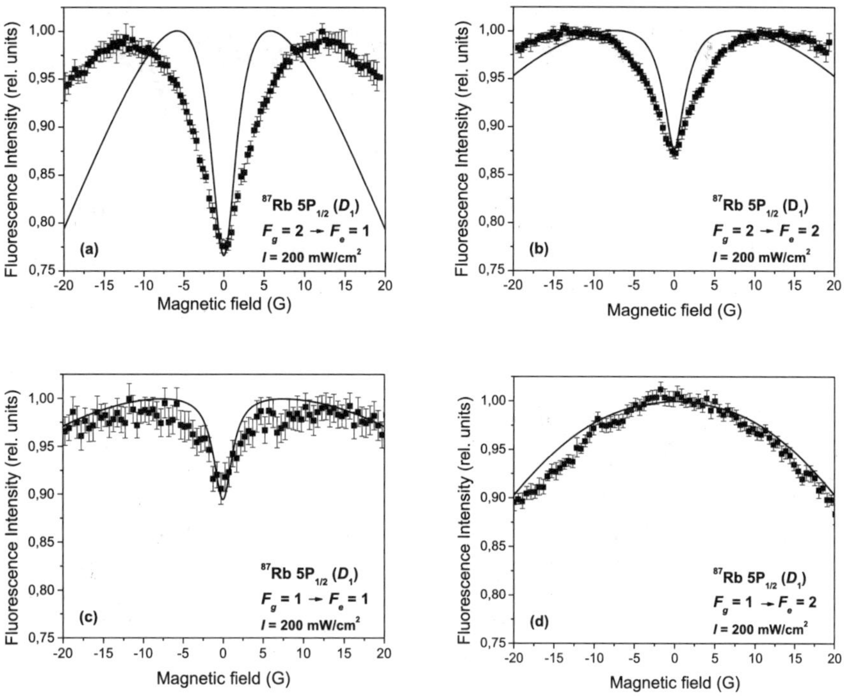

Fig. 5. Fluorescence intensity versus magnetic field for ^{87}Rb at D_1 excitation. Filled squares, experiment; solid line, theory. The excited state, total angular momentum of the ground F_g and excited states F_e of the transition and laser power density I are given in each panel. Wall separation $L = \lambda$, the wavelength of the light.

to obtain the best values for these parameters, a number of theoretical calculations were run with various parameter sets, and the one that gave the best correspondence for all the hyperfine transitions and both isotopes was chosen for the final results. It is worth noting that our aim was to use the same parameters for all transitions while for each particular transition better consistency could be achieved with slightly different parameters. The parameters chosen for the final results were as follows: the laser power density was related to the Rabi frequency as $I = \Omega_R^2$, the transit/collision rate γ=1.44 MHz, the FWHM for the residual Doppler profile was 60 MHz, and the background signal in the experimental results was 50%, while the spectral linewidth of laser was assumed to be 10 MHz. Once established, the same parameters were used for all calculations.

Figure 6 shows the transition $F_g = 1 \rightarrow F_e = 1$ for ^{87}Rb at various laser power densities. The filled squares represent experimental results while the

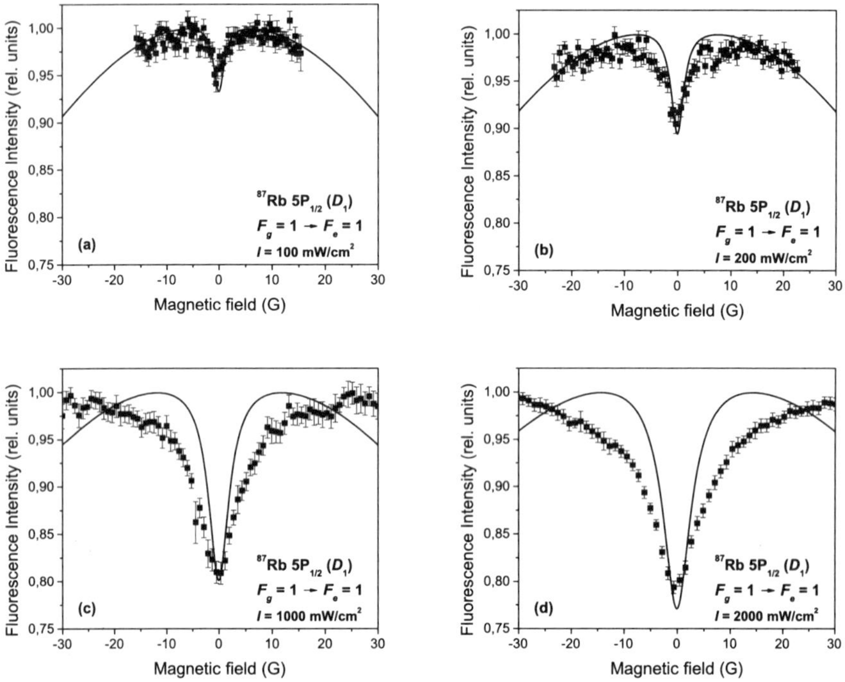

Fig. 6. Resonance signals for ^{87}Rb at the $F_g = 1 \rightarrow F_e = 1$ transition for different laser power densities I. Filled squares, experiment; solid line, theory. $L = \lambda$.

solid line denotes the results of theoretical calculations. Similar to the case of ordinary vapor cells, discrepancies between theory and experiment arise when the laser power density (and, hence, the Rabi frequency) is increased. The discrepancies mostly affect the shape of the resonances, especially at larger field values, while the contrast is still described very well even at high laser power densities (Fig. 7). The reason of this most likely is the exponential transit/collision relaxation mechanism assumed in the model is not valid at higher Rabi frequencies, and some nonlinear mechanism should be considered.

As was seen in the case of the ordinary cell (see Fig. 6 in Ref. 8) the resonances are sensitive to detuning, in the case of an ETC one would expect a much larger sensitivity as the Doppler profile is significantly reduced. Indeed, as can bee seen in Fig. 8, there is a rapid drop in the resonance contrast when the detuning the laser frequency changes by just a few tens of megahertz. In fact, the contrast drop in an ETC at 20 MHz detuning roughly equals the contrast drop in an ordinary cell when laser

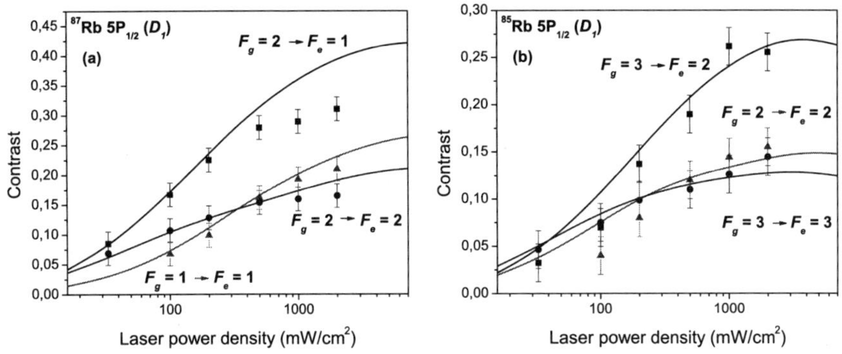

Fig. 7. Resonance contrast as a function of laser power density for (a) ^{87}Rb and (b) ^{85}Rb dark resonances. Markers, experiment; solid line, theory. $L = \lambda$.

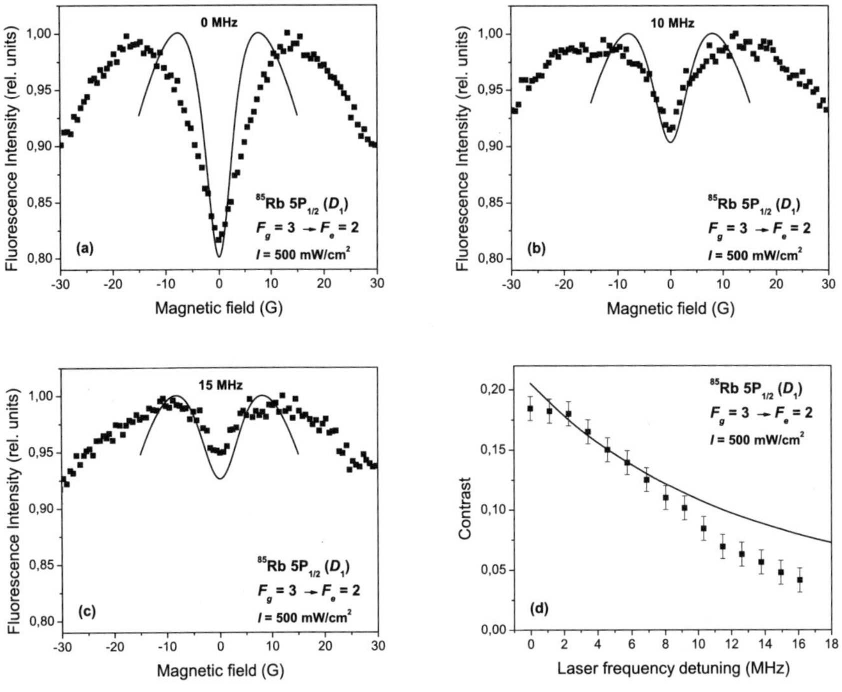

Fig. 8. Resonance signals for ^{87}Rb at the $F_g = 1 \rightarrow F_e = 1$ transition for different laser power densities I. Filled squares, experiment; solid line, theory. $L = \lambda$.

is detuned by 300 MHz. These results also are in agreement with results of the dark resonance formation in an ETC in a Λ-system with two lasers (^{85}Rb $5S_{1/2}$,$F_g = 2 \rightarrow 5P_{3/2} \rightarrow 5S_{1/2}$,$F_g = 3$).[17,18] Namely, as the coupling laser was detuned from the resonance with an atomic transition, a strong increase of the resonance width of the electromagnetically induced transparency and a worsening of the contrast were recorded. In ordinary cm-size cells, effectively the opposite behavior would be observable.

Due to the wedge form of the ETC it was possible to obtain resonance signals for various wall separation distances. The separation could be changed and controlled simply by changing the position of incident beam on the cell and observing the intensity of transmitted light, which obeyed the Fabri-Perot rules and produced maximal signal when wall separation equals an integer times half the wavelength. We have measured resonances for L equal to $\lambda/2$, $3\lambda/4$, λ, and $3\lambda/2$, and the results are shown in Fig. 9. In order to model these results we assumed that transit/relaxation time γ^{-1} and the residual Doppler profile are linearly proportional to the cell thickness, decreasing if the separation is decreased and vice versa. As can be seen in Fig. 9, this does a pretty good job in describing the contrast changes, although it fails to predict the width dynamics of the resonances. In this case we did not explore the dependence on wall separation in more detail, but some nonlinear effects might be responsible for the severe change in resonance shape while changing wall separation.

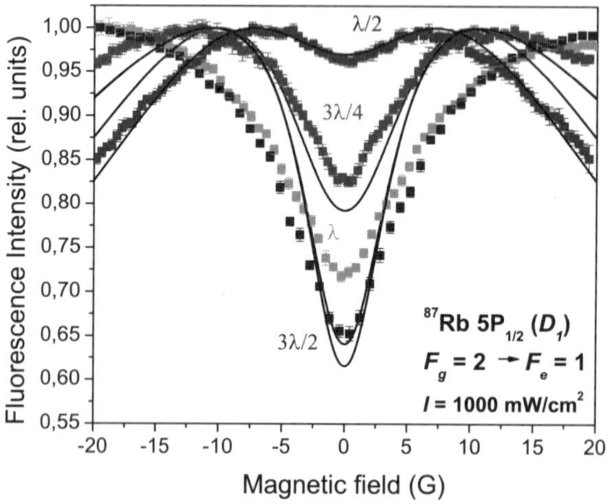

Fig. 9. Resonance signals for ^{87}Rb at the $F_g = 1 \rightarrow F_e = 1$ transition for different laser power densities I. Filled squares, experiment; solid line, theory. $L = \lambda$.

In general the agreement between the experimental data and calculations was satisfactory over a wide range of different experimental conditions and for four distinct hyperfine transitions in each of two isotopes, though it was not nearly as good as in case of an ordinary cell.[8] Several reasons can be named that could be responsible for the discrepancies in the case of ETC. First of all, as a result of the sensitivity of the resonance shapes to the laser detuning (Fig. 8), in the case of the ETC, small inevitable drifts in the laser frequency could cause significant changes in measurements, whereas in the case of ordinary cell, such drifts would make very little difference. Secondly, since the signals are sensitive to the cell thickness (Fig. 9) and since the cell thickness is a nonlinear function of position and is changing rapidly, the actual laser beam (0.4 mm in diameter) covered a region of varying cell thickness.

5. Conclusions

Nonlinear magneto-optical resonances have been measured for all hyperfine transitions of the D_1 line of ^{85}Rb and ^{87}Rb in an ETC under a wide variety of experimental conditions, which included different laser power densities, laser detunings, and ETC thicknesses. As expected, six dark resonances have been observed on hyperfine transitions with $F_g \geq F_e$. The expected bright resonances ($F_g < F_e$) where not observed, in agreement with the theoretical model. The theoretical model was based on the optical Bloch equations and was essentially the same as the one used to describe the resonances in ordinary cells, although some parameters where changed to adapt the model to the case of the ETC. These parameters included the non-coherent transit/collision relaxation rate γ, which was increased considerably, and the residual Doppler profile. which was decreased, to meet the conditions atoms face in the thin space between the cell windows. The last parameter to change was the conversion factor between Rabi frequency and laser power density. With all these parameters changed, satisfactory agreement between theory and experiment was achieved using the same parameter set for all hyperfine transitions in both isotopes. Overall, the resonances in the ETC are much broader and have lower contrast when compared to resonances in ordinary cells. The observations are strongly supported by the results of theoretical model and our understanding of physical processes present in the described conditions.

The theory does a pretty good job at describing the resonances at low power densities and the overall dynamics of the dark resonance contrasts while changing the laser power density. Problems arise when describing the shape of the resonance at higher laser power densities, especially when predicting the shape of the resonance "wings", i.e., the part of the signal at higher magnetic field values after the maximum in the fluorescence signal has been reached. It is worth pointing out the high sensitivity of the resonances to laser detuning from the exact hyperfine transition frequency, which is very well described by the theoretical model. Another interesting result concerns changes in resonance shape with changing wall separation. However, this result still lacks a full theoretical understanding and should be subject to further investigations.

Acknowledgments

The Riga group would like to thank Maris Tamanis for assistance with the experiments and Christina Andreeva for useful discussions and to acknowledge support from the Latvian National Research Programme in Material Sciences Grant No. 1-23/50 and the Latvian Science Council Grant No. LZP 09.1196. The work in Ashtarak was supported in part by the INTAS South-Caucasus Grant 06-1000017-9001. L. K. acknowledges support from the ESF project Nr. 2009/0138/1DP/1.1.2.1.2./09/IPIA/VIAA/004, and F. G. acknowledges support from the ESF project Nr. 2009/0223/ 1DP/1.1.1.2.0./09/APIA/VIAA/008.

References

1. D. Sarkisyan and A. Papoyan, *Optical processes in micro- and nanometric thin cells containing atomic vapor* (Nova Science Publishers, 2009), vol. 263 of *Horizons in World Physics*, chap. 3, pp. 85124.
2. A. Sargsyan, G. Hakhumyan, A. Papoyan, D. Sarkisyan, A. Atvars, and M. Auzinsh, Applied Physics Letters **93**, 021119 (2008).
3. Y. Dancheva, G. Alzetta, S. Cartalava, M. Taslakov, and C. Andreeva, Optics Communications **178**, 103 (2000).
4. G. Alzetta, A. Gozzini, L. Moi, and G. Orriols, Il Nuovo Cimento B **36**, 5 (1976).
5. R.W. Schmieder, A. Lurio,W. Happer, and A. Khadjavi, Physical Review A **2**, 1216 (1970).
6. C. Andreeva, A. Atvars, M. Auzinsh, K.Bluss, S. Cartaleva, L. Petrov, and D. Slavov, Physical Review A **76**, 063804 (2007).
7. M. Auzinsh, R. Ferber, F. Gahbauer, A. Jarmola and L. Kalvans, Physical Review A **78**, 013417 (2008), ISSN 10941622.
8. M. Auzinsh, R. Ferber, F. Gahbauer, A. Jarmola and L. Kalvans, Physical Review A **79**, 053404 (2009), ISSN 10941622.
9. D. A. Steck, *Rubidium 85 D line data* (2009), (revision 2.1.2, 12 August 2009), URL http://steck.us/alkalidata.

10. D. A. Steck, *Rubidium 87 D line data* (2009), (revision 2.1.2, 12 August 2009), URL http://steck.us/alkalidata.
11. D. Sarkisyan, D. Bloch, A. Papoyan, and M. Ducloy, Optics Communications **200**, 201 (2001).
12. C. Andreeva, S. Cartaleva, Y. Dancheva, V. Biancalana, A. Burchianti, C. Marinelli, E. Mariotti, L. Moi, and K. Nasyrov, Physical Review A **66**, 012502 (2002), ISSN 10941622.
13. S. Stenholm, *Foundations of Laser Spectroscopy* (Dover Publications, Inc., Mineole, New York, 2005).
14. N. G. van Kampen, Physics Reports **24**, 171 (1976).
15. G. Breit and I. I. Rabi, Physical Review **38**, 2082 (1931), ISSN 15366065.
16. E. B. Aleksandrov, M. Chaika and G. I. Khvostenko *Intereferntce of Atomic States* (Springer Verlag, Berlin, 1993).
17. A. Sargsyan, D. Sarksiyan and A. Papoyan, Physical Review A **73**, 033803 (2006), ISSN 10941622.
18. Y. Pashayan-Leroy, C. Leroy, A. Sargsyan, A. Papoyan and D. Sarkisyan, J. Opt. Soc. Am. B **24**, 1829 (2007).

LASER ISOTOPE SEPARATION IN ATOMIC VAPOUR. PHOTO-CHEMICAL METHOD VS. PHOTO-IONIZATION ONE

P. A. BOKHAN, N. V. FATEYEV and D. E. ZAKREVSKIY
Institute of Physics of Semiconductors, RAS, Russia

V. V. BUCHANOV
Institute of Chemical Physics, RAS, Russia

M. A. KAZARYAN*
Institute of Chemical Physics, RAS, Russia
**E-mail: kazar@sci.lebedev.ru*

Two methods of laser isotope separation in atomic vapour are compared. The first of them is a well developed Photo-ionization method. The other method is based on isotope-selective excitation of long-living atomic states and subsequent chemical reaction of excited atom with special reagents. It is shown that this method has some principal advantages compared to Photo-ionization method.

Keywords: Rydberg levels; isotope separation; spectroscopy; atomic vapour; photoionization.

1. Introduction

The development of optics and in particular laser spectroscopy makes it possible to obtain comprehensive information about the structure and shift of spectral lines due to isotope effect. Achieved in recent years, advances in laser physics, methods of adjustment, control and stabilization of laser frequency converted laser sources of scientific and laboratory instruments to industrial devices. This greatly simplifies the use of laser isotope separation and opened up new opportunities for both isotopically or chemically pure substances.

Laser methods are compared favorably with those of other high selectivity of the process (up to 10^3)(other than electromagnetic), and, potentially, the most effective for the production of many products.

Development of laser isotope separation methods is generally carried out in substantial programs in many countries worldwide, primarily in the United States, France, Japan and other. Dominant part of the work was devoted to the method of selective Photo-ionization, what had been in these programs called AVLIS (Atomic Vapour Laser Isotope Separation). Currently, however, has matured the need to develop more efficient and competitive methods of laser isotope separation. Here we include the method of Photo-chemical separation of isotopes, having economic prospects for large-scale production.

The aim of the paper is to briefly compare the two methods of isotope separation. Size of the article does not allow to do it fully, so it marks only the main points. For more information, numerous references can be found here[1] and in the monograph the of authors.[2]

2. Coherent and Incoherent Interaction Radiation with Atoms

In practice, studies of laser isotope separation found two physically distinct cases. In the first case there is incoherent interaction of radiation with atoms, in another case, on the contrary, the interaction is essentially a coherent character. Incoherent model of interaction allows us to consider radiation as a stream of photons, which in the collision with the atoms excite or destroy atomic states. Like a stream of particles the radiation is described by the equations of transfer. This approximation is widely used in the calculation of laser kinetics. Equations incoherent approximation described in detail in many papers, and in particular, the review.[3]

Physics of the excitated levels may vary depending on line width and pulse duration. In the early stages of research, as a rule, fairly broadband radiation was used. The line width of this radiation is higher than the inverse time of the pulse duration or of the radiative lifetime. In this case, is acceptable to use the interaction of radiation with atoms in the analysis of incoherent approximation.

Incoherent interaction of light with atoms occurs when the coherence time of radiation or transverse relaxation time levels is much smaller than the characteristic time for changes in the populations of atomic levels (in practice this time comparable with the pulse duration). Otherwise, the interaction is coherent in nature.

When the isotope shifts correspond to the Gigahertz range of frequencies of radiation, the selectivity of excitation is provided by radiation with a sufficiently large width of the emission line (0.5 – 1) GHz. Such a line width in accordance with $\Delta\nu\tau_c \sim 1$ ($\Delta\nu\tau_c$ - effective width of the line, τ_c - characteristic coherence time) corresponds $\tau_c \sim (1-2)$ ns. Actual pulse duration tunable radiation in the visible or ultraviolet radiation lie in the range $(5-50)$ ns. Thus, the coherence time is much less than the duration of the pulse and can use the description of the process of isotope separation in the approximation of incoherent interaction of radiation with atoms. If the isotope shifts of less than 1 GHz, the approximation of incoherent interaction can be applied only in rare cases.

In connection with the development of laser technology extremely narrow line widths were reached, limited only by the pulse duration. The typical duration of the pulses in the powerful sources of tunable radiation lies in the range $(5-50)$ ns and becomes comparable with the characteristic times of radiative decay of levels. In this case you must use a coherent model of the interaction of radiation with atoms. The model is based on the equation of Liouville for the density matrix of a multilevel system:

$$i\hbar\left(\frac{\partial}{\partial t}+\hat{\Gamma}\right)\hat{\rho}=\left[\hat{H}_0+\hat{V},\hat{\rho}\right], \tag{1}$$

where $\hat{\Gamma}$ - operator describing the relaxation of the system, $\hat{H}_0$ - the Hamiltonian of the unperturbed system, $\hat{V}$ - operator describing the interaction with the electric field radiation In the coherent approximation, the excitation of the upper third level is no longer seen as a sequential excitation through intermediate level. The interaction of a multilevel system with radiation occurs as a single process. Under optimum conditions, virtually 100% excitation of the upper level with an extremely narrow linewidth of the excitation can be achieved. In the coherent case, by detuning the frequency can be achieved that the intermediate level did not really affect the efficiency of excitation and the line width of radiation absorption.

3. Photo-Ionization Method

The method of isotope separation has had been a long history of development (over 30 years). It description has an extensive literature, a list of which would amount to tens of pages of text. This chapter covers only the main features of the method.

The method is based on multistage isotope-selective excitation of atoms. The uppermost excited level can be ionized, generally speaking, in different ways, but as a rule, it uses laser radiation. Isotope-selective excitation of atoms has been made possible by obtaining tunable laser with a line

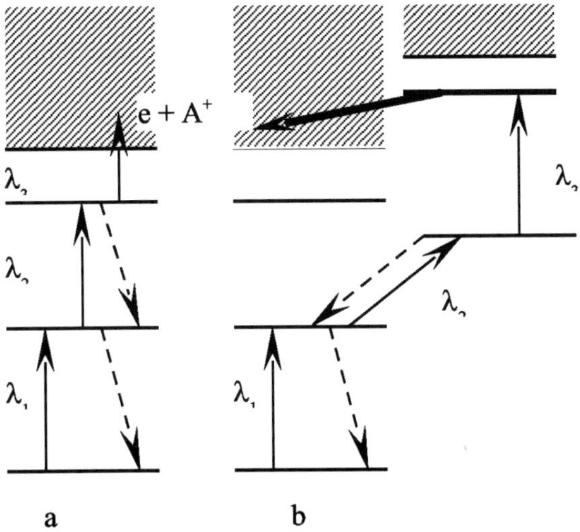

Fig. 1. Scheme stepwise photo-ionization: a – direct photo-ionization; b – through auto-ionization state (marked by a solid line). Dotted arrows indicate the radiative decays of levels; solid arrows marked decay of auto-ionization on the electron and ion.

width less than the characteristic frequency shift of the absorption lines of radiation by different isotopes.

One of the most important tasks of the method consists in selecting a suitable scheme of photo-ionization. Scheme can be two-level, three-level, as in Fig. 1, and in some cases, and a four-level.

The characteristic time of excitation to a higher level is defined by formula $\tau = (\sigma\Phi)^{-1}$, where σ is cross-section of absorption, Φ - flux quanta. Energy per pulse per unit area $W = h\nu\Phi\tau = h\nu/\sigma$. Even in the Doppler broadening not less than 10^{-15} cm^{-2}. If you chouse to assess the characteristic quantum energy 2 eV, then $W \sim 320$ "$\mu J/cm^2$. When pulse repetition rate is 10 kHz (typical pulse repetition frequency of pumping), we obtain the required average intensity of radiation at the level of 3.2 W/cm^2. Achieved to date, the radiation power of laser sources may exceed that amount in the ultraviolet, visible and infrared spectrum.

For direct photo-ionization is necessary to use energies several orders of magnitude higher. This is due to low photo-ionization cross-sections, which constitute $(10^{-17} \div 10^{-18})$ cm^2. Average radiated power under the conditions described above, is estimated at the level of $(100 \div 1000)$ W/cm^2. Low section of direct photo-ionization caused respectively weak absorption of ionizing radiation. This creates a serious problem of reconciling the optical paths at different stages in the process of photo-ionization. Energy process of isotope separation will be mainly determined by photo-ionization process.

Instead of direct photo-ionization one can use ionization via decay of auto-ionization levels (Fig. 1,b). These levels are above the first ionization potential. The emission lines from the auto-ionization states were found in the majority of chemical elements. At the present there are quite intensive research works done on the identification of auto-ionization levels, clarifying their position, and, determining their lifetime. The cross section for excitation of these levels may reach $(10^{-15} \div 10^{-16})$ cm^2 and the above problem is largely mitigated. Therefore, finding suitable auto-ionization levels is one of the most important tasks of spectroscopy.

Another way of photo-ionization is the excitation of long-lived Rydberg states, closely adjacent to the ionization border. Relatively high ionization cross section, the proximity to the border of the ionization makes it possible to use powerful infrared lasers (CO_2, CO, etc.).

In addition, the application of the electrostatic field shifts the Rydberg levels. When the electric field high-lying state can be above the ionization limit and can decay into an electron and ion. Disadvantage of the method of ionization through Rydberg levels is the high sensitivity of levels to weak external induced electric fields, which "dilute" these levels and hinder their excitation.

AVLIS method, in addition to issues related to the photo-ionization requires solutions and other important issues. In general, the process of isotope separation comprises the following steps: evaporation of the material, creating a collimated beam of atoms, the extraction of ions and the accumulation of certain isotopes in the reservoirs of the extractor, the photo-ionization of atoms collected and atoms, which are not subjected to photo-ionization (the system blade and the gas pump working chamber) (Fig. 2).

In order to be evaporated by the elements, a source of vapor crucible type, in which the substance evaporates in a closed volume and in a vapor goes through the slit. With high pressure in the evaporation chamber may form a gas-dynamic jet with the formation of clusters, which can scatter and absorb radiation. At low pressure collisionless gas escape is used. The advantage of the crucible method of evaporation is its simple design and low voltage heating elements.

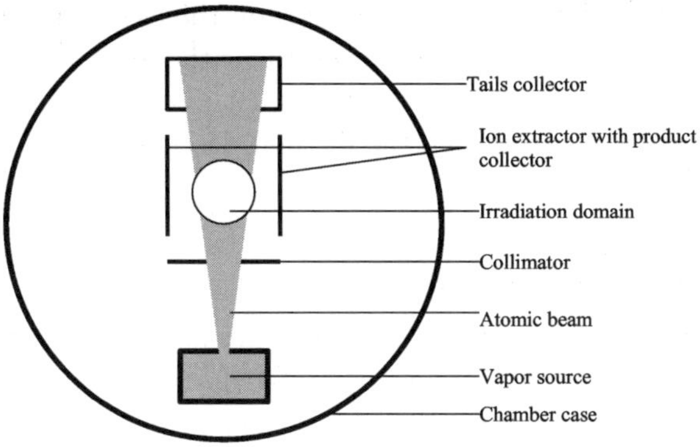

Fig. 2. Schematic diagram of installation in cross-section view.

Electron-beam evaporator may be an alternative source of vapors. The principle of his actions is that the electron beam bombards the surface of the evaporated material and heats to high temperatures. The result is a gas-dynamic flow of steam. While expanding the steam cools and this leads to low temperatures in the radiation zone (~ 200 K). This temperature can reduce the Doppler broadening and the concentration of low-lying metastable states of atoms. The disadvantages of the method include high voltage, high coefficient of reflection of electrons from the surface and the unwanted ionization of vapor.

In many atoms of elements isotopic structure partially or completely is hidden by the Doppler contour. To reduce the width of the Doppler broadening is used by collimation of atomic beam. To do this, a system of plates with slits is mounted. With this out of the stream is allocated fairly narrow beam of atoms with a small spread of velocities in the direction of the beam of radiation. Atoms that have not gone through the slits have deflected the system of gas extraction. The big drawback of collimation is that in the process of isotope separation only a fraction of the flow of atoms from the evaporator is useful to use. The remaining atoms create an undesirable background that reduces the selectivity of isotope separation because of the falling atoms on the collectors for the desired isotope and requires a rather cumbersome system of pumps to maintain vacuum at low pressure ($10^{-6} - 10^{-7}$ Torr).

Extractor is used to extract ions from the stream of atoms. There are various methods of extraction of ions. The most common method is the electrostatic extraction. Applied voltage is on the electrodes of the extractor. It provides the extraction of ions on the cathode. High-density plasma is quasi-neutral and the electric field does not penetrate. Limiting density of the plasma in the extraction, the method described above, is estimated at 10^{10} cm^{-3}.

4. Photo-Chemical Method

The rapid development of photo-chemistry revealed a large number of light-induced reactions can be assumed that for the majority of chemical elements can be found such a reagent, which binds the excited atom, forming a molecule. In this case an atom in the ground state does not have enough energy to break chemical bonds, and it does not enter into a chemical reaction with the reagent. Formed molecules can be released from the mixture of gases with the help of known physical and chemical methods.

Laser technology has reached such a level that made isotope-selective excitation of almost any atomic level possible. However, for the purposes of laser photo-chemical separation is preferable to use long-lived (metastable) states for the following reasons. It is clear that the probability of chemical interaction between the excited atom must exceed the probability of the radiative decay, and this means that

$$k^* N > 1/\tau, \tag{2}$$

where k^* - reaction rate, τ - radiative lifetime, N - concentration of molecules of the reagent. The concentration of molecules is limited to approximately the level of 10^{17} cm^{-3}. At higher concentrations affect unwanted collision line broadening and the reaction of isotopic exchange

$$A_1 R + A_2 = A_2 R + A_1, \tag{3}$$

where A_1 and A_2 - two different isotopes, R - molecular group. Even at a fairly high rate constant $k^* = 10^{-10}$ cm^3/s, we find that $\tau \gg 100$ ns. Therefore, the effective photo-chemical separation can be acheaved when the excited states radiative is more than lifetime of not less than 1 μs.

Possible schemes of excitation of long-lived levels are shown in Fig. 3. The simplest scheme is the one-photon excitation of long-lived level (Fig. 3,a). If the isotopic level structure is hidden by the Doppler contour, the isotopic selectivity is achieved by tuning the frequency of radiation from the center of the Doppler contour (see below).

Application of laser radiation in the photo-chemical method is effective for a number of elements, even if their isotopic structure is hidden by the

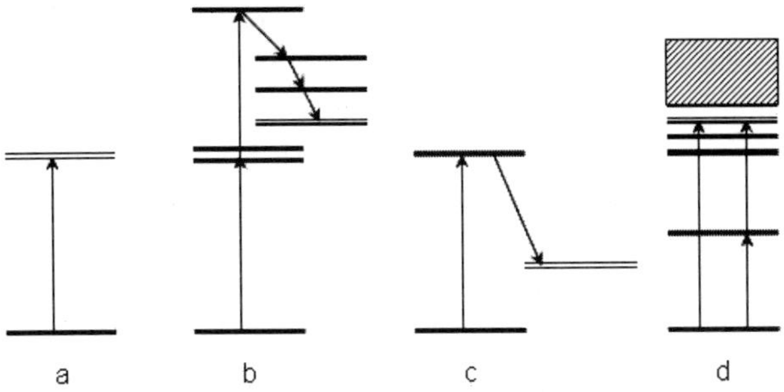

Fig. 3. Photo-excitation of long-living states (marked with double line).

Doppler contour. In this case, the method uses the property of the Doppler contour, which increases detuning frequency simultaneously on the frequencies of lines of all isotopes improves the selectivity of excitation. Fig. 4 shows the Doppler spectra of two isotopes, shifted relative to each other by half the width of the Doppler contour ($\Delta\nu_D$). Such isotopic shifts occur quite often. As seen from the graphs, the ratio of the excitation probabilities of atoms of different isotopes by the line radiation with a narrow range increases with the detuning. Thus, when the detuning is of half the width of the Doppler contour, this ratio is about 10. It is interesting to note that for Lorentzian contour, on the contrary, the selectivity asymptotically decreases.

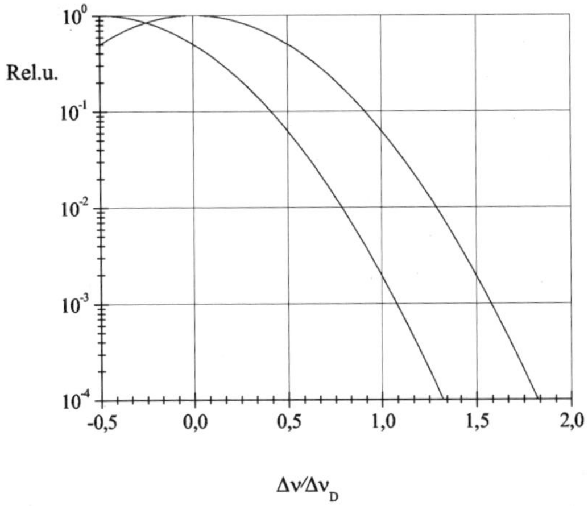

Fig. 4. Shifted Doppler profiles of the excitation probability for atoms.

Typically, the laser beam has a narrower emission line than the inhomogeneous Doppler width of the transition. During one pulse radiation with a narrow range, especially if the impact is carried out on the edge of the Doppler contour, is excited by a small fraction of the atoms. Nevertheless, for a multiple number of pulses it can be fully arouse and "burn" one isotope, which has a position of the line the closest to the frequency of radiation. You can also create conditions under which the concentration of atoms of other isotopes will decrease slightly. To ensure that repeated radiation exposure to the atoms, one uses a longitudinal or transverse mode relatively weak pumping a mixture of original isotope, gas-reagent and a buffer of inert gas. In many cases, due to chemical reaction a single pass gas mixture through the active zone at the same time can "burn out" several isotopes, while maintaining the desired isotopes.

The choice of radiation energy for the effective separation of isotopes must meet the following requirements. The energy of radiation does not exceed the level of saturation of the environment along the length of the active zone. In this case the medium becomes transparent and the radiation is no longer excites the atoms of the desired isotope. The energy of radiation is not simply lost, but also affects the selectivity, arousing more unsaturated atoms of undesired isotopes. The lower is the energy of radiation, the higher the selectivity of excitation. On the other hand the requirement of high productivity leads to increase the radiation energy at the expense of selectivity. In each case, it is necessary to choose a compromise between two contradictory requirements to improve the selectivity and efficiency.

Another scheme is based on two-photon excitation of high lying levels with subsequent decay of long-lived state (Fig. 3,b). The decay may take place due to spontaneous transitions, and as a result of superradiance from the upper level. Using colliding beams of radiation, one can significantly reduce the width of the Doppler contour in two-photon excitation. These schemes are used for the separation of isotopes of zinc (see below).

In many cases, the metastable levels are below the resonance (Pb, Ba, Cu, Au, Mn, Eu, etc.). Then the scheme shown in Fig. 3c is applied. By their nature, long-lived Rydberg levels exhibit high chemical activity due to the remoteness of the valence electron from the nucleus. This can be used in circuits with single-photon and multiphoton excitation of Rydberg levels (Fig. 3,d). Single-photon scheme was implemented for the separation of isotopes of rubidium (see below).

In the separation of isotopes by a photo-chemical method the concentration of atoms of isotopes is limited to the process of resonant energy transfer from atom to atom of one isotope of another. This process is known decrease on the downside the selectivity of excitation. In typical conditions, the cross section of energy transfer from the resonant level, when it is in the range of $10^{-12} - 10^{-13}$ cm^{-2}.[1,2] Assuming for the evaluation of the cross section of energy transfer from long-lived states at the level of 10^{-13} cm^{-2}, we obtain an upper limit for the concentration of atoms 10^{13} cm^{-3}. This value is by three orders of magnitude higher than the limit concentration of atoms allowed in the photo-ionization isotope separation. Size of the camera when using the photo-chemical method is dramatically reduced. It becomes possible direct excitation of long-lived levels, including the metastable. It is possible to have isotope-selective excitation of atoms with detuning from the central part of the absorption of several Doppler widths (see below). The rate of gas flow rate is reduced to $1 - 10$ m/s and is provided by easy flow through the chamber, as transverse pumping and longitudinal with respect to the laser beam.

5. Chemical Reactions of Atoms in Excited States

The rate of these processes depends on the energy difference ΔE between the initial and final states and has a pronounced resonance character. Almost an exact match corresponds to very large cross section ($\sim 10^{-13}$ cm^2). Moderate cross-section ($\sim 10^{-16}$ cm^2) obtained by the energy difference of a few kT (k - Boltzmann constant, T - gas temperature), and for the energy difference exceeding a few tenths of eV, the cross sections are much smaller.

Chemical reactions of electronically excited atoms with molecules to form stable complex play an important role in laser isotope separation. They may be excluded from the process one or two stages of excitation. This allows to simplify and cheapen the process of isotope separation.

When the atom M is faced with a molecule of XY, then there may be chemical reaction scheme:

$$M, M^* + XY \longrightarrow M^+(XY)^- \longrightarrow MX + Y; k_1, k_1^* \quad (4)$$

where M, M^* - atom in the ground the and electronically excited state, respectively, XY - molecule, k_1, k_1^* - rate constants of reactions in the ground and excited state. Long-lived reaction products may be deposited on the walls of the reaction chamber. At the same time unreacted atoms can be collected elsewhere. If the rate of reaction (4) the excited atoms exceeds it for unexcited ($k_1^* \gg k_1$) and the selectivity of excitation is provided, it is possible to implement the separation of isotopes.

Basic requirements for the reaction (4), used for separating of isotopes are as follows:

(1) Interactions with unexcited atoms (background reaction) is small, i.e. satisfy the condition $k_1 N \gg 1/t$, where N - the concentration of molecules XY, t - residence time of atoms in the collision zone;
(2) All the excited atoms should react during their lifetime $k_1^* N > 1/\tau$, where τ - lifetime of the excited state;
(3) The reaction products MX should be well accommodated on the walls of the reaction chamber. Radicals Y should be less reactive with atoms M.

The choice of molecules of gas-reagent for a particular atom, satisfying these conditions, and providing the necessary selectivity is a difficult task, because for complex molecules in the reaction chamber is implemented in numerous variety of secondary physicochemical processes that leads to loss of excitation selectivity and separation, as well as loss of product. However, as shown in experiments, for most of the atoms it is possible to choose an appropriate gas-reagent providing the necessary selectivity of excitation. As will be shown below, the majority of photo-chemical reactions have cross-sections (10 – 100) Å^2, i.e. significantly exceed the gas-kinetic. Such high velocities have been explained in the framework of the "harpoon model".[4] The basis of this mechanism is the notion that the approach of the atom and molecule make possible nonadiabatic transition: the atom M "throws" its valence electron (harpoon) to an atom of the molecule a very close and then pulls him to her by the Coulomb attraction. I.e. reaction (4) occurs through the formation of an intermediate complex $M^+(XY)^-$.

In other more accurate models, the process of electron transfer to molecules includes a large number of covalent and ionic electronic - vibrational potential surfaces. Each surface is characterized by the vibrational quantum numbers of molecules XY and XY^-. This surfaces intersect, forming a lattice with a large number of nodes. As a result, of intersection, the collision process $M + XY$ moves the system at a rather complicated trajectory, passing a large number of intersections, at least until the transition of electrons on the surface corresponding to the energetically most favorable. However, depending on the energy of reaction, products may be in the excited electronic or vibrational states.

Resonant collisions are called processes occurring on the scheme:

$$A^{**} + BC \longrightarrow A^* + BC^* + \Delta E \tag{5}$$

In this case, the excitation energy from one of the colliding partners transfers to another. This type of collision occurs under condition $\Delta \leqslant kT$ (the energy defect of reaction must be small or comparable with the relative energy of the colliding particles). As a BC partner with the collision of electronically excited atom A may be the same atom or the other isotope. An important property of resonant processes is the possibility of realization cross-sections of about 5×10^{-14} cm^2 (Ref. 5) which are much greater than gas-kinetic. This process limits the concentration of atoms and molecules of the gas-reactant in the interaction region, and thus limit the performance of laser isotope separation.

6. Two-Photon Excitation of Lead Atoms

This section discusses the possibility of removing the radioactive isotope ^{210}Pb by photo-ionization. Lead and lead-tin alloys with low radioactivity are used for the production of microchips with a clock frequency of above 500 MHz. Natural lead contains the isotope ^{210}Pb, emerging from the dissolution of ^{238}U. In its subsequent decay, ^{210}Pb becomes ^{210}Po, which decays to ^{206}Pb via α-decay. Fast *alpha*-particle penetrate into the p-n junction. Because of this reason the chip malfunctions. ^{210}Pb can be removed only by means of lead isotope separation. For laser isotope separation, this problem is almost perfect. The concentration of the isotope in lead is very small $\sim 10^{-13}$%. In this case, the absorption of radiation by the atoms ^{210}Pb is negligible and mitigates the problems associated with self-focusing of radiation and the spatial and temporal alignment of radiation pulses. As a result, it is necessary to use a rather extended region of interaction of radiation with matter (a few tens of meters).

Fig. 5 shows the scheme of atomic levels of lead, which can be used in the process of photo-ionization. The solid lines show the possible channels of the photo-ionization, dashed lines indicate the transitions caused by spontaneous decay of states. The first transition from the ground to the resonant level is determined almost uniquely, because transitions to higher levels demands too small wavelengths. On the second transition few options are possible. One of the most appropriate level is 8p $^3P_1$ is excited by use of rhodamine-6G dye laser. The choice of photo-ionizing radiation largely depends on the structure of auto-ionization levels and their lifetimes. With the laser system providing a wide range of tunable radiation were selected auto-ionization levels, excitation cross section of which is 10^{-15} cm^{-2}.

The main purpose of the calculation was to define the conditions for the optimal system performance, taking into account the restrictions on the

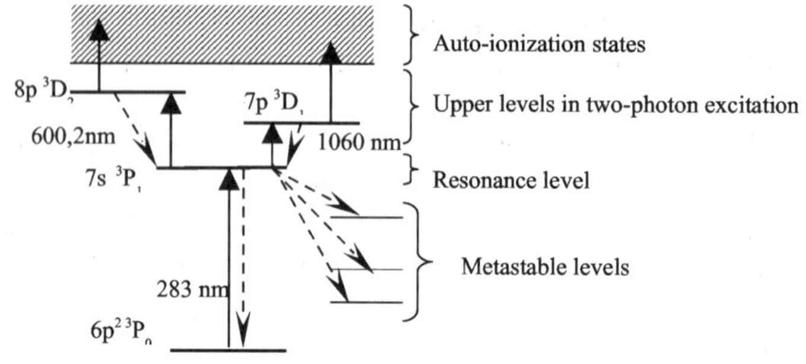

Fig. 5. Transitions in the lead atom.

parameters of radiation.It is necessary to find the conditions under which during one pulse in the whole active volume the maximum possible number of isotopes ^{210}Pb ions at a given degree of ionization of this isotope (not less than 0.9) can be produced.

Calculations were made for ideal conditions of a collimated atomic beam. It was found that the main limitation to system performance was due to the absorption of radiation by atoms of all isotopes, except the allocated ^{210}Pb, which is negligible in the mixture. Minimizing the absorption leads to strict requirements on the magnitude of the radiation frequency detuning from the exact resonance. The latter depends on the radiation energy and pulse durations. Each of the radiation energy has its own optimal detuning frequency. At energies of 50 μJ/cm^2 it is in the range of 30 – 60 GHz, depending on the form of pulses.

Fig. 6 shows the results of calculations with optimal detuning frequency and delayed ionizing pulse for the transitions 6p^2 ^{3}P$_0$ $\rightarrow$ $7s$ ^{3}P$_1$ $\rightarrow$ $8p$ ^{3}D$_2$. In addition to the time profiles of pulses, the graph shows the curves for the relative population of the third level (n_3), the metastable levels (n_m) and concentration of ions (n_i) isotope ^{210}Pb. For this purpose, the following conditions should be fulfilled:

(1) pulse energy at the first and second conversion - 50 μJ/cm^2;
(2) pulse duration - 15 ns;
(3) average probability of photo-ionization of the third level - 10^9 s^{-1};
(4) photo-ionization pulsewidth - 10 ns;
(5) detuning of the resonant frequency - 47 GHz.

As can be seen in this case the degree of ionization isotope ^{210}Pb is greater

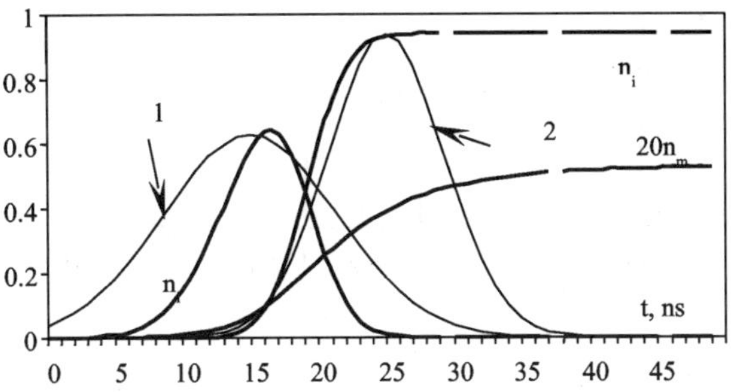

Fig. 6. Calculated profiles of (1) the radiation pulses and (2) photo-ionization pulses; temporal dependence of the population of the upper (n_3) and metastable (n_m) levels and the ion concentration (n_i)

than 0.9. The degree of ionization of the remaining isotope was close to 0.001.

When one uses 50% of the radiation power of the first transition the limiting system performance is 1 g/s (about 2 tons per month of isotopically modified product). It is already commercially viable large-scale production.

7. Photo-Chemical Separation of Isotopes of Zinc and Rubidium on Single-Photon Scheme

The beginning of this section presents some results of calculations on the laser separation of isotopes of zinc, which reveal the features of the separation of isotopes by one-photon scheme.

Natural mixture of zinc consists of even isotopes - ^{64}Zn (48.6%), ^{66}Zn (27.9%), ^{68}Zn (18.8%), ^{70}Zn (0.6%), ^{70}Zn (0.6%), and odd - ^{67}Zn (4.1%) isotope. Isotope shifts of the main isotopes are shown in Fig. 7. Doppler width of absorption lines significantly exceeds the characteristic isotopic shift. Therefore, the isotopic structure is hidden by the Doppler contour.

The calculations were performed for both transverse (Fig. 8), and for the longitudinal flow of gas and gave similar results. Results are presented in Ref. 2.

In the case of transverse pumping the gas flow is a gradual "enlightenment" of the medium due to "burning" of the atoms of 64 isotope. Preferential excitation of this isotope takes place because the frequency of the radiation detuning from the line of the 64 isotope at 2.1 GHz, and the line of 66 isotope is at 2.76 GHz (the width of the emission spectrum - 500 MHz).

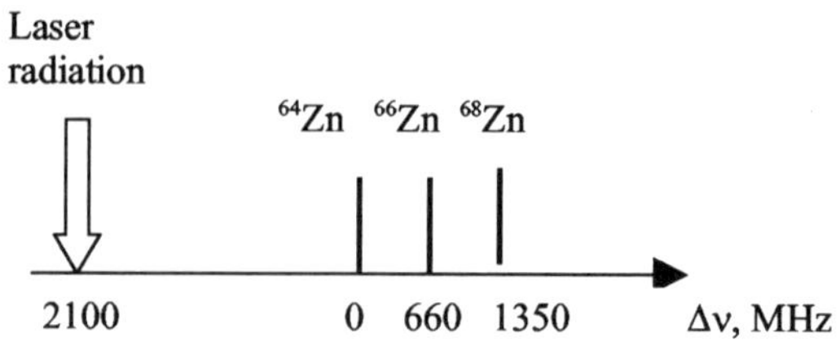

Fig. 7. Isotope shifts of the main isotopes of zinc and shift center line of radiation from the resonant frequency of the isotope ^{64}Zn

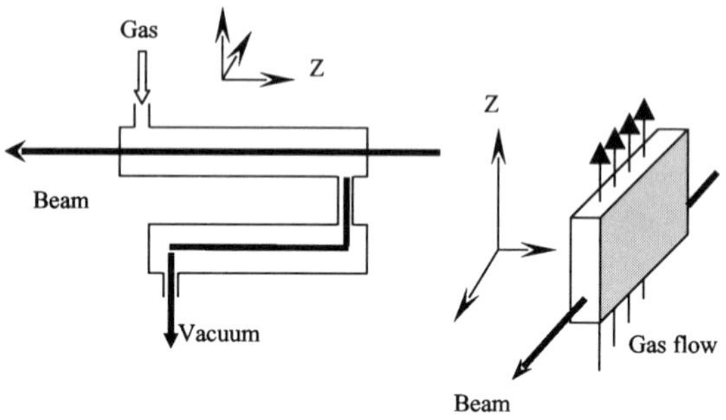

Fig. 8. Longitudinal and transversal circulation of gas mixture.

The main parameters and results of calculations of one of the options are presented in Table 1. The leading parameter in the table is N_0L (product of the concentration of atoms at the inlet flow on the length of the active medium). The width of the irradiated zone of 2 cm corresponds flow rate of 0.5 m/s.

In the case of high concentrations resonant energy transfer and exchange chemical reactions are possible. These processes reduce the efficiency of isotope separation. It is therefore necessary to reduce the concentration of atoms with a simultaneous increase in the length of the active zone. Specific values of concentration of atoms depend on the of the rate constants of these reactions and the desired degree of separation of isotopes. When the concentration is 5×10^{13} cm^{-3}, the length of active zone is 6 m.

Let us consider the longitudinal gas flow. The main parameters and results of calculations of one of the options are presented in Table 2 and the missing parameters presented in Table 1. The calculations showed that

Table 1. Parameters of the calculation of the separation of isotopes of zinc in the cell with a transverse flow.

Parameter	Value [a]
The radiation power	3 W
Pulse repetition rate	10 kHz
Width at half maximum of the emission line	500 MHz
Cross-sectional area of the irradiated zone	4.5 cm^2
Parameter N_0L	3×10^{16} cm^{-2}
Pollution degree of 66-isotope by 64-isotope and 67-isotope	0.0099
The number of pulses per time of transition through active zone	500
The number of quanta per atom of 66-isotope	20.0
The absorption in the active zone	52%
Productivity, atoms/s	2.3×10^{17}

Table 2. Parameters of the calculation of the separation of isotopes of zinc in the cell with a longitudinal flow.

Parameter	Value [a]
The radiation power	2 W
Pollution degree of 66-isotope by 64-isotope and 67-isotope	0.029
The number of pulses per time of transition through active zone	3100
The number of quanta per atom of 66-isotope	6.75
The absorption in the active zone	71%
Productivity, atoms/s	4.6×10^{17}

at the end of the core, the 64-isotope content is a strong decrease of almost two orders of magnitude compared with the content in the original mixture. On the other hand, the 66 isotope content decreased only by half, i.e. there is significant enrichment of zinc by 66 isotope.

The calculation results showed high efficiency of "burning" of isotopes, whose lines are closest to the emission line. Thus, when the output power is 1 W, in order to reduce the concentrations of 68 isotope factor of ten, 2.5 photon per atom is required, and the concentration of the 66-isotope decreased by 30% and 64 isotope - 5%.

Thus, for example, such "inconvenient" for laser separation of elements like zinc, it is shown that the proposed method has an acceptable energy efficiency and could easily compete with other methods.

In Ref. 2 reported on the experimental realization of laser separation of zinc atoms single-photon method. The separation was done by heating using and external furnace, a quartz tube 60 cm long, 3 cm diameter. It was used mixture of zinc atoms with argon atoms and molecules of the gas-reagent. Pumping rate of the mixture was 0.5 l/s, at the pressure of argon (1 ÷

2) Torr, and the concentration of reagent atoms was at 10^{16} cm^{-3}. Under these experimental conditions, the drift of atoms through the interaction region is about 1 second. The atoms of the isotope arrived in the region of separation from the reservoir, which is heated by a separate furnace with temperature, providing the necessary concentration of atoms. The radiation power of the laser propagated along the flow providing a "burning" of the excited atoms by chemical reaction. Thus formed products were deposited on the chamber walls of separation. Vapors of the remaining isotopes were condensed on the walls of the chamber, as the walls were kept at room temperature. Useful photo-chemical reaction was provided by collisions of atoms of zinc in the 4p $^3P_1$ state, excited by laser radiation with molecules of diethyl ether $(C_2H_5)O$. The rate constant of this reaction was $k = 1.61 \times 10^{-9}$ cm^3/s. The upper limit of the constant rate for the atoms in the unexcited state is 10–14 cm^3, which is 5 orders of magnitude less than for 4p $^3P_1^0$ state. This value was determined from the results of experiments on registration of the concentration of zinc atoms in the collector chamber in the presence or absence of gas-reagent.

Average power laser radiation propagating in the cell separation was ~ 2 W, beam diameter ~ 1 cm, repetition rate 12 kHz, pulse width 10 ns, line width $\Delta\nu = 36$ MHz.

When the frequency of the radiation power of the laser tuned to the center of the Doppler contour of zinc atoms, recorded almost 100% "burning" of the atoms in the separation cell. This confirms the efficiency excitation of zinc atoms in the separation cell and the efficiency of chemical reactions:

$$Zn^* + (C_2H_5)_2O \longrightarrow ZnO + (C_2H_5)_2 \tag{6}$$

The product of this reaction ZnO deposited on the walls of the chamber. Effective chemical reactions are also observed with several other molecules, but the highest degree of selectivity is obtained with molecules of diethyl ether $(C_2H_5)_2O$.

The achieved accumulation of the product was ~ 1 g of the substance at a three-hour exposure by UV radiation with an average power of 2 watts and the speed of pumping$\cdot$100 l/s.

Laser isotope separation of rubidium atoms by conventional methods AVLIS runs into difficulties. First, there is no effective scheme for the photoionization by recently developed effective tunable laser sources. Second, a large vapor pressure at room temperature leads to loss of selectivity.

In Ref. 6 experimentally was investigated a single-photon laser isotope separation of rubidium, based on the "burning" selectively excited

Rydberg states of rubidium atoms in a state $11P_{3/2}$ due to chemical reaction. The natural mixture of rubidium atoms consists of two odd isotopes ^{85}Rb (72.17%) and ^{87}Rb (27.83%). The isotopic shift for the rubidium atoms is small, but one can find the absorption lines belonging to different isotopes and separated by a value greater than the Doppler broadening. The oscillator strength of transition ($5S_{1/2}$-$11P_{3/2}$) is small, therefore requires a large power density of laser radiation to saturate the transition. Absorption cross-section can be calculated from the known oscillator strength.

Excitation of atoms on transition ($5S_{1/2}$-$11P_{3/2}$) by narrow-band tunable pulsed radiation with $\lambda = 311$ nm was carried out from the same source, which was used for laser separation of zinc. Average power in the interaction region was ~ 0.6 W with a beam diameter of 1 cm in the same cell was injected gas-reagent.

Experimental results showed that when the frequency of the exciting radiation tuned to the absorption peak of the isotope ^{87}Rb, at the end of the chamber of interaction content of this isotope was reduced more than 90% of the initial value. Through condensation of atoms that came from the zone of interaction, can be carried out accumulation the product of weighted quantities. Estimated productivity was $\sim 10^{-2}$ g/day at the experimental parameters. Increasing the concentration of atoms, the pumping rate and the laser power can significantly exceed this value.

8. Photo-Chemical Separation of Isotopes of Zinc on the Two-Photon Scheme

The method is based on two-photon excitation of Zn at 6s $^3S_1$, by the absorption of two colliding photons ($\lambda_1 = 307.6$ nm and $\lambda_2 = 303.6$ nm) (Fig. 9). The proximity of the energy of these photons in the two-photon process ensures the reducing of the Doppler broadening to a value of 16 MHz, which is considerably less than the isotopic shift. The state $6s^3S_1$ then decays into 4p $^3P_{0j}$ in the chain of spontaneous and induced transitions through intermediate levels as 5p $^3P^0_{0,1,2}$ and 5s $^3S_1$, and directly (Fig. 9). Interchange transitions between the systems of singlet and triplet levels of the Zn atoms are very weak, so after two-photon excitation due to rapid processes (for a time less than 30 ns), all atoms are in a position 4p $^3P_{0j}$. Optimal detuning frequency of the radiation on the first interchange transition was $\delta\nu_1 = 4 \div 5\,\Delta\nu_D$. To separate the desired isotope was used exothermic photo-chemical reaction:

$$Zn^*(4p\,^3P_{0,1,2}) + CO_2 \longrightarrow ZnO + CO + 1.3eV \qquad (7)$$

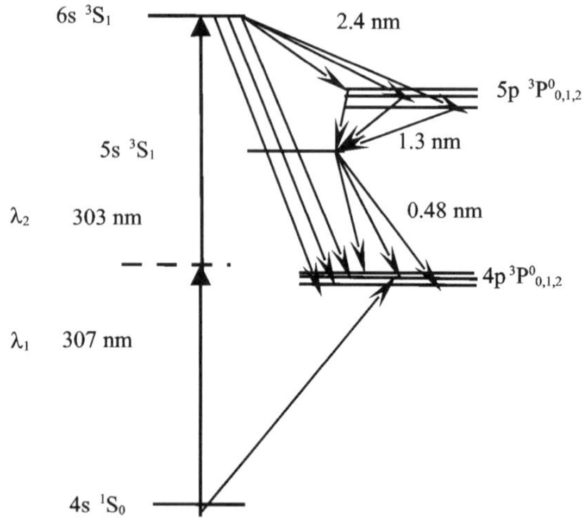

Fig. 9. Diagram of transitions in zinc atom

Measurements of the rate constant of reaction (7) was carried out in a pumped gas cell by the rate of accumulation of CO. The constant within the error (20%) is equal to $k = 2.5 \times 10^{-10}$ cm^{3}s^{-1}. The rate constant of reaction of zinc atoms in the ground state with the molecules of $\mathrm{CO_2}$, is described by the formula:

$$k(cm^2/s) = 7.9 \times 10^{-8} \times (-10886/T), T(K) \tag{8}$$

At the operating temperatures ($T \approx 350°$) it is 4-5 orders of magnitude lower than the rate of reaction of the excited atom. This ensures high selectivity of the process. At CO_2 gas pressure 0.1 Torr the reaction rate is an order of magnitude higher than the probability of spontaneous decay 4p $^3\mathrm{P}^0_j$ state (10^5 s^{-1}). The quantum efficiency exceeds 50%.

Product of chemical reaction (7), as well as several others, is a stable molecular ZnO, which is deposited on the collector. Formed CO molecules are pumped from the separation zone. Isotope separation was carried out in different patterns of longitudinal and transverse pumping CO_2 with Zn at relatively high pressure of CO_2 ($\sim$ 1 Torr). In the experiments mainly used two-flow pumping of the working mixture along a radiation beam at a speed of $\sim$ 2 l/s and zinc across the beam from the evaporator to the cold wall. There was a collector for collecting the product in the form of isotopically modified zinc oxide. Chemical methods are then used to extract the isotope - enriched zinc. Practical implementation of laser isotope separation showed that at two-photon excitation of zinc atoms nonlinear dependence of the absorption of radiation on the concentration of atoms is observed. The reason for this dependence is the appearance of cascade superradiance. Experimentally were found several lines at 2.4 μm, 1.3 μm and 0.48 μm. The measured power at 1.3 μm was about 100 mW. The generation of radiation leads to an undesirable broadening of the absorption of pump radiation. As a result, the effectiveness of two-photon excitation falls several times. The observed effect forced to use relatively low concentrations of atoms. As a result of two-photon excitation the excitation energy dissipates through the spontaneous and induced transitions. Fig. 9 shows 12 possible transitions. On one part of the transitions forms a population inversion and run the superradiance, while the other transitions inversion does not occur.

With the aim of theoretical interpretation of experimental results and evaluation of the separation ability it was carried out numerical simulation of the dynamics of population levels. In accordance with the calculations using the colliding beams of radiation can be achieved by a high degree of excitation (0.7) of zinc atoms 64 isotope in the long-lived states 4p $^3\mathrm{P}^0_{0,1,2}$. This state is formed due to radiative transitions, starting from the upper level.

The calculations showed that the degree of contamination in 1% 64-isotope by other isotopes is achieved when the absorption coefficient does not exceed 0.004 cm^{-1}. Experimental measurements of the absorption coefficient gave close values.

Requirements for efficiency and selectivity contradict each other, so for each task one needs to find a compromise between specific performance, efficiency and selectivity of the separation process. For example, if the selectivity of 50$\div$100, the radiation energy density E_1 and E_2 must not exceed 350 μJ/cm^2, and the concentration of zinc atoms should be not higher than $N = 2 \times 10^{13}$ cm^{-3}. Under these conditions, an equivalent absorption coefficient for the isotope ^{66}Zn $k_\nu \sim (3 \div 4) \times 10^{-3}$ cm^{-1}, which allows efficient (50%) use the pump radiation at moderate absorption length 150$\div$200 cm. Production range for this case is 0.6 g/hour. Required selectivity achieved by choice of appropriate experimental parameters.

Thus, the use of two-photon excitation of zinc, with a small detuning from the intermediate state in conjunction with the photo-chemical reaction allows the effective separation of isotopes for the case where the lines of resonant transitions of the isotopes are disposed within the Doppler absorption contour.

9. Conclusion

After years of intensive research the following disadvantages of the photo-ionization method of laser isotope separation can be mentioned:

(1) need for big separation volume because of the low density of the desired isotope;
(2) need to maintain a high vacuum, to avoid the background density of atoms;
(3) need to create complex and bulky optical systems for guidance of the beams with optical path length in the tens or hundreds of meters;
(4) need to create complex systems collimation of atomic beams and ion extraction;
(5) need to use gas flows with transonic flow rate, leading to the fact that the atoms are irradiated only once.

These shortcomings deprived by photo-chemical method of laser isotope separation, although it has its drawbacks. Historically, however, preference was given to the photo-ionization laser method of isotope separation. Only in the last few years, interest to this method is markedly increased, and is now determined its prospects. Although the study of isotope separation by laser photo-chemical method for large-scale installation in Russia began only few years ago, however, these results suggest that this method could compete with the traditional photo-ionization method.

References

1. V. Yu. Baranov, Isotopes. Properties, Production and Application (IzdAT, Moscow, 2000).
2. P. A. Bokhan, V. V. Buchanov and N. V. Fateev, M. M. Kalugin, M. A. Kazaryan, A. M. Prokhorov, Laser isotope separation in atomic vapor (Wiley-VCH Verlag GmbH Co. KGaA, 2006).
3. S. I. Yakovenko, Quant. Electron.**28**, 945 (1998).
4. D. Gershbah, Scattering of molecular beams, accompanied by the reaction, (Mir, Moscow, 1969).
5. G. Massey, E. Barhop, Electronic, Ionic Collisions, accompanied by the reaction, (IL, Moscow, 1958).
6. P. A. Bochan, D. Ed. Zakrevskii, N. V.Fateev, JETP Lett.**75**, 202 (2002).

TWO-DIMENSIONAL CONFINED TERAHERTZ WAVE PROPAGATION IN GAP PLASMON WAVEGUIDE FORMED BY TWO CYLINDRICAL SURFACES

Yu. H. AVETISYAN*, A. H. MANUKYAN, H. S. HAKOBYAN
and T. N. POGHOSYAN

Department of Radiophysics, Yerevan State University,
Yerevan, 1 Alek Manoogian 0025, Armenia
**E-mail: yuriav@ysu.am*

Undistorted 2D-confined THz pulse propagation through plasmon waveguide formed by two metal cylindrical surfaces is demonstrated. Simple theoretical model is developed to calculate effective index and field distribution of the fundamental TM_0 -mode in waveguide formed by gold, doped Si, and polyvinylidene fluoride (PVDF) materials. It is shown that doped Si and PVDF are good candidates for plasmon THz waveguides, especially in case of the narrow-band applications. Metal waveguides are preferable when negligible dispersion of propagating waves is required in wide bandwidth.

Keywords: Terahertz, waveguide, THz time-domain spectroscopy.

1. Introduction

Guiding of terahertz (THz) wave within a subwavelength cross section has recently been attracting a great deal of attention because of ever increasing demands for miniaturization of THz circuits. Compact integration of THz functionality could benefit a number of applications ranging from spectroscopy[1] to communication.[2] However, large wavelength scale in the THz frequency range ($\lambda = 300$ μm at 1 THz) has hindered compact dielectric-based THz waveguides due to the diffraction properties of electromagnetic waves. Typical THz device features, on the order of a half wavelength or greater, are at least three or four orders of magnitude larger than the nanoscale electronic counterparts.

In optical wavelength region (visible and near-infrared radiations), the plasmon waveguides have attracted recently much attention because they have the potential to guide light in a region beyond the so-called diffraction limit. The mode guided along the metallic interface is known as the surface plasmon-polariton (SPP).[3] Various geometries have been proposed to achieve sub-wavelength confinement of the SPP in the plane transverse to the propagation direction.[3–6] Among these proposals, the air gap plasmon waveguide formed by a two metal cylindrical surfaces is particularly interesting due to an opportunity of the 2-dimensional (D) field confinement and easiness of fabrication. For these reasons, we focused our attention on such waveguides to obtain SPP-assisted 2D THz-wave propagation.

It should be noted that idea of SPP is not easily transformable from optics to THz range. The strong field localization occurs only for SPPs with frequencies close to the plasma frequency, which is defined by $\nu_p^2 = Ne^2/4\pi^2 m_e \varepsilon_0$, where N is the electron density, e electron charge, ε_0 is the vacuum permittivity and m_e electron mass. Since the free electron density in metals is typically on the order of 10^{23} cm^{-3}, the plasma frequency is limited to the visible and ultraviolet regions. At frequencies significantly below the plasma frequency (like the THz range), large negative permittivity strongly prohibits electromagnetic fields from penetration inside a metal, and plasmon excitation on the metal-dielectric interface becomes challenging. Thus efficient plasmonic excitation at THz frequencies requires materials with lower plasma frequencies.

Since long it is known,[7] that periodic arrays of metal elements can be used to form composite media with low plasma frequencies. More recently[8] it was demonstrated that a periodic lattice of thin metallic wires could exhibit a plasma frequency even in microwave. However, many period of such a structure are required that is a limiting factor to the use of such media in practical applications. Besides, the periodic lattice of thin metallic wires is difficult and expensive to produce. An alternative composite media are structured metal surfaces with periodical grooves or holes.[9] This medium is characterized by an effective permittivity that has a plasmon form with a plasma frequency dictated by the geometry of the hole or the groove. However, effective permittivity of the periodical structures is strongly frequency dependent that limit operation bandwidth of the plasmonic devices.

In this paper we are focusing on media having plasma frequency well below that of metals, which do not depend on the use of components with dimensions related to the frequency of interest. The doped semiconductors are good alternatives to metals for the excitation of low-frequency plasmon.[10] As the carrier densities in semiconductors are much lower than those in metals, the plasma frequency is much smaller, being typically at mid- or far-infrared frequencies. Therefore, the permittivity of semiconductors at THz is comparable to that of metals at optical frequencies. A decisive advantage of semiconductors is that their carrier density and mobility, and consequently the SPPs, can be easily controlled by thermal or optical excitation of free carrier. However, SPP propagation length in doped semiconductors is usually short due to free-carriers THz-wave absorption.

It has been known that low frequency plasmons can be observed in intrinsically conducting polymers.[11] Quite recently porous fiber with ferroelectric polyvinylidene fluoride (PVDF) polymer was proposed as surface plasmon resonance (SPR) sensor.[12] The PVDF film (usually used in piezoelectric devices) is widely commercially available and it can be easily bonded to different substrates including metals. Moreover a use of PVDF gives an opportunity to design a fully polymeric THz plasmonic structure, which potentially makes the fabrication process highly suitable for the industrial scale-up. However, strong frequency dependence of PVDF permittivity hinders its applications in broadband devices. Thus, above mentioned materials have a number of pros and cons in THz plasmonics.

The first objective of this paper is to present the characteristics of THz-wave propagation in gap Plasmon Waveguide formed by two Cylindrical Surfaces (PWCS), made of metal, semiconductor, and PVDF (Fig. 1). The analytic description SPP modes in gap PWCS with metal surfaces has been recently obtained in optical regime.[6] In theoretical part the simple approach to calculate fundamental TM_0 mode size and its dispersion in THz region is presented. Waveguide is considered as parallel plate waveguide (PPWG) with adiabatically varying distance between plates $h(y)$. It is known[13] that effective refractive index at TM_0 mode in plasmon PPWG is approximately inversely proportional to the distance between plates, $N_{\text{eff}} \sim 1/h(y)$. Therefore model of graded-index dielectric waveguide[14] can be used to describe THz-wave confinement along y-axis.

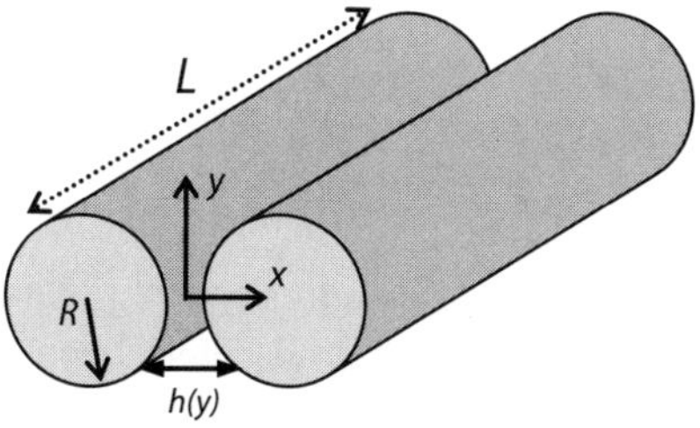

Fig. 1. Schematic view of the waveguide.

The second objective of this paper is to demonstrate undistorted 2D-confined THz-pulse propagation in PWCS. The low-loss, undistorted, and strongly confined THz-pulse propagation in waveguides is a priority for many practical applications.[2] The experimental part of paper describes the application of THz time domain spectroscopy (TDS) to measure transmission of THz-pulses through PWCS having cylinders made of Au-coated aluminium. The choice of metal (gold) is caused by necessity of both the low losses and negligible dispersion of the propagating waves in wide frequency region 0.1 - 3 THz.

2. Theoretical Model

2.1. *Complex permittivity materials in THz range*

In order to model the gap plasmon waveguide, the complex permittivity of cylindrical surfaces (made of metal or doped semiconductors) was assumed to be Drude-like, which corresponds to the approximation of a single energy independent scattering time, τ.[15] This quantity is related to the mobility μ, by $\mu = e\tau/m^*$. Then the frequency-dependent complex permittivity $\dot{\varepsilon}(\nu) = \varepsilon'(\nu) + i\varepsilon''(\nu)$ is

$$\dot{\varepsilon}(\nu) = \varepsilon_\infty - \frac{\nu_p^2}{\nu(\nu + i\gamma)}, \tag{1}$$

where ν_p is the plasma frequency, γ is the damping rate (inverse scattering time) of the carriers, and ε_∞ is the permittivity at extremely high frequencies that is unity for metals or permittivity of undoped material for semiconductors.

When material of cylindrical surfaces is PVDF film, the dispersion of complex permittivity is determined by relationship

$$\dot{\varepsilon}(\nu) = \varepsilon_{\mathrm{opt}} + \frac{(\varepsilon_{\mathrm{dc}} - \varepsilon_{\mathrm{opt}})\nu_0^2}{\nu_0^2 - \nu^2 - i\gamma\nu}, \tag{2}$$

where ν_0 is the frequency of the transverse-optical mode of the material, $\varepsilon_{\mathrm{opt}}$ and $\varepsilon_{\mathrm{dc}}$ are the optical and low-frequency dielectric constants respectively.

Using Eq. (1) with known values of ν_p and γ,[16] the complex permittivities of different high-conductivity metals have been calculated in 0.1 - 3 THz region. In all cases the negative real part of $\dot{\varepsilon}$ is an order of 10^5. Because THz region is far enough from plasma frequency ν_p the value of ε' is practically independent of frequency. For instance, in gold $-\dot{\varepsilon} = 1.14 \times 10^5$

at 0.1 THz and it is slowly reduced to $-\dot{\varepsilon} = 0.94 \times 10^5$ at 3 THz. Such small dispersion is very attractive for plasmonic waveguides operating in wide frequency region. The imaginary part of the permittivity in metals (which is related to dissipation) is usually 3-10 times bigger than $|\varepsilon'|$, whereas in optical region $\varepsilon'' \ll |\varepsilon'|$. For gold the ratio - $\varepsilon'/\varepsilon''$ is 0.16 for 1 THz, whereas it is approximately 13 for 300 THz ($\lambda = 1$ μm).

To illustrate a behavior of permittivity of the doped semiconductors the doped Si was chosen. The silicon's characteristics in THz range are well established[15] and it is widely available and cheapest semiconductor material. In addition, Si already used in many studies related to SPP-assisted THz-wave transmission through array of holes.[17,18] For n-doped Si the dependencies of the real and imaginary parts of permittivities on frequency were calculated with Eq. (1). Because scattering time of the carriers is dependent of its density, the values of γ have been assumed $\gamma = 1.35$ THz for free carriers concentration $N = 10^{17}$ cm^{-3} and $\gamma = 4.1$ THz for $N = 10^{18}$ cm^{-3} in accordance to Baccarani-Ostaja model.[19] The calculated dependencies of $\varepsilon'(\nu)$ and $\varepsilon''(\nu)$ for doped Si with $N = 10^{18}$ cm^{-3} and $N = 10^{17}$ cm^{-3} are presented in Fig. 2.

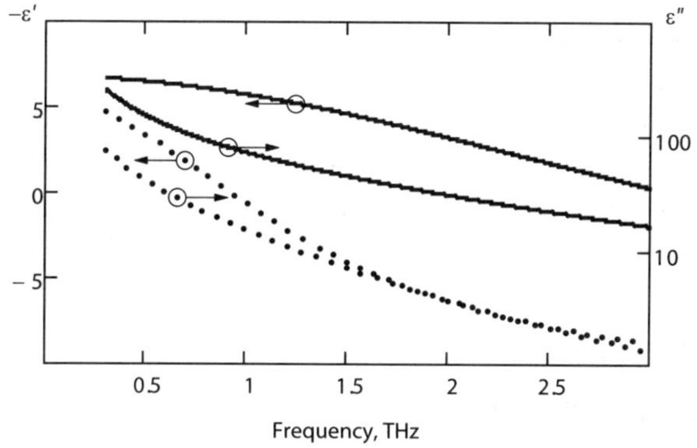

Fig. 2. Dependencies of negative real and imaginary parts of doped Si on frequency for carrier concentrations $N = 10^{18}$ (solid lines) and $N = 10^{17}$ (dotted lines).

It is seen that lightly doped silicon ($N = 10^{17}$ cm^{-3}) has negative ε' only for $\nu < 0.9$ THz and therefore it cannot support SPP-mode excitation at interface air-Si for higher frequencies. The doping with $N = 10^{18}$ cm^{-3} is quite sufficient to have negative real part of $\dot{\varepsilon}$ as far as 3 THz, which is close to highest frequency of the common TDS systems.

The calculated dependencies of the real and imaginary parts of PVDF permittivity on frequency are plotted in Fig. 3. The following values $\varepsilon_{\text{opt}} = 2$, $\varepsilon_{\text{dc}} = 50$, $\nu_0 = 0.3$ THz, and $\gamma = 0.1$ THz have been used in Eq. (2) accordingly to published data.[19]

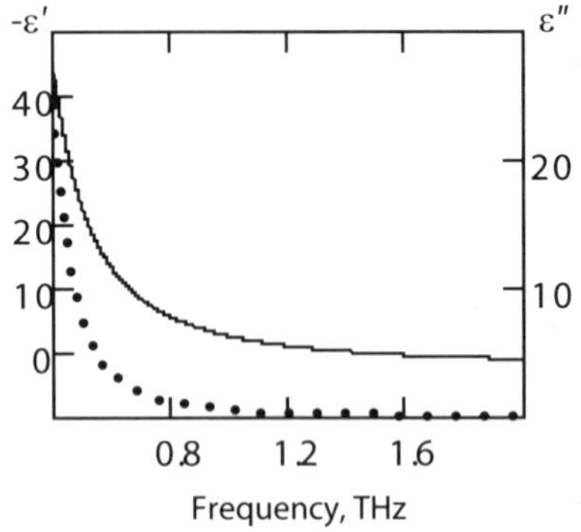

Fig. 3. Dependencies of negative real (solid line) and imaginary (dotted line) parts of PVDF on frequency.

The Fig. 3 shows that plasmon-like SPP excitation may be carried out in frequency range from $\sim$ 0.4 THz to $\sim$ 1.4 THz. The upper limit is determined by condition $\varepsilon' < 0$ and lower limit to necessity of the relatively small value of ε'', i.e. low absorption. Thus the use of PVDF in plasmon devices is limited by narrowband applications in range of approximately 0.4 - 1.4 THz.

2.2. *Effective index of PPWG*

In this section we calculate the effective refractive index $Re(\dot{n}) = c/u_{\text{ph}}$ (u_{ph} is the phase velocity, c is the velocity of light) of PPWG formed by two infinitive parallel plates in YZ plane. The plate's separation h_0 is chosen equal to the minimal distance between cylindrical surfaces $h(y{=}0)$ of PWCS (see Fig. 1). Here obtained results will be used in the next section to calculate propagation characteristics of real PWCS.

The dispersion of SPP-mode in PPWG filled with air is given by[3]

$$\tanh\left[\frac{k_0 h_0}{2}\sqrt{\dot{n}^2 - 1}\right] + \frac{\sqrt{\dot{n}^2 - \dot{\varepsilon}}}{\left(\sqrt{\dot{n}^2 - 1}\right)\dot{\varepsilon}} = 0, \tag{3}$$

where $k_0 = 2\pi\nu/c$ is the wavenumber in air, $\dot{\varepsilon}$ is the complex permittivity of the material of plates. Using above calculated dependencies of complex permittivities for doped Si, PVDF, and gold, the numerical solution of (3) can be obtained. The dependencies of the detuning of effective refractive index $\Delta = Re(\dot{n}) - 1$ and absorption coefficient $\alpha = k_0 \cdot Im(\dot{n})$ on frequency for PPWG with $h_0 = 0.1$ mm in cases when plates are made of the doped Si ($N = 10^{18}$ cm^{-3}) and gold are presented in Fig. 4. For clarity of the picture the values corresponding to gold covered PPWG are increased 50 times.

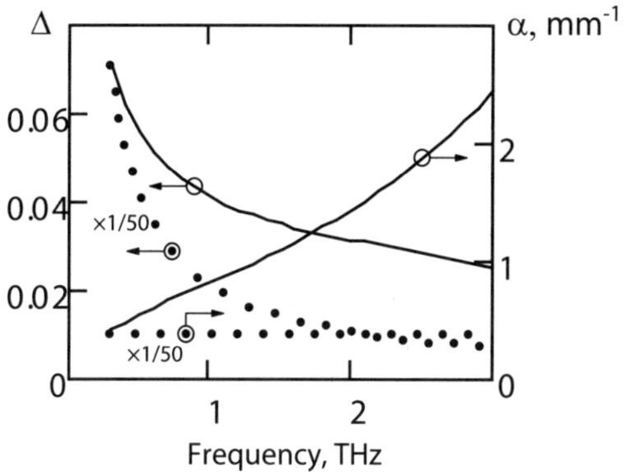

Fig. 4. Dependencies of Δ and absorption coefficient α on frequency for PPWG made of doped Si with $N = 10^{18}$ cm^{-3} (solid lines) and gold (dotted lines).

While the absorption coefficient in gold-PPWG is small $\alpha = 7.6 \cdot 10^{-3}$ mm^{-1}, effective refractive index is very close to 1. As we shall be convinced below, the relatively large value of Δ is favorable to have strong field confinement in lateral direction (along y-axis) of proposed PWCS. In PPWG with plates made of doped Si the effective refractive index n more appreciably differs from 1, but absorption losses are higher. Nevertheless the use of doped Si is suitable for chip-scale waveguiding (a few of millimeters).

The Fig. 5 illustrates dependencies of Δ and α on frequency for PVDF covered PPWG. It is seen that attenuation of THz-wave is strongly enhanced with increase of the frequency since nearly 1 THz. Hence it is expedient to limit operation bandwidth of PVDF-based PPWG by frequencies about 0.4 - 1 THz. In this region the value of $Re(\dot{n})$ is significantly higher than that of waveguide made of doped Si. So the main advantage of PVDF using is an opportunity to obtain small mode size in lateral direction of PWCS.

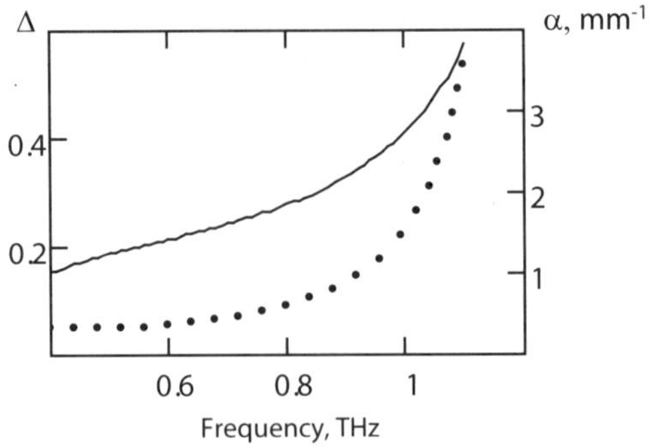

Fig. 5. Dependency of Δ (solid line) and α absorption coefficient (dotted line) on frequency for PPWG made of PVDF.

2.3. *Effective index of PWCS*

Effective refractive index $N_{\text{eff}} = c/V_{\text{ph}}$ (V_{ph} is the phase velocity in PWCS) is calculated by considering PWCS as PPWG with adiabatically varying plate's separation. Using designations in Fig. 1, the distance between cylindrical surfaces of PWCS can be written as

$$h(y) = h_0 + \frac{y^2}{R}, \tag{4}$$

where R is the radius of cylinders and condition $|y| \ll R$ is used.

Based on results of the previous section we can conclude that in all cases $\Delta \ll 1$. Therefore the Eq. (3) can be simplified by using approximate expression $\tanh(x) \approx x$ for sufficiently thin waveguide $h_0 \ll 1/k_0\sqrt{\Delta}$. Then by replacing h_0 with $h(y)$ and using condition $\dot{\varepsilon} \gg 1$, the following expression for real part of refractive index can be obtained

$$Re(\dot{n}) = 1 + \frac{1}{k_0 h_0 \sqrt{|\dot{\varepsilon}|}(1 + y^2/Rh_0)} \cos\frac{\varphi}{2}, \tag{5}$$

where $\varphi = \tan^{-1}(\varepsilon''/|\varepsilon'|)$.

The Eq. (5) shows that refractive index is maximal in middle of waveguide (y=0) and it is decreased with increase of $|y|$. Therefore THz beam can be confined in y-direction by similar way that of graded-index dielectric waveguide. The theory of such waveguides is well developed. The analytical expressions for effective refractive index and field distribution were obtained for some profiles of the refractive index, including $n(y) \propto 1/\cosh^2(y)$.[14]

The last profile corresponds to our case because approximate expression $1+y^2/Rh_0 \approx \cosh^2(y/\sqrt{Rh_0})$ is satisfied if $y^2 \ll Rh_0$. Based on results for $\cosh^{-2}(y)$ profile of refractive index,[14] it is easy to obtain that fundamental TM_0-mode field distribution along y-axis $E_x(y)$ and effective index N_{eff} of PWCS are given by

$$E_x \propto \cosh^{-s}\left(\frac{y}{\sqrt{Rh_0}}\right), \tag{6}$$

$$N_{\text{eff}} = \sqrt{1+\frac{s^2}{Rh_0k_0^2}}, \tag{7}$$

where

$$s = 0.5\left(\sqrt{1+\frac{8Rk_0\cos(\varphi/2)}{\sqrt{|\dot{\varepsilon}|}}}-1\right). \tag{8}$$

From Eq. (6) and (8) follows that small $|\dot{\varepsilon}|$ of the cylinders material is favorable for strongly confined TM_0-mode propagation in waveguide. The TM_0-mode intensity distributions (plotted accordingly to (6) and (8)) for PWCS made of gold and doped Si ($N = 10^{18}$ cm^{-3}) for radius of cylinders $R = 5$ cm, minimal distance $h_0 = 0.1$ mm at frequencies 0.4 THz and 1 THz are presented in Fig. 6(a) and 6(b). The field distribution in PVDF-based waveguide is nearly same that of PWCS with doped Si and it is not shown in Fig. 6(a).

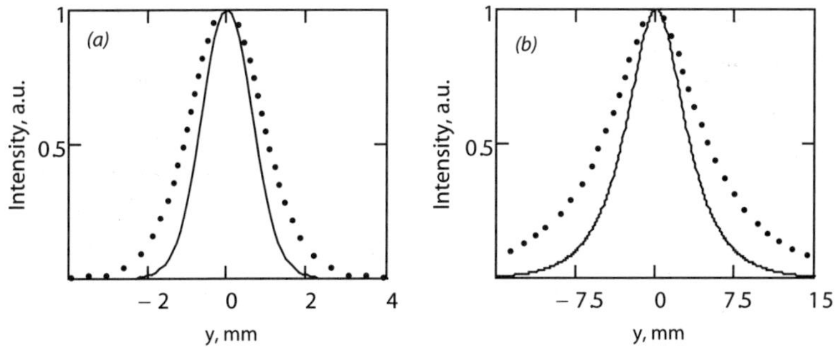

Fig. 6. Fundamental mode intensity distributions in PWCS made of doped Si (a) and Au (b) at frequencies 1 THz (solid lines) and 0.4 THz (dotted lines).

It is seen that field distributions are frequency dependent as it was generally expected. In case of Si-based PWCS the electrical field is more strongly concentrated in a vicinity of the middle of waveguide (y=0) than that of Au-based PWCS. It is related to relatively small $|\dot{\varepsilon}|$ for Si material.

The dispersion of waveguide is very important for applications of PWCS with broadband THz radiation, such as THz-pulses. The calculated dependencies $N_{\rm eff} = N_{\rm eff}(\nu)$ are nearly same that of dispersion $n = n(\nu)$ of corresponding PPWG. For Au-based PWCS the value of $N_{\rm eff}$ is very close to 1 independently of frequency. For Si-based PWCS the value of $N_{\rm eff}$ is 1.076 at 0.1 THz and it is reduced to $N_{\rm eff} = 1.026$ at 3 THz. Even such small dispersion of the waveguide limits its application with broadband THz-pulses. Indeed, after propagating 3 mm distance the delay of spectral components at 0.1 THz and 3 THz is 0.5 ps, which is close to duration of the THz-pulses usually used in practice.

3. Experiment

Here an opportunity of the undistorted 2D-confined THz-pulse propagation in plasmon waveguide formed by two gold-covered cylindrical surfaces is demonstrated. To measure transmission of THz-pulses through PWCS we used TDS technique[21,22] similar to previously used for PPWG with slightly curved plates.[23,24]

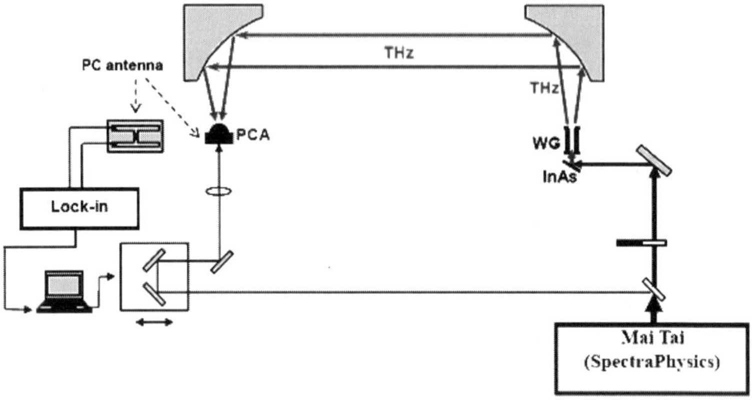

Fig. 7. Schematic view of experimental setup.

Experimental setup is schematically illustrated in Fig. 7. A pulse train derived from a mode-locked Ti:sapphire femtosecond laser is split into excitation (pump) and detection (probe) pulses. The excitation pulses impinge on an InAs surface field emitter which leads to the emission of THz-pulse.[25] The THz radiation after propagation through the studied PWCS is collected with off-axis parabolic mirror. Another mirror is used to focus THz radiation onto a detection photoconductive antenna. This antenna is based on an Auston switch,[26] which is a piece of low-temperature grown GaAs, placed in between two metal contacts. The photoconductive switch is gated by probe Ti:Sapphire beam. This enables the time domain detection of the THz field amplitude by scanning the time delay between pump and probe pulses with a motorized translation stage. In all measurements the THz beam was polarized parallel to the x-axis to excite fundamental TM_0 mode in PWCS.

The studied PWCS had the radius of cylinder $R = 5$ cm and minimal distance between them was $h_0 = 0.1$ mm. The full length of the waveguide was $L_\Sigma = 7.3$ cm, where $L = 5$ cm corresponds to inherent PWCS and $L_{\rm in} = 2.3$ cm to 10°-tapered section. In tapered part of the waveguide the metal planes separation is gradually changed from 4 mm at input to $h_0 = 0.1$mm at the output (Fig. 8). Such input adapter allows to increase coupling efficiency of the incident THz beam (having aperture a few mm) with really PWCS.

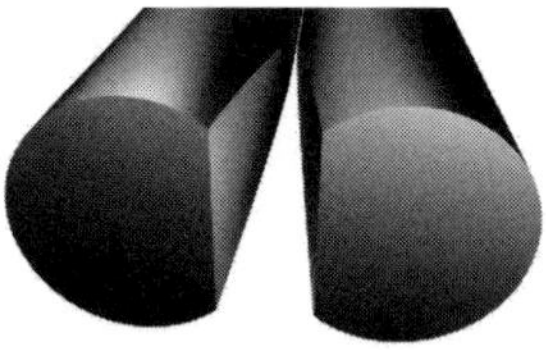

Fig. 8. The tapered part of the waveguide.

Figure 9(a) shows the waveform of THz pulse propagating in ambient air. The oscillations after main pulse are related to THz absorption in water vapor. In the same Fig. 9(a) the pulse form after propagating through the PWCS is presented as well. In the last case the vertical and horizontal offsets are applied to clarity picture. The comparison of the waveforms indicates nearly undistorted pulse propagation in the PWCS. It is confirmed by the corresponding Fourier amplitude spectrum, which gives no indication of a low-frequency cutoff (Fig. 9(b)).

To estimate the THz beam confinement along y-axis the knife edge test[27] was used. The movable metal plate was placed close to the waveguide output. Similar measurements were carried out in the same position of the

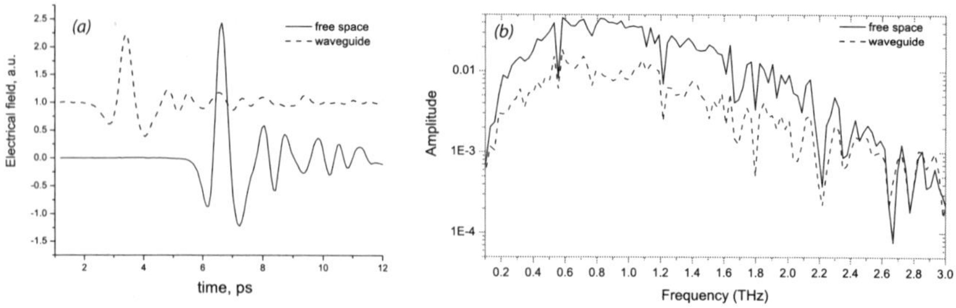

Fig. 9. THz pulses waveforms (a) and corresponding spectra (b) for free space and waveguide propagation.

knife in absence of the waveguide as well. The beam intensity distributions for both cases (with and without waveguide) at frequencies 0.4 THz and 1 THz are presented in Fig. 10.

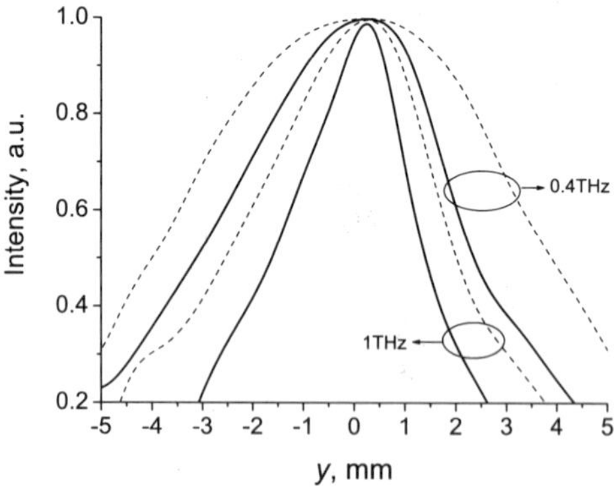

Fig. 10. THz beams intensities distributions in case of use WG (solid lines) and without WG (dotted lines) for frequencies 0.4THz and 1THz.

It is seen that THz waves were indeed guided inside the gap region with a satisfactory mode quality over a wide frequency range. For frequency of 1 THz the full-width at half-maximum (FWHM) beam size is nearly 3 mm that indicates on lateral confinement in the waveguide. Note that according to calculation (see Fig. 6(b)) the beam size about 4 mm has to be expected. The main reason of this discrepancy is probably related to inaccuracy of measurement by knife edge method caused by strong diffraction of THz-wave.

4. Conclusion

Undistorted 2D-confined THz pulse propagation through plasmon waveguide formed by two metal cylindrical surfaces is demonstrated. A simple theoretical model is developed to calculate effective index and field distribution of the fundamental TM_0-mode in waveguide formed by gold, doped Si, and PVDF materials. It is shown that doped Si and PVDF are good candidates for plasmon THz waveguides, especially in case of the narrow-band applications. In Si-covered waveguide the estimated beam spot area is nearly 0.1 mm^2 (at 1 THz) that indicates opportunity of strongly confined THz-wave propagation.

Acknowledgments

We acknowledge the support of the Ministry of Science and Education of Armenia and ANSEF Grant EN-2202.

References

1. T. Ohkubo, M. Onuma, J. Kitagawa, Y. Kadoya, Appl. Phys. Lett. **88**, 212511 (2006).
2. M.Tonouchi, Nature Photon. **1**, 97 (2007).
3. S. Bozhevolnyi, Plasmonic Nanoguides and Circuits (Pan Stanford, Singapore, 2008).
4. K. Tanaka and M. Tanaka, Appl. Phys. Lett. **82**, 1158 (2003).
5. J. Kim, Opt. Lett. **32**, 3405 (2007).
6. S. I. Bozhevolnyi, Kh. V. Nerkararyan, Opt. Express **17**, 10327 (2009).
7. R. N. Bracewell, Wireless Engin. **31**, 320 (1954).
8. J. B. Pendry, A. J. Holden, W. J. Stewart, I. Youngs, Phys. Rev. Lett. **76**, 4773 (1996).
9. J. B. Pendry, L. M. Moreno, F. Garcia-Vidal, Science **305**, 847 (2004).
10. J. G. Rivas, M. Kuttge, H. Kurz, P. H. Bolivar, J. A. Sánchez-Gil, Appl. Phys. Lett. **88**, 082106 (2006).
11. R. Kohlman et al., Handbook of Conducting Polymers, Ed. by Skotheim, Elsenbaumer and Reynolds (Marcel Dekker, New York, 1998).
12. A. Hassani, M. Skorobogatiy, CLEO/QELS-2009 (Optical Society of America, Washington, DC, 2009), CM16.
13. B. Ghamsari, H. Majedi, J. Appl. Phys. **104**, 083108 (2008).
14. T. Tamir (Ed.), Guided-Wave Optoelectronics (Springer-Verlag, Berlin, 1990).
15. T.-I. Jeon, D. Grischkowsky, Phys. Rev. Lett. **78**, 1106 (1997).
16. M. A. Ordal, R. J. Bell, R. W. Alexander, Jr. L. Long, M. R. Querry, Appl. Opt. **24**, 4493 (1985).
17. J. G. Rivas, C. Schotsch, P. H. Bolivar, H. Kurz, Phys. Rev. **B 68**, 201306 (2003).

18. E. Hendry, F. J. Garcia-Vidal, L. Martin-Moreno, J. G. Rivas, M. Bonn, A. P. Hibbins, M. J. Lockyear, Phys. Rev. Lett. **100**, 123901 (2008).
19. G. Baccarani, P. Ostoja, Solid-State Electron. **18**, 579 (1975).
20. T. Furukawa, Phase Transitions, **18**, 143 (1989).
21. P.R. Smith, D.H. Auston, M.C. Nuss, IEEE J. Quantum Electron. **24**, 255 (1988).
22. R. Mendis, D. Grischkowsky, Opt. Lett. **26**, 846 (2001).
23. Yu. H. Avetisyan, A.H. Makaryan, K. Khachatryan, A. Hakhoumian, Arm. J. Phys. **2**, 122-128 (2009).
24. Yu. Avetisyan et al., Tech. Dig. of CLEO'09, (Baltimore, USA, May 31 – June 5, 2009), paper CThQ2.
25. P. Gu, M. Tani, S. Kono, K. Sakai, and X.-C. Zhang, J. Appl. Phys. **91**, 5533 (2002).
26. D. H. Auston, K. P. Cheung, P. R. Smith, Appl. Phys. Lett. **45**, 284 (1984).
27. G. Brost, P. D. Horn, A. Abtani, Appl. Opt. **24**, 38 (1985).

BROADBAND SIMILARITON: FEATURES AND APPLICATIONS

A. ZEYTUNYAN, A. MURADYAN, G. YESAYAN, and L. MOURADIAN*

Ultrafast Optics Laboratory, Faculty of Physics, Yerevan State University, 1, Alex Manoogian Street, Yerevan 0025, Armenia

**E-mail: lmouradian@ysu.am*

F. LOURADOUR and A. BARTHÉLÉMY

Département Photonique, XLIM Institut de Recherche, Faculté des Sciences, 123 ave. A. Thomas, F-87060 Limoges, France

Self-similarity in optics, along with the soliton physics, recently attracts interest of the researchers of ultrafast and nonlinear fiber optics.[1] Both the parabolic similariton of active fibers, and nonlinear-dispersive similariton generated in passive fiber are of interest, especially for applications in ultrafast optics, such as pulse compression and shaping, similariton-based temporal lensing and spectrotemporal imaging, spectral interferometry, etc. However, the signal analysis-synthesis problems in a few femtosecond time scale demand the generation and study of broadband similaritons of ~100 nm bandwidth.

We generate the broadband similariton, and characterize it experimentally to reveal its nature and distinctive properties, and to describe it mathematically. We carry out the complete characterization of the broadband similariton by means of the chirp measurement through the technique of spectral compression and frequency tuning in the sum-frequency generation process. Our studies are of interest in view of applications of similariton to the signal analysis and synthesis problems in ultrafast optics, particularly for similariton-induced temporal lensing and similariton-based spectral interferometry. Our developed method of similariton chirp measurement can serve also for the fiber characterization.

Keywords: Femtosecond, fiber, similariton, chirp, pulse measurement, spectral interferometry, spectral compression, imaging.

1. Introduction

In the beginning of 90s Anderson and co-authors predicted a parabolic solution for the nonlinear Schrödinger equation with dispersion and high nonlinearity, that is, the self-similar propagation of a pulse with the parabolic temporal, spectral and phase profiles.[2] Generation of such kind of pulses in practice is possible in active fibers such as rare-earth-doped fiber amplifiers,[3] Raman fiber amplifiers,[4] and also in fibers with distributed dispersion.[5] Such a similariton can be generated in a laser resonator[6] as well, or in a tapered fiber with decreasing normal dispersion, using either passive dispersion-decreasing fiber or a hybrid configuration with Raman amplification.[7] The temporal and spectral profiles of the similariton generated in a fiber amplifier are independent of the input pulse profile; they are determined only by the input pulse energy and amplifier parameters.[8] The linear chirp of parabolic similariton is independent of the input pulse energy and depends only on the gain and dispersion of the amplifier.[3] All above mentioned studies are related to the class of parabolic similaritons. Recently, a new type of similariton was generated in a passive fiber (without gain) under the combined impacts of Kerr-nonlinearity and dispersion.[9,10] The spectral interferometric study of this nonlinear-dispersive similariton[10] showed the linearity of its chirp (parabolic phase), with a slope given only by the fiber dispersion. This property leads to the spectrotemporal similarity and self-spectrotemporal imaging of nonlinear-dispersive similariton, with the accuracy given by the spectral broadening and pulse stretching together. The bandwidth of similariton generated in passive fiber is given by the input pulse power only.[10] Both the parabolic similariton of active fiber and nonlinear-dispersive similariton of passive fiber are of interest for applications in ultrafast optics, especially for the pulse compression[11] and shaping,[12] similariton-referencing temporal lensing and spectrotemporal imaging,[13] spectral interferometry,[14] etc.

The signal analysis-synthesis problems in a few femtosecond time scale, however, demand the generation and study of broadband similariton. Particularly, the resolution of femtosecond oscilloscope, based on spectrotemporal imaging in the similariton-induced parabolic lens, is given by the bandwidth of similariton.[13] For the similariton-based spectral interferometry, the application range is as large as broadband the similariton-reference is. The pulse compression ratio is also as high as the spectral broadening factor is. Thus, there is a need to characterize and study the broadband similariton. The spectral interferometry permitted the complete characterization and study of the nonlinear-dispersive similariton of up to 10 nm bandwidth.[10] The same study for the characterization of the $\sim$100 nm bandwidth similariton is technically difficult, as a generator of a known broadband reference pulse (of the bandwidth $\geq$100 nm) is specially required.

The subject of our work is the generation of broadband similariton (of ~100 nm bandwidth), and its experimental and numerical studies. The aim of our studies is the revealing of the nature and distinctive properties of broadband similariton in view of applications to the signal analysis and synthesis problems in ultrafast optics, particularly for similariton-induced temporal lensing and similariton-based spectral interferometry.

The outline of our work is the following: first, we present the results of our numerical modeling of the process under study based on the solution of the generalized nonlinear Schrödinger equation, afterwards, we describe our experiment of similariton chirp measurement, based on the technique of spectral compression and frequency tuning in the sum-frequency generation process, discuss the results leading to the mathematical description of the broadband similariton, and finally, present the comparative study of similariton-based spectral interferometry and temporal lensing - spectrotemporal imaging as two applications of broadband similariton.

2. Chirp of Broadband Similariton

The nonlinear-dispersive similariton of the bandwidths $\Delta\lambda \leq 10$ nm (at $\lambda \sim 800$ nm), according to its spectral-interferometric characterization,[10] is described with its slowly varying amplitude $A_f(t)$ and parabolic phase $\varphi_f(t)$ as

$$A_f(t) \propto \tilde{A}_f^*(\omega)|_{\omega=\gamma t} = |\tilde{A}_f(\omega)| \exp(\imath\beta_2 f\omega^2/2)|_{\omega=\gamma t}, \tag{1}$$

$$\varphi_f(t) = -\phi_f(\omega)|_{\omega=\gamma t} = \beta_2 f\omega^2/2|_{\omega=\gamma t} = \gamma t^2/2, \tag{2}$$

where $\tilde{A}(\omega) \equiv FT[A(t)]$ is the spectral amplitude, $\gamma = (\beta_2 f)^{-1}$ - chirp factor, β_2 - coefficient of group velocity dispersion (GVD), t - time, ω - frequency, and f - fiber length. For a pulse pure dispersive propagation in the temporal Fraunhofer zone, we have a "spectron" pulse:

$$A_d(t) = FT^{-1}[\tilde{A}_0(\omega)\exp(-\imath\beta_2 f\omega^2/2)] \propto \tilde{A}_0(\omega)\exp(\imath\beta_2 f\omega^2/2)|_{\omega=\gamma t}. \tag{3}$$

Thus, the phase of nonlinear-dispersive similariton coincides with the dispersion-induced phase of spectron $\varphi_f(t) = \gamma t^2/2 = t^2(2\beta_2 f)^{-1}$. The applications of similariton demand to check and generalize this key peculiarity.

To have the complete physical pattern and reveal the distinctive peculiarities of the generation and propagation of broadband similaritons in the active and passive fibers, we carry out the numerical modeling of the processes under study. Our mathematical description is based on the generalized nonlinear Schrödinger equation, which describes the broadband parabolic similariton generally, nonlinear-dispersive similariton in the absence of gain, and the spectron pulse of pure dispersive propagation in the absence of gain and nonlinearity. Our numerical analysis, based on the solution of generalized nonlinear Schrödinger equation using the split-step Fourier method, shows the impact of the key physical factor of third-order dispersion (TOD) for the broadband pulses. The fiber TOD is expressed additively in the $\phi_f(\omega)$ and $\varphi_f(t)$ parabolic spectral and temporal profiles of the phase (and chirp) due to its small impact as compared to the GVD, and its value is the same for the parabolic and nonlinear-dispersive similaritons, coinciding also with the one for the spectron pulse of pure dispersive propagation:

$$\Delta\varphi_f(t) \approx \Delta\phi_f(\omega)|_{\omega=\frac{t}{\beta_2 f}} = \frac{1}{6}\beta_3\omega^3 f|_{\omega=\frac{t}{\beta_2 f}} = \frac{1}{6}\frac{\beta_3}{\beta_2^3 f^2}t^3, \tag{4}$$

where β_3 is the TOD coefficient. Thus, the chirp measurement of a broadband similariton becomes urgent, since it gives the TOD of fiber and permits to generalize the description for these three types of broadband pulses.

In our experiment, we generate a broadband nonlinear-dispersive similariton of up to 100 nm bandwidth (FWHM) in a piece of passive fiber. Afterwards, we carry out the chirp measurement through frequency tuning in the sum-frequency generation (SFG) and spectral compression process.[13] Fig. 1 shows the schematic of the experimental setup. We use the radiation of Coherent Verdi V10 + Mira 900F femtosecond laser system with the following parameters of radiation: 100 fs pulse duration, 76 MHz repetition rate, 1.6 W average power, 800 nm central wavelength. A beam-splitter (BS)

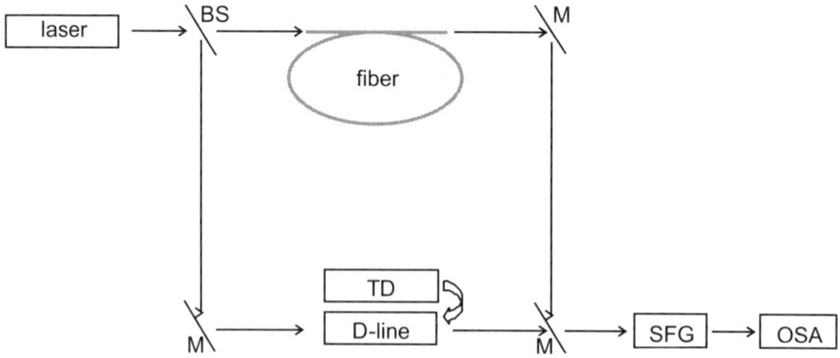

Fig. 1. Experimental setup: BS - beam splitter, M - mirrors, TD - temporal delay, D-line - dispersive delay line, SFG - BBO crystal, OSA - optical spectrum analyzer.

splits the laser radiation into the low- and high-power parts (80% + 20%). We direct the low-power pulse to the dispersive delay line (D-line) with anomalous dispersion (3.5 m-separated pair of SF11 prisms with the reverse mirror), and dispersively stretch it 22 times, resulting in 3.1 ps autocorrelation duration. In the second path, we inject the high-power pulse into the standard single-mode fiber (1.65 m Newport F-SPF PP@820 nm) by means of a 10× microscope objective and generate a broadband nonlinear-dispersive similariton. Afterwards, we direct these two pulses to the nonlinear β-barium borate (BBO) crystal (type 1 - ooe, 800 nm operating wavelength) to have SFG-spectral compression. The spectrum is recorded by an optical spectrum analyzer (OSA Ando 6315). The temporal delay between these two pulses leads to the frequency shift of the SFG-signal according to the concept of the temporal lens.[13] For SFG-spectral compression, we have up- and down-chirped pulses with the central frequencies at the centers of the temporal profiles. A temporally delayed pulse gates from the similariton a neighbor frequency to the central one, and the SFG-signal becomes frequency shifted, relevantly. We experiment this by shifting the reverse mirror of the D-line and recording the relevant SFG-compressed spectra. We obtained ± 20 nm frequency tuning for 23 times SFG-spectrally compressed pulses (down to 0.12 nm at the 400 nm central wavelength). This technique can be considered as a version of cross-correlation frequency-resolved optical gating (X-FROG),[15,16] modified by spectral compression, providing a sharper spectral signal in a larger spectral (and temporal) range. To check this modification, we carry out the experiment of frequency tuning with D-line for spectral compression and without it, replacing the D-line with a simple temporal delay (TD). The relevant 3D frequency tuning patterns are shown in Fig. 2(b) and (c).

Fig. 2 shows the spectrum of broadband similariton (a) representing its spectrotemporal image, and 3D frequency tuning patterns with spectral compression (b) and without it (c) for the chirp measurement. A temporal delay in the range of ± 16 ps results in a ± 20 nm wavelength (± 0.24 PHz frequency) shift for SFG-spectrally compressed signal, corresponding to chirp measurement of similariton at 800 nm central wavelength in the range of 160 nm.

Figure 3 shows the derivative of spectral phase of broadband similariton obtained from the frequency tuning data of Fig. 2(b). The circles in Fig. 3 are the measured experimental points, the dashed curve is a linear fit, and the solid curve is a parabolic fit. The slope of the chirp linear component gives the GVD coefficient, while the parabolic component is re-

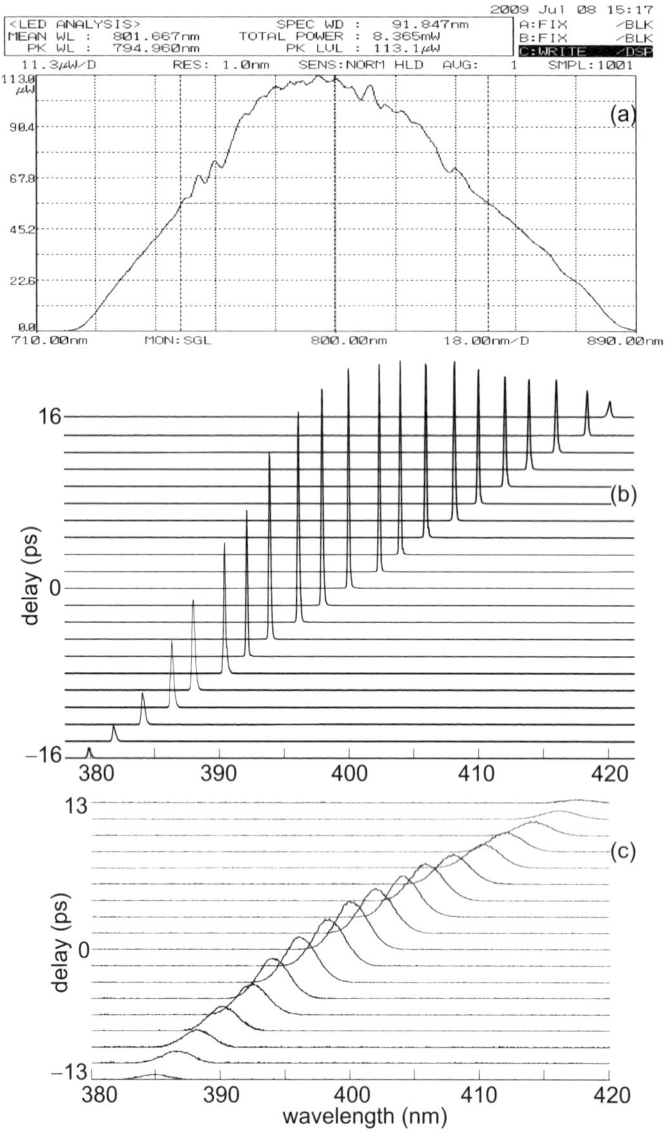

Fig. 2. The measured spectrum of broadband similariton (a), and 3D frequency tuning patterns with spectral compression (b) and without it (c) for the chirp measurement: the 40.307 nm (0.475 PHz) frequency tuning at 400 nm is adequate to the 161.228 nm spectral range of similariton at 800 nm for the 32.34 ps range of temporal delay between SFG-interacting pulses.

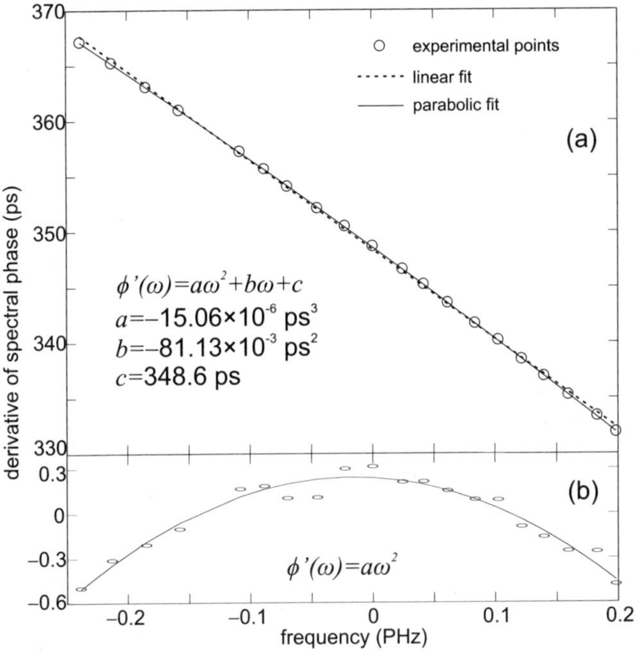

Fig. 3. (a) Derivative of spectral phase of broadband similariton obtained from the frequency tuning data of Fig. 2(b), and (b) the quadratic component of the spectral phase derivative, separately.

sponsible for TOD. We extract the dispersion coefficients after a relevant processing of the Fig. 3 data and get the following values for GVD and TOD: $\beta_2 = 49.17$ fs^2/mm and β_3=18.25 fs^3/mm. The ~1% difference from the linear fit of the similariton chirp in the range of ~100 nm gives the range of applications for aberration-free similariton-based spectrotemporal imaging[13] and spectral interferometry.[14]

Thus, we have generated nonlinear-dispersive similariton of up to 100 nm bandwidth, and carried out its complete characterization through the chirp measurement, using the technique of frequency tuning in the SFG-spectral compression. Our studies state that only fiber dispersion determines the phase (chirp) of broadband nonlinear-dispersive similariton resulting in its following description:

$$A_f(t) \propto |\tilde{A}_f(\omega)| \exp[\frac{\imath}{2}(\beta_2\omega^2 + \frac{1}{3}\beta_3\omega^3)f]|_{\omega=\frac{t}{\beta_2 f}}. \quad (5)$$

The fiber TOD results in the same additional phase for all broadband spectron, nonlinear-dispersive and parabolic similaritons, of the following value:

$$\Delta\varphi_f(t) = \frac{1}{6}\frac{\beta_3}{\beta_2^3 f^2}t^3. \tag{6}$$

The ~1% accuracy of the linear fit for the chirp of the 100 nm-bandwidth similariton gives the range of applications for aberration-free similariton-based spectrotemporal imaging[13] and spectral interferometry.[14]

3. Similariton-Based Spectral Interferometry and Femtosecond Oscilloscope

Based on our studies on broadband similariton, we develop and demonstrate a method of similariton-based spectral interferometry for the femtosecond pulse characterization in comparison with the pulse spectrotemporal imaging in the similariton-induced temporal lens. Application of similariton to the reference-based methods upgrades them up to the self-referencing ones, substantially improving their performance due to the enlarged application range along with the simplicity of the principle and configuration. Our comparative experimental study, carried out together with the theoretical check, evidences the quantitative accordance and high precision of the both similariton-referencing methods of spectral interferometry and spectrotemporal imaging.

The classical method of spectral interferometry (SI) is based on the interference of the signal and reference beams spectrally dispersed in a spectrometer, with the spectral fringe pattern caused by the difference of the spectral phases.[17] The constant (or known) spectral phase of the reference permits retrieving the spectral phase of signal, and, together with the spectrum measurement, recovering the complex temporal amplitude of the signal through Fourier transformation. The setup of classic SI is rather simple, and the measurement is accurate as any interferometric one, but its application range is restricted by the bandwidth of the reference. The SI characterization of a signal underwent a nonlinear interaction with medium requires a special broadband reference to fully cover the broadened signal spectrum. To avoid this restriction, the self-referencing methods of spectral shearing interferometry, such as the spectral phase interferometry for direct electric field reconstruction (SPIDER)[18] and spectral shearing interferometry resolved in time (SPIRIT),[19] are developed. This improvement promotes the SI to the class of the most popular and commercialized methods of accurate measurements at the femtosecond time scale, up to the competition with the FROG,[20] but the optical arrangement becomes much

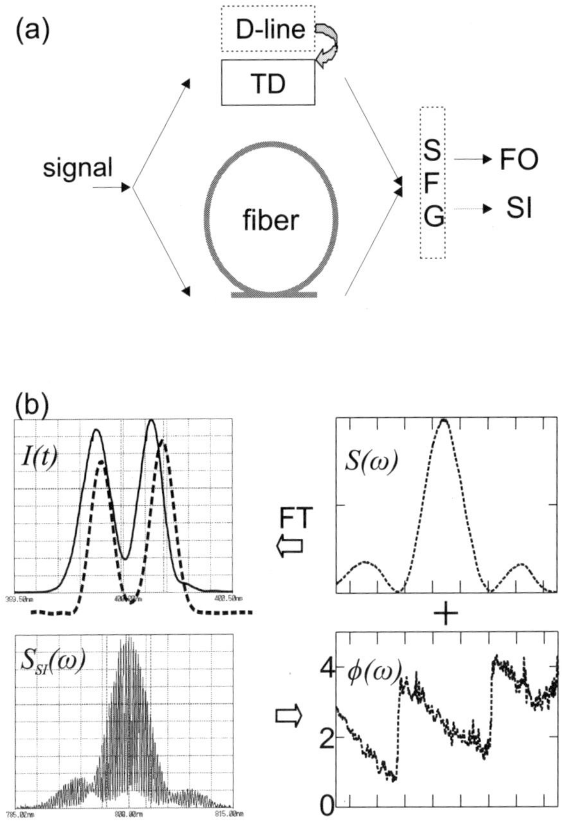

Fig. 4. Schematic of the setup and performance for the comparative experiment on the similariton-based SI and FO: (a) The signal beam is split; the powerful part generates NL-D similariton in a passive fiber, which is afterwards coupled with the low-power part (passed through TD) giving the SI-fringe pattern $S_{SI}(\omega)$ in OSA (the scheme is modified to FO by replacing the TD by D-line and placing BBO crystal for SFG at the input of OSA); (b) The SI-fringe pattern registered ($S_{SI}(\omega)$, left bottom) gives the signal spectral phase ($\phi(\omega)$, right bottom), which together with the signal spectrum recorded ($S(\omega)$, right top), via Fourier transformation (FT), gives the pulse temporal ($I(t)$, left top, dashed), which is compared with the SFG-spectrotemporal image of FO ($S_{SFG}(\omega) \propto I(t)$, left top, spectrogram).

more complicated. Our proposed method of similariton-based SI, along with its self-referencing performance, keeps the simplicity of the principle and configuration of the classic SI.

The schematic of similariton-based SI is shown in Fig. 4. Splitting the signal beam, its part is injected into a fiber to generate the nonlinear-dispersive similariton-reference, with the complex spectral amplitude

$\tilde{A}_f(\omega) = |\tilde{A}_f(\omega)| \exp(-\imath\beta_2^f f\omega^2/2)$, and the parabolic phase $\phi_f(\omega)$ and chirp factor $\gamma_f = -(\ddot{\phi}_f)^{-1} = (\beta_2^f f)^{-1}$, given by the fiber length f and dispersion β_2^f.[10,13] The residual part of the signal, with the complex spectral amplitude $\tilde{A}(\omega) = |\tilde{A}(\omega)| \exp[\imath\phi(\omega)]$, passing an optical time delay (TD), is coupled with the similariton in a spectrometer. The spectral fringe pattern $S_{SI}(\omega) = 2|\tilde{A}(\omega)||\tilde{A}_f(\omega)| \cos[\phi(\omega) - \phi_f(\omega)] = 2|\tilde{A}(\omega)||\tilde{A}_f(\omega)| \cos[\phi(\omega) + \beta_2^f f\omega^2/2]$, on the background of the signal and similariton spectra, completely covers the signal spectrum $S(\omega) = |\tilde{A}(\omega)|^2$, and the whole phase information is available for any signal. The known spectral phase of the similariton-reference allows to retrieve the signal spectral phase $\phi(\omega)$, and by measuring also the signal spectrum, to reconstruct the complex temporal amplitude $A(t)$ of the signal through Fourier transformation. Thus, the method of similariton-based SI joins the advantages of both the classic SI[17] and spectral shearing interferometry,[18,19] combining the simplicity of the principle and configuration with the self-referencing performance. Experimentally examining the similariton-based SI, we compare the measurements with the ones carried out by a prototype of the femtosecond oscilloscope (FO) based on the pulse spectrotemporal imaging in the similariton-induced temporal lens in the process of sum-frequency generation (SFG). Our comparative study, involving also autocorrelation check, along with the demonstration of the similariton-based SI serves also for the inspection of the prototype of the similariton-based FO, the measurements of which previously were compared with the autocorrelation only.[13]

Method of SFG-spectrotemporal imaging for the direct femtosecond scale measurements is based on the conversion of temporal information to the spectral domain in a similariton-induced parabolic temporal lens.[13] The setup of similariton-based SI is modified to FO by replacing the temporal delay (TD) with a dispersive delay line (D-line) and placing a nonlinear crystal for SFG at the system output (Fig. 4). In the spectral domain, the dispersive delay works as a parabolic phase modulator, and the signal $\tilde{A}(\omega)$ passed through is described as $\tilde{A}_d(\omega) = \tilde{A}(\omega) \exp(\imath\ddot{\phi}_d\omega^2/2)$ with the given coefficient $\ddot{\phi}_d \approx -\gamma_d^{-1}$. In the fiber arm, we have nonlinear-dispersive similariton with the known parameters as in the case of similariton-based SI. In both arms of the setup, we have linearly chirped pulses, and the temporal and spectral complex amplitudes repeat each other in the temporal Fraunhofer zone, i.e. spectron pulses are formed:[10,13] $A_d(t) \propto \tilde{A}_d(\omega)$, and $A_f(t) \propto \tilde{A}_f(\omega)$ with $\omega = \gamma_{d,f}t$. Under the conditions of the opposite and same value chirps $\gamma_f = -\gamma_d \equiv \gamma$, and constant similariton spectrum throughout the signal spectrum, the output temporal SFG-signal repeats the input spectral amplitude: $A_{SFG}(t) \propto A_d(t) \times A_f(t) \propto \tilde{A}(\omega)$. Accordingly, the output spectral and input temporal amplitudes repeat each other $\tilde{A}_{SFG}(\omega) \propto A(t)$, and the output SFG-spectrum displays directly the input temporal pulse: $S_{SFG}(\omega) = |\tilde{A}_{SFG}(\omega)|^2 \propto |A(t)|^2 = I(t)$, with the scale $\omega = \gamma t$. The resolution of such a similariton-based FO is given by the transfer function of the similariton's spectrum,[13] and FO with a similariton-reference of the bandwidth of a few tens of nanometres provides the direct measurement of temporal pulse in a spectrometer, exceeding the resolution of the recent achievement of silicon-chip-based ultrafast optical oscilloscope[21] by an order of magnitude.

In the experiment, we use a combined setup for implementing both methods of the similariton-based SI and SFG-spectrotemporal imaging, modifying the setup of Fig. 4 by removing the D-line for SI-measurements. As a pulse generator we use standard Coherent Verdi V10 + Mira 900F femtosecond laser system with the following parameters of radiation: 100 fs pulse duration, 800 nm central wavelength, 1.6 W average power at a 76 MHz repetition rate. We shape different amplitude- and phase-modulated pulses at the setup input and split the signal radiation by a beam-splitter (80%+20%). We direct the low-power part to the TD or D-line (a pair of SF11 prisms with a reverse mirror) for similariton-based SI and spectrotemporal imaging, respectively. In the second path, we inject the high-power pulse (with average power ~10–100 mW) into a standard single-mode fiber (1.65 m Newport F-SPF PP@820 nm) by a microscope objective (10×) and generate a nonlinear-dispersive similariton (of 20-80 nm bandwidth). For the similariton-based SI-measurements, we couple these two pulses directly into the spectrometer (OSA Ando 6315) and register the SI fringe pattern, and signal spectrum. To retrieve the spectral phase, we use the Fourier-transform algorithm of the fringe-pattern analysis.[22] For the FO-measurements, we place a nonlinear SFG-crystal (1mm-thick BBO, ooe) at the input of OSA, and register directly the SFG-spectrotemporal image. We carry out the similariton-based SI and FO measurements together with the checking by a standard APE PulseCheck autocorrelator.

First, we test the similariton-based SI for the laser pulses stretched and chirped in SF11 glasses of different thickness, comparing the results with the autocorrelation measurements (Fig. 5). For the stretched pulses, we have the following coefficients for the dispersion-induced parabolic spectral phases: $\ddot{\phi}$=1.94×10^{-3}, 4.94×10^{-3}, 6.34×10^{-3}, and 10.78×10^{-3} ps^2 for the 0, 2, 3, and 5 cm-glasses, respectively. The SI-reconstructed pulses, correspondingly, have durations of 108, 197, 252, and 365 fs, in a good ac-

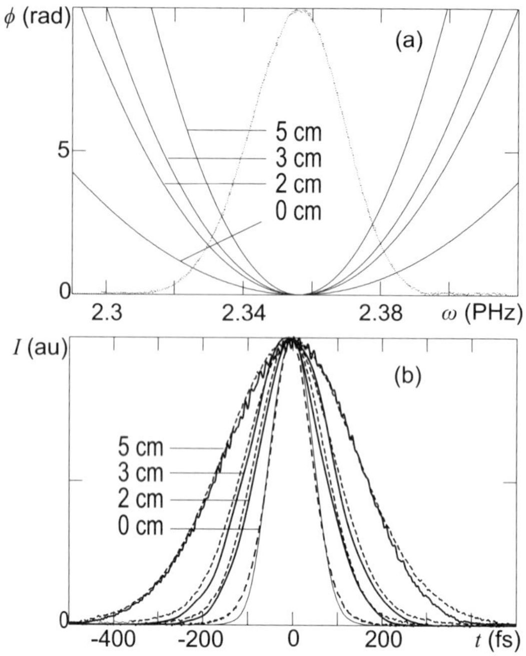

Fig. 5. Similariton-based SI for the pulses dispersively stretched and chirped in the SF11 glasses of different thickness: (a) retrieved spectral phases with the measured spectrum, and (b) SI-reconstructed temporal pulses (solid) in comparison with autocorrelation measurements (dashed).

cordance with the autocorrelation measurements of 110, 210, 261 and 381 fs (in assumption of Gaussian pulses).

To compare the measurements of the similariton-based SI and FO, we shape double-peak signal pulses with the spectral domain amplitude- and phase-modulation given by the peaks' temporal distance T and their proportion μ. The temporal amplitude $A(t) = A_0(t) + \mu A_0(t+T)$ corresponds to the complex spectral amplitude $\tilde{A}(\omega) = \tilde{A}_0(\omega)\rho(\omega)\exp[\imath\phi(\omega)]$, with the $\rho(\omega) = \sqrt{1+\mu^2+2\mu\cos(\omega T)}$ amplitude- and $\phi(\omega) = \arctan\{\sin(\omega T)/[\mu^{-1}+\cos(\omega T)]\}$ phase-modulation. To shape such double-peak pulses, we expand the laser beam and place a thin glass plate in its part. The beam part passed through the plate obtains a time delay with respect to the free-propagated part. Moving the plate along the vertical axis, we adjust the power proportion between the peaks. The thickness of the plate gives the time delay between the peaks; a 0.09 mm thick glass plate gives a 150 fs delay according to autocorrelation check. Us-

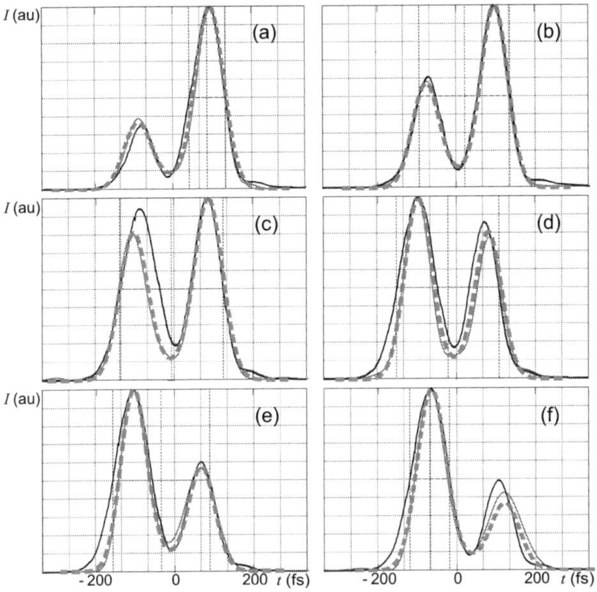

Fig. 6. Comparison of similariton-based SI and FO for different double-peak signal pulses: black solid curves are related to the FO-measurement, gray - to the similariton-based SI-characterisation, gray dashed - to the theory.

ing the double-peak signal pulse, we experiment comparatively the FO and similariton-based SI: the spectrotemporal images of the signal are compared with the SI-reconstructed pulses, according to the experimental procedures described above (Fig. 4). Fig. 6 illustrates this experiment by the results for a set of double-peak signal pulses. Fig. 7 is related to the best result of Fig. 6 (b): the quantitative accordance of the measured spectrum (a) and retrieved spectral phase (b) with the theoretical curves (gray dashed) leads to an accurate pulse reconstruction through similariton-based SI (c, gray solid). An accurate spectrotemporal imaging (c, black spectrogram) is ensured by the similariton of the bandwidths of $\geq$40 nm. The differences between these independent SI- and FO-measurements and theoretical curve are hardly seen, evidencing the accurateness of both the similariton-based methods.

Thus, we have experimentally demonstrated the methods of similariton-based SI and spectrotemporal imaging as two applications of similariton. Reference-based methods become self-referencing by the use of similariton. Results of our measurements carried out by similariton-based SI and spectrotemporal imaging are in quantitative accordance. The method of

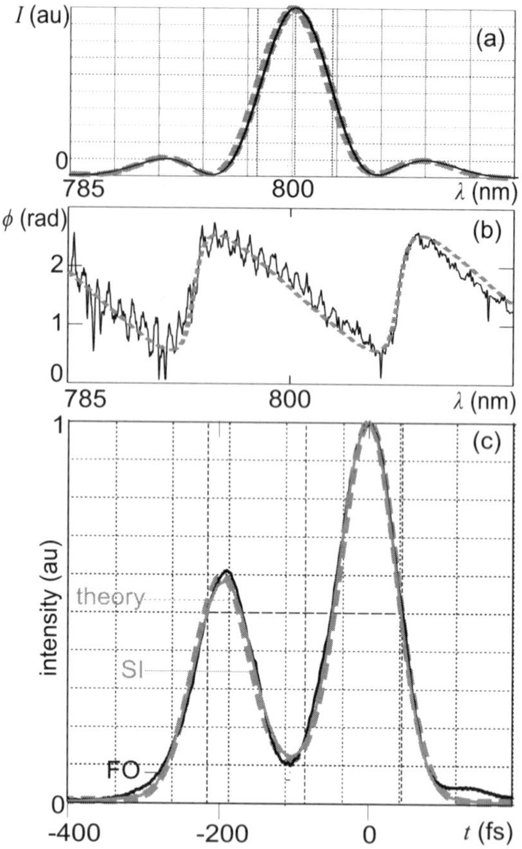

Fig. 7. An accurate registration of a double-peak signal pulse through the similariton-based SI and FO: (a) measured spectrum, (b) retrieved spectral phase, and (c) pulse. The gray dashed, gray solid and black curves are for the theory, similariton-based SI and FO, respectively.

similariton-based spectrotemporal imaging has an advantage of direct pulse measurement leading to the development of a femtosecond optical oscilloscope, but it does not give phase information without additional interferometric measurement. The method of similariton-based SI provides complete (amplitude and phase) characterization of femtosecond signal.

4. Conclusion

We have generated a broadband similariton of ~100 nm bandwidth, and experimentally studied it by means of its chirp measurement through the technique of frequency tuning in the SFG-spectral compression process. The applied technique allows measuring the coefficients of fiber GVD (β_2) and TOD (β_3). The fiber TOD impacts additively on the phases of parabolic and nonlinear-dispersive similaritons with the value $\Delta\varphi_f(t) = \beta_3\beta_2^{-3}f^{-2}t^3/6$, the same as for the pulse dispersive propagation. In a comparative study, the similariton-based methods of spectral interferometry and spectrotemporal imaging for accurate femtosecond scale temporal measurements are demonstrated as two applications of broadband similariton in ultrafast optics.

Acknowledgments

This work was carried out within the framework of the Centre National de la Recherche Scientifique (CNRS), France - State Committee of Science (SCS), Armenia collaborative programme project IE007. A. Zeytunyan acknowledges SCS and National Foundation of Science and Advanced Technology (NFSAT) for financial support in the framework of the Early Career Support Program (grant numbers A-16 and ECSP-09-50).

References

1. J. M. Dudley, C. Finot, G. Millot, and D. J. Richardson, Nat. Phys. **3**, 597 (2007).
2. D. Anderson, M. Desaix, M. Karlson, M. Lisak, and M. L. Quiroga-Teixeiro, J. Opt. Soc. Am. B **10**, 1185 (1993).
3. V. I. Kruglov, A. C. Peacock, J. D. Harvey, and J. M. Dudley, J. Opt. Soc. Am. B **19**, 461 (2002).
4. C. Finot, G. Millot, C. Billet, and J. M. Dudley, Opt. Express **11**, 1547 (2003).
5. T. Hirooka, and M. Nakazawa, Opt. Lett. **29**, 498 (2004).
6. F. Ö. Ilday, F. W. Wise, and F. X. Kärtner, Opt. Express **12**, 2731 (2004).
7. C. Finot, B. Barviau, G. Millot, A. Guryanov, A. Sysoliatin, and S. Wabnitz, Opt. Express **15**, 15824 (2007).
8. C. Finot, G. Millot, and J. M. Dudley, Opt. Lett. **29**, 2533 (2004).
9. G. Yesayan, K. Palanjyan, T. Mansuryan, A. Zeytunyan, L. Mouradian, P. Kockaert, and P. Emplit, in *Nonlinear Photonics*, OSA Technical Digest (CD) (Optical Society of America, 2007), paper JWA18.
10. A. Zeytunyan, G. Yesayan, L. Mouradian, P. Kockaert, P. Emplit, F. Louradour, and A. Barthélémy, J. Europ. Opt. Soc. Rap. Public. 09009, Vol. **4** (2009).
11. V. I. Kruglov, D. Méchin, and J. D. Harvey, J. Opt. Soc. Am. B **24**, 833 (2007).
12. C. Finot, and G. Millot, Opt. Express **12**, 5104 (2004).

13. T. Mansuryan, A. Zeytunyan, M. Kalashyan, G. Yesayan, L. Mouradian, F. Louradour, and A. Barthélémy, J. Opt. Soc. Am. B **25**, A101 (2008).
14. A. Zeytunyan, G. Yesayan, L. Mouradian, F. Louradour, and A. Barthélémy, in *Frontiers in Optics*, Technical Digest (CD) (Optical Society of America, 2009), paper FWI5.
15. D.T. Reid, P. Loza-Alvarez, C.T.A. Brown, T. Beddard, and W. Sibbett, Opt. Lett. **25**, 1478 (2000).
16. J.M. Dudley, X. Gu, L. Xu, M. Kimmel, E. Zeek, P. O'Shea, R. Trebino, S. Coen, and R.S. Windeler, Opt. Express **10**, 1215 (2002).
17. J. Piasecki, B. Colombeau, M. Vampouille, C. Froehly, and J. A. Arnaud, Appl. Opt. **19**, 3749 (1980).
18. C. Iaconis and I. A. Walmsley, Opt. Lett. **23**, 792 (1998).
19. V. Messager, F. Louradour, C. Froehly, and A. Barthélémy, Opt. Lett. **28**, 743 (2003).
20. D. J. Kane and R. Trebino, Opt. Lett. **18**, 823 (1993).
21. M. A. Foster, R. Salem, D. F. Geraghty, A. C. Turner-Foster, M. Lipson, and A. L. Gaeta, Nature **456**, 81 (2008).
22. M. Takeda, H. Ina, and S. Kobayashi, J. Opt. Soc. Am. **72**, 156 (1982).

INDEX